Waterbirds in the UK 2004/05
The Wetland Bird Survey

Alex Banks, Mark Collier, Graham Austin,
Richard Hearn and Andy Musgrove

WeBS

Published by

British Trust for Ornithology, Wildfowl & Wetlands Trust,
Royal Society for the Protection of Birds and Joint Nature Conservation Committee

November 2006

BTO WWT RSPB JOINT NATURE CONSERVATION COMMITTEE

ISBN 1-904870-77-5
ISSN 1353–7792

This publication should be cited as:
Banks, A.N., Collier, M.P., Austin, G.E., Hearn, R.D. & Musgrove, A.J. 2006. *Waterbirds in the UK 2004/05: The Wetland Bird Survey.* BTO/WWT/RSPB/JNCC, Thetford.

Published by: BTO/WWT/RSPB/JNCC

Cover: Eider at dawn by Julian Novorol. Julian has been a WeBS counter at Hamford Water since 1967.

Photographs: Dawn Balmer, John Bowers, Mark Collier, Tommy Holden and Richard Vaughan.

Produced by the British Trust for Ornithology.

Printed by Crowes Complete Print, 50 Hurricane Way, Norwich, NR6 6JB.

Typeset in Times New Roman, Arial and Trebuchet MS fonts.

Available from: BTO, The Nunnery, Thetford, Norfolk IP24 2PU, UK, and Natural History Book Service, 2-3 Wills Road, Totnes, Devon TQ9 5XN, UK.

This report is provided free to all WeBS counters and those who participate in the other national waterbird surveys, none of whom receive financial reward for their invaluable work. Further feedback is provided to counters through the annual WeBS Newsletter. For further information please contact the WeBS Office at the BTO.

ACKNOWLEDGEMENTS

This book represents the twenty-forth report of the Wetland Bird Survey and comprises information from WeBS and complementary national and local surveys, *e.g.* goose censuses. It is entirely dependent on the many thousands of dedicated volunteer ornithologists who supply the data and to whom we are extremely grateful. The Local Organisers who coordinate these counts deserve special thanks for their contribution.

We are also grateful to the following people for providing technical assistance, supplementary information and additional data, and comments on draft texts:

Helen Baker, Niall Burton, Dave Butterfield, Peter Cranswick, Olivia Crowe, Emma Glaister, Colette Hall, Martin Heubeck, Steve Holloway, Baz Hughes, Rowena Langston, Ilya Maclean, Heidi Mellan, Margaret Morris, Mark Rehfisch, Lucy Smith, David Stroud, Chris Waltho and Jenny Worden. Many amateur observers also provide reports of their studies; these are acknowledged within the text.

Grateful thanks to all and apologies to anyone who has inadvertently been missed.

The WETLAND BIRD SURVEY

Organised and funded by

British Trust for Ornithology
The Nunnery, Thetford, Norfolk IP24 2PU
www.bto.org

Wildfowl & Wetlands Trust
Slimbridge, Gloucestershire GL2 7BT
www.wwt.org.uk

Royal Society for the Protection of Birds
The Lodge, Sandy, Bedfordshire SG19 2DL
www.rspb.org.uk

Joint Nature Conservation Committee
Monkstone House, City Road, Peterborough PE1 1JY
www.jncc.org.uk

CONTACTS

WeBS National Coordinator: Andy Musgrove
WeBS Core Counts: Mark Collier
WeBS Low Tide Counts: Alex Banks
WeBS Counter Coordinator: Steve Holloway

General queries: webs@bto.org

WeBS Office
British Trust for Ornithology
The Nunnery
Thetford
Norfolk IP24 2PU
UK
Tel: 01842 750050
Fax: 01842 750030
E-mail: firstname.surname@bto.org
or webs@bto.org
www.bto.org/survey/webs

NATIONAL GOOSE CENSUSES

Organised and funded by the Wildfowl & Wetlands Trust and the Joint Nature Conservation Committee.
Contact: Jenny Worden
E-mail: Jenny.Worden@wwt.org.uk
or monitoring@wwt.org.uk

Wildfowl & Wetlands Trust
Slimbridge
Glos GL2 7BT
UK
Tel: 01453 891225
Fax: 01453 891901
www.wwt.org.uk/research/monitoring

OTHER NATIONAL WATERBIRD SURVEYS

Details of and contacts for many of the other waterbird surveys used in this report, and of forthcoming surveys, can be obtained via the web sites of the four WeBS partner organisations.

ERRATA TO 2003/04 REPORT

Please note the following corrections to data presented in the 2003/04 WeBS annual report:

- p26-7 Icelandic Greylag Geese during May to September should read 0 and counts from these months are added to the Re-established population, which become; 5738 May, 12743 Jun, 14794 Jul, 18483 Aug, 28858 Sep.
- p49 The 2001/02 peak of Black-necked Grebe in Langstone Harbour was a Low Tide count of 15; the 2003/04 five-year mean peak was therefore 14.
- p59 Bewick's Swan indices are incorrect see species text in this report for details.
- p60 Whooper Swan indices are incorrect see species text in this report for details.
- p61 The 2002/03 Whooper Swan peak at Nene Washes should read 143 (standard Core Count), giving five-year mean peak of 97.
- p68 Counts of Greylag Geese for Machrihanish and Moine Mhor are presented as being from the Northwest Scottish Population. Birds at these sites are now considered to be from the Icelandic Population, however, five-year mean peaks at these sites were below the threshold for this population and would not have appeared in the table.
- p69 The table header for Re-established Greylag Goose states 'Sites with mean peak counts of 300 or more…' this should read '…400 or more…'.
- p78 The 2000/01 peak of East Canadian High Arctic Light-bellied Brent Goose at the Dee Estuary should read 39 consequently giving a 2003/04 five-year mean peak of 35.
- p118 The GB maximum count for Avocet was during December.
- p130 Only two Pectoral Sandpipers were present at Cliffe Pools.

CONTENTS

Summary

The Wetland Bird Survey and Waterbirds in the UK

The Wetland Bird Survey (WeBS) is a joint scheme of the British Trust for Ornithology (BTO), the Wildfowl & Wetlands Trust (WWT), Royal Society for the Protection of Birds (RSPB) and Joint Nature Conservation Committee (JNCC) to monitor non-breeding waterbirds in the UK. The principal aims of the scheme are to identify population sizes, to determine trends in numbers and distribution, and to identify important sites for waterbirds. WeBS Core Counts are made annually at around 2,000 wetland sites of all habitats; estuaries and large still waters predominate. Monthly coordinated counts are made mostly by volunteers, principally from September to March, with fewer observations during summer months. Data from other sources, *e.g.* roost counts of grey geese, are included in this report where relevant.

This report presents total numbers counted for all species in the most recent year in Great Britain and Northern Ireland. Annual indices are provided for the more numerous species, as are monthly indices showing relative abundance during the winter.

Species accounts provide yearly maxima for all sites supporting internationally and nationally important numbers. Sites with changed status are highlighted and significant counts are discussed. Counts are placed in an international context where possible, and relevant research is summarised. Waterbird totals are provided for all sites meeting criteria for international importance and species occurring in internationally important numbers on each are identified.

WeBS Low Tide Counts are made on selected estuaries to determine the distribution of birds during low tide and to identify important feeding areas that may not be recognised during Core Counts, which are made mostly at high tide. A summary of results for these estuaries, and distribution maps for selected species, are provided.

Waterbird totals recorded by the Irish Wetland Bird Survey, a similar scheme operating in the Republic of Ireland, are also included.

The 2004/05 year

This report summarises counts during 2004/05 and previous years (since 1960 for wildfowl, 1969 for waders and the early 1980s or 1990s for other species). During 2004/05, WeBS counters covered 3,300 count sectors at around 2,000 count sites, during the crucial 'winter' period of September to March. At least 1,500 were counted in any one of these months and almost 1,100 were covered continually throughout this period. This represents a fantastic effort all around and a huge thank you must go to all those involved.

Bewick's and Whooper Swans continue to use the Ouse Washes in ever-increasing numbers, in the latter case leading to a dramatically increasing British wintering population.

Nationally, the number of wintering Pink-footed Geese rose slightly following a successful breeding season. Numbers of European White-fronted Geese recorded in Britain continued to fall and reached their lowest level since records began, although this is likely to be due to increasing numbers remaining in continental Europe throughout the winter period rather than a reflection on the overall population. Greenland White-fronted Geese showed further declines and in 2004/05 were present in their lowest numbers for 15 years. Unlike European Whitefronts, this decline has been witnessed across the entire population and is thought to be linked to poor breeding success.

Increases in the numbers of Icelandic Greylag Geese were identified. The numbers of birds remaining in Iceland throughout the autumn has also increased. The increase in Greenland Barnacle Geese was attributed to a successful breeding season; numbers of this species have risen at the principal site, Islay, as well as other key sites. Despite numbers of Svalbard Barnacle Geese remaining stable, productivity over the past five years has remained low. Numbers of Dark-bellied Brent Geese showed their first increase for over ten years. The proportion of young birds was the highest for five years, however, this remained below the level of the late 1990s. October 2004 saw the highest ever total of East Canadian High Arctic Light-bellied Brent

Geese at Strangford Lough; this was despite many birds still being present in Iceland. By mid-winter most birds had dispersed from the main sites, this was reflected by a decline in the annual index. Lower than average numbers of Svalbard Light-bellied Brent Geese were present throughout the winter. As usual Lindisfarne held the largest numbers.

There were mixed fortunes for Shelduck, which continued to decline in Britain and increase in Northern Ireland. Wigeon were present in slightly lower numbers than the past couple of years; however, the British wintering population remains healthy. Numbers of wintering Gadwall reached their highest-ever level in Britain and yet at the same time numbers in Northern Ireland were one of the lowest ever recorded. As usual Teal numbers were subject to a high level of fluctuation, although the underlying trend remains stable. Mallard continued their long-term decline, reaching record low levels in both Britain and Northern Ireland. Pintail is another species that has showed long-term declines, but since 2000/01 numbers in both Britain and Northern Ireland have risen steadily. Despite there being little change in the numbers of Shoveler in Britain numbers reached a record low in Northern Ireland.

A key feature of recent WeBS counts has been the dramatic decline of some diving duck species in Northern Ireland, due largely to trends in numbers using Loughs Neagh & Beg. Following recent crashes, data for 2004/05 showed sustained low numbers of Pochard and Tufted Duck. However, Scaup peaked at their highest ever level, showing that concerns over this decline proved thankfully unnecessary. Numbers of these three species have been less erratic in Britain with a trivial decline for Pochard, a rise for Tufted Duck and no change for Scaup. In the last seven years Goldeneye have shown a decline of over 15% in Britain, this decline is much more pronounced in Northern Ireland; however, numbers here have risen slightly for the second year running. The British wintering population of Red-breasted Mergansers rose by around 10% in 2004/05, whereas Goosander numbers fell by almost 20% in the same period.

Counted maxima for Red-throated Diver were the lowest to date, while those for Black-throated and Great Northern Divers were well above average. The British trends for both Little and Great Crested Grebes show steady increases. Little Egret again continued to expand its range and population, although less rapidly than in recent years. In Britain both Moorhen and Coot numbers have remained relatively stable. However, Coot numbers in Northern Ireland have fallen dramatically over the past five years, largely due to the declines at Loughs Neagh and Beg.

Oystercatcher numbers remained fairly stable, numbers at the top site, Morecambe Bay, being the highest for seven years. Avocet have shown a steady increase over the past 15 years and 2004/05 data followed the pattern of recent years. There were mixed fortunes for Lapwing numbers, which rose in Britain and fell in Northern Ireland. Ringed Plover numbers continued to fall; this was despite a counted maximum similar to those of the past 15 years. Numbers of wintering Golden Plover increased in Northern Ireland and reached record numbers in Britain. Conversely, Grey Plover have continued to decline in both Britain and Northern Ireland. Knot showed further signs of increase; numbers on The Wash were the highest for over ten years. Wintering numbers of both Sanderling and Dunlin continued to decline. There was also a considerable fall in Bar-tailed Godwit numbers. While Redshank and Turnstone numbers remained similar to the past few years, there was a definite increase in the number of wintering Curlew in 2004/05. Additionally, record peaks of wintering Black-tailed Godwit and Greenshank were documented.

As in 2003/04, further declines were recorded in most of the main gull species, the exception being Herring Gull, which rose by almost 70% on the previous year. Summed maxima of Sandwich Tern were higher than in the previous year. However, Little Tern and Common Tern numbers were lower than in 2003/04. As with all gulls and terns the optional coverage of these groups during WeBS counts means that numbers recorded will largely be dependent on coverage at each site.

Introduction

The UK is of outstanding international importance for waterbirds. Lying on some of the major flyways for Arctic-nesting species, large numbers of waterbirds are attracted, especially during winter, by the relatively mild climate and extensive areas of wetland, notably estuaries. The UK thus has both moral and legal obligations to conserve both these waterbirds and the wetlands upon which they depend.

As a signatory to a number of international conservation conventions, and as a member of the EU, the UK is bound by international law. In particular, the 'Ramsar' Convention on Wetlands of International Importance especially as Waterfowl Habitat, the EC Birds Directive and the EU Habitats and Species Directive, between them, require the UK to identify important examples of wetland and other habitats and sites important for birds and designate them for protection. Implicit in these obligations is the need for regular monitoring to identify and monitor such sites. These instruments also lay particular significance on the need to conserve migratory populations, and consequently most of the waterbird populations in the UK.

The UK has ratified the Agreement on the Conservation of African-Eurasian Migratory Waterbirds (AEWA) of the 'Bonn' Convention on the Conservation of Migratory Species of Wild Animals. AEWA entered into force in 1999. It is a specific Agreement requiring nations to take coordinated measures to conserve migratory waterbirds given their particular vulnerability due to their migration over long distances and their dependence on networks that are decreasing in extent and becoming degraded through non-sustainable human activities. Article three of the Agreement requires, among other things, that sites and habitats for migratory waterbirds are identified, protected and managed appropriately, that parties initiate or support research into the ecology of these species, and exchange information and results. Explicit in this Agreement is that adequate monitoring programmes are set in place to fulfil these objectives and the Action Plan to the Agreement specifically requires that nations endeavour to monitor waterbird populations.

AIMS AND OBJECTIVES OF WeBS

The Wetland Bird Survey (WeBS) aims to monitor all non-breeding waterbirds in the UK to provide the principal data on which the conservation of their populations and wetland habitats is based. To this end, WeBS has three main objectives:

- to assess the size of non-breeding waterbird populations in the UK;
- to assess trends in their numbers and distribution; and
- to assess the importance of individual sites for waterbirds.

These results also form the basis for informed decision-making by conservation bodies, planners and developers and contribute to the sustainable and wise use and management of wetlands and their dependent waterbirds. The data and the WeBS report also fulfil some of the objectives of the Conventions and Directives listed above. WeBS also provides UK data to Wetlands International to assist their function to coordinate and report upon waterbird monitoring at an international scale.

Structure and organisation of WeBS

WeBS is a partnership scheme of the British Trust for Ornithology (BTO), Wildfowl & Wetlands Trust (WWT), Royal Society for the Protection of Birds (RSPB) and the Joint Nature Conservation Committee (JNCC), the last on behalf of English Nature (EN) (now Natural England), Scottish Natural Heritage (SNH) and the Countryside Council for Wales (CCW), and the Environment and Heritage Service in Northern Ireland (EHS).

WeBS continues the traditions of two, long-running count schemes which formed the mainstay of UK waterbird monitoring since 1947 (Cranswick *et al.* 1997). WeBS Core Counts are made at a wide variety of wetlands throughout the UK. Synchronised counts are conducted once per month, particularly from September to March, to fulfil all three main objectives. In addition, WeBS Low Tide Counts are undertaken on selected estuaries with the aim of identifying key areas used during the low tide period, principally by feeding birds; areas not otherwise noted for their importance by Core Counts which are normally conducted at high tide.

The success and growth of these count schemes accurately reflects the enthusiasm and dedication of the several thousands of volunteer ornithologists who participate. It is largely due to their efforts that waterbird monitoring in the UK is held in such international high regard.

Aim of this report

This report presents syntheses of data collected between April 2004 and June 2005 (see *The WeBS Year*), and in previous years, in line with the WeBS objectives. Data from other national and local waterbird monitoring schemes, notably annual goose censuses, are included where WeBS data alone are insufficient to fulfil this aim, so that the report provides a single, comprehensive source of information on waterbird status and distribution in the UK. All nationally and internationally important sites for which data exist are listed.

Changes since the 2003/04 report

The arrangement of species largely follows the order recently adopted by the British Ornithologists' Union (Knox *et al.* 2002). This broadly means that swans through ducks can be found before divers. The rest of the species order remains largely unchanged.

Previous WeBS reports have presented data from April to March inclusive, meaning that any spring data (April to June) effectively concerned birds from the previous winter period. As WeBS is predominantly concerned with non-breeding waterbirds, whose peak numbers typically occur during the winter months and can still be present through into much of the spring, it is more meaningful to present spring data along with that of the preceding winter.

This report presents data from July through to June, thus including the autumn and spring data from the adjoining wintering period. Data for April to June 2004 are presented in Appendix 3.

WEATHER IN 2004/05

This summary of UK weather is drawn from the Meteorological Office web site at www.metoffice.gov.uk. Figures in brackets following the month refer to the Core Count priority date for the month in question. Arctic breeding conditions for birds that winter in the UK are summarised from information collated by Soloviev & Tomkovich at the web site www.arcticbirds.ru.

United Kingdom

Although **April** (25) temperatures were above average across the whole of Britain and Northern Ireland, sunshine was in short supply and the month was dominated by wet conditions. The exception, however, was north Wales and northwest England, which experienced lower than average rainfall.

Higher than average rainfall continued into **May** (23) in southeast England, although for the rest of the UK conditions were somewhat drier. Most areas experienced conditions that were sunnier and warmer than average, particularly eastern Scotland.

June (20) saw the return of wet weather to northern England and Scotland, whilst further south sunny and dry conditions prevailed. Temperatures were above average despite being cold in the second half of the month.

Although temperatures, rainfall and sunshine in **July** (18) were all around average for the UK, the were large variations between regions. Eastern England received twice its average monthly rainfall whilst in contrast some areas of Scotland only received half of the expected.

August (22) was dominated by rain with many areas experiencing double their monthly rainfall. Much of northeast England and southern Scotland experienced low amounts of sunshine, although further north Shetland basked in unusually high levels of sunshine. Average temperatures were normal for the time of year.

As the year moved on wet weather persisted in western areas although for the rest of Britain **September** (19) saw the return of drier, sunnier conditions. Temperatures were slightly above average across the country. By the end of the second week, conditions became more unsettled with showers persisting through most of the second half of the month. Scotland and some western areas experienced strong winds. The month finished with eastern areas experiencing thundery conditions.

October (17) began wet and windy in most areas with some localized thunderstorms in the second week. High pressure brought a series of fronts that brought wet and sometimes windy conditions to most of England and Wales. Scotland also experienced bands of heavy rain and like the rest of the country ended with a few days of dry weather at the end of the month.

Dry conditions continued into **November** (14) with high pressures dominating early on in the month. Temperatures started mild with few frosts, however, a sudden fall in temperature in the middle of the month was a result of northerly winds, which also brought heavy rain and widespread frosts and snow to higher ground, and even to some areas of southern England. Milder conditions returned at the end of the month with cloudy and rainy weather to most areas.

Despite starting widely mild and sunny **December** (12) turned drizzly across most of England and Wales by the start of the second week. This was short lived as high pressure helped keep much of eastern England dry, before southeasterly winds brought cold and misty conditions. Southwesterly winds brought milder, albeit wetter and windier conditions in the middle of the month. Scotland experienced mainly mild and cloudy conditions until strong winds in the middle of the month brought change. Showers persisted through the third week of December although these turned wintry bringing snowy conditions across much of the country.

The unsettled weather continued into the first two weeks of **January** (16), which saw heavy rain and strong blustery winds across most of the country. These conditions persisted into the third week, except in eastern England, which experienced clear skies bringing sunshine and frosts in the middle of the month. Rain soon returned to most of the country with some wintry spells especially in northern areas. Temperatures finally rose on the last day of the month with sunshine in some areas.

February (13) started mainly dry and although by the start of the second week a cold front brought unsettled and wintry conditions across much of England and Wales that persisted for the rest of the month. Scotland experienced some milder and wetter weather towards the middle of the month before colder, wintry conditions prevailed.

Cold, wintry conditions continued into the first half of **March** (13) with snowy and icy conditions throughout much of England, Wales, Scotland and Northern Ireland. These cold temperatures gave way to milder, cloudier and wetter conditions in the middle of the month, which continued for the following couple of weeks. Mild temperatures were a feature across most of the country.

April (10) began mild with alternative spells of wet and settled periods typical for the time of year. Some higher areas experienced brief snow showers in the middle of the month. Temperatures remained high towards the end of the month with many areas, especially Scotland and Northern Ireland experiencing increasing periods of sunshine.

Changeable weather remained into **May** (8) with rain easing during the middle of the month. Temperatures remained cool and rain returned for the remainder of the month. The final few days saw heavy rain at times.

Rain continued into the first week of **June** (26) before drier warmer conditions intervened. High temperatures and fine weather continued in England and Wales for much of the month, however, there were heavy showers across much of England during the final week. Scotland and Northern Ireland experienced cooler damper conditions during the second half of the month.

Table 1. The percentage of inland count units (lakes, reservoirs, gravel pits, rivers and canals) in the UK with any ice and with 75% or more of their surface covered by ice during WeBS counts in winter 2004/05 (England divided by a line drawn roughly between the Humber and the Mersey Estuaries).

Region	Ice	S	O	N	D	J	F	M
Northern Ireland	>0%	0	0	0	0	0	0	0
	>74%	0	0	0	0	0	0	0
Scotland	>0%	0	0	1	8	9	7	5
	>74%	0	0	<1	5	3	2	<1
N England	>0%	0	0	2	2	1	2	4
	>74%	0	0	<1	0	<1	<1	0
S England	>0%	0	0	1	<1	<1	1	<1
	>74%	0	0	<1	<1	<1	0	<1
Wales	>0%	0	0	0	1	0	1	0
	>74%	0	0	0	0	0	0	0

Arctic Breeding Conditions 2004

Much of the Russian Arctic and Fennoscandia experienced cold conditions during June and July. Despite a late spring and cold June in the western Taymir temperatures quickly rose during late June and July. Rodent numbers were low across Fennoscandia and average on the Kola Peninsula. Higher numbers were found in northwest Europe, the Yamal and Taymir Peninsulas.

SURVEY METHODS

The main source of data for this report is the WeBS scheme, providing regular monthly counts for most waterbird species at the majority of the UK's important wetlands. In order to fulfil the WeBS objectives, however, data from a number of additional schemes are included in this report. In particular, a number of species groups necessitate different counting methodologies in order to monitor numbers adequately, notably most geese and seaducks, and the results of other national and local schemes for these species are routinely included.

The methods for these survey types are outlined below and more detail can be found in Gilbert *et al.* (1998). It should be noted that site definition is likely to vary between these surveys (see *Interpretation of Waterbird Counts*).

WeBS Core Counts

WeBS Core Counts are made using so-called 'look-see' methodology (Bibby *et al.* 2000), whereby the observer, familiar with the species involved, surveys the whole of a predefined area. Counts are made at all wetland habitats, including lakes, lochs/loughs, ponds, reservoirs, gravel pits, rivers, freshwater marshes, canals, sections of open coast and estuaries. Numbers of all waterbird species, as defined by Wetlands International (Rose & Scott 1997), are recorded. In the UK, this includes swans, geese, ducks, divers, grebes, cormorants, herons, Spoonbill, rails, cranes, waders and Kingfisher. Counts of gulls and terns are optional.

In line with the recommendations of Vinicombe *et al.* (1993), records of all species recorded by WeBS, including escapes, have been published to contribute to the proper assessment of naturalised populations and escaped birds. Following Holmes & Stroud (1995), non-native species which have become established are termed 'naturalised'. These species are categorised according to the process by which they became established: naturalised feral (domesticated species gone wild); naturalised introduction (introduced by man); naturalised re-establishment (species re-established in an area of former occurrence); or naturalised establishment (a species which occurs, but does not breed naturally, *e.g.* potentially Barnacle Goose in southern England). With the exception of vagrants, all other non-native species have been classed as 'escapes'. The native range is given in the species account for naturalised species, escapes and vagrants.

Most waterbirds are readily visible. Secretive species, such as snipes, are generally under-recorded. No allowance is made for these habits by the observer and only birds seen or heard are recorded. The species affected by such biases are well known and the problems of interpretation are highlighted individually in the Species accounts. Most species and many subspecies are readily identifiable during the counts. Categories may be used, *e.g.* unidentified scoter species, where it is not possible to be confident of identification, *e.g.* under poor light conditions.

Species present in relatively small numbers or dispersed widely may be counted singly. The number of birds in large flocks is generally estimated by mentally dividing the birds into groups, which may vary from five to 1,000 depending on the size of the flock, and counting the number of groups. Notebooks and tally counters may be used to aid counts.

Counts are made once per month, ideally on predetermined 'priority dates'. This enables counts across the whole country to be synchronised, thus reducing the likelihood of birds being double counted or missed. Such synchronisation is imperative at large sites, which are divided into sectors, each of which can be practicably counted by a single person in a reasonable amount of time. Local Organisers ensure coordination in these cases due to the high possibility of local movements affecting count totals. The priority dates are pre-selected with a view to optimising tidal conditions for counters covering coastal sites at high tide on a Sunday (see *Coverage*). The dates used for individual sites may vary due to differences in the tidal regime around the country. Coordination within a site takes priority over national synchronisation.

Counts suspected to be gross underestimates of the true number of non-secretive species present are specifically noted, *e.g.* a large flock of roosting waders only partially counted before being flushed by a predator, or a distant flock of seaduck in heavy swell. These counts may then be treated differently when calculating site totals (see *Analysis*).

Data are input by a professional data input company. Data are keyed twice by different people and discrepancies identified by computer for correction. Any particularly unusual counts are checked by the National Organisers and are confirmed with the counters if necessary.

WeBS Low Tide Counts

This survey aims to assess numbers of waterbirds present during low tide on estuaries, primarily to assess the distribution of feeding birds at that time (Musgrove *et al.* 2003; see the section *Low Tide Counts* for a full explanation of methods).

This survey occasionally provides higher counts for individual sites than Core Counts, for example, where birds feed on one estuary but roost on another. These data are validated before being used for site assessment against 1% thresholds.

Supplementary daytime and roost counts

Supplementary counts are made at some sites where WeBS counts are known to under-represent the true value of the site. In particular, some species occur in much larger sites when using the site as a nighttime roost, *e.g.* geese, Goosander and gulls, that are not present during WeBS daytime counts. Some sites are also counted more frequently than once monthly by some observers.

Supplementary counts are collected by counters familiar with the site for WeBS survey, thus employing the same site definition and, for daytime counts, the same counting methods, and are submitted on standardised recording forms adapted from those used for WeBS Core Counts.

Goose roost censuses

Many geese (*Anser and Branta spp*) spend daylight hours in agricultural landscapes, and are therefore missed during counts at wetlands by WeBS. These species are usually best counted as they fly to or from their roost sites at dawn or dusk since these are generally discrete wetlands and birds often follow traditional flight lines approaching or leaving the site. Even in half-light, birds can generally be counted with relative ease against the sky, although they may not be specifically identifiable at mixed species roosts.

In order to produce population estimates, counts are synchronised nationally for particular species (see *National totals* below), though normally only one or two such counts are made each year. The priority count dates are determined according to the state of the moon, since large numbers of geese may remain on fields during moonlit nights. Additional counts are made by some observers, particularly during times of high turnover when large numbers may occur for just a few days.

In some areas, where roost sites are poorly known or difficult to access, counts of birds in fields are made during the daytime. As with WeBS Core Counts, the accuracy of the count is noted.

Additional counts

Additional, *ad hoc*, data are also sought for important sites not otherwise covered by regular monitoring, particularly open coast sections in Scotland, whilst the results of periodic, coordinated surveys - such as the non-estuarine coastal waterbird survey (NEWS), International Greenland Barnacle Goose Census and International Whooper & Bewick's Swan Census - are included where the data collected are compatible with the presentation formats used in this report.

The accuracy of counts of waterbirds on the sea is particularly dependent on prevailing weather conditions at the time of or directly preceding the count. Birds are often distant from land, and wind or rain can cause considerable difficulty with identifying and counting birds. Wind not only causes telescope shake, but even a moderate swell at sites without high vantage points can hamper counts considerably. The need to count other waterbirds in 'terrestrial' habitats at the site often precludes the time required for an accurate assessment of seaducks. Many sites may be best covered using aerial surveys, though this technique has been little used in the UK historically. Consequently, the best counts of most divers, grebes and seaduck at open coast and many estuarine sites are made simply when conditions allow; only rarely will such conditions occur by chance during WeBS counts. Synchronisation between different sites may be difficult or impossible to achieve, and thus coordination of most counts to date has occurred at a regional or site level, *e.g.* within the Moray Firth and within North Cardigan Bay.

The extensive use of aerial survey methods in nearshore marine waters in recent years means that data are available for a number of sites. However, the boundaries of such sites frequently do not correspond to those counted for WeBS Core Counts, and indeed the area surveyed from the air can vary between years. As a result, such aerial surveys are now tabulated separately within the relevant species accounts. These surveys employ a 'distance sampling' methodology (see Buckland *et al.* 2001, 2004), whereby only a proportion of birds is counted, and the missed proportion estimated by statistical means. Some published reports from these surveys provide only the counted number, whilst others include the calculated estimates also (which often have relatively wide confidence intervals).

Some data are provided directly by individuals (for example, reserve wardens), often undertaking counts for site survey purposes, but whose data are not formally published in a report.

A significant point is that these additional data are taken from published sources, from surveys with the specific aim of monitoring waterbirds, and where methods have been published - or where data have been collected by known individuals, usually undertaking site-based surveys, and are provided directly for use in *Waterbirds in the UK*. Casual records and data from, *e.g.* county bird reports, where the methods and/or site boundaries used are not documented, are not included. Reports and data for important sites from surveys that the authors know to have taken place in recent years are actively sought for inclusion in this report, but it is likely that other sources of suitable data are overlooked. The inclusion of additional data for some species and sites does not, thus, indicate that the tables in the Species accounts include all such suitable data.

Irish Wetland Bird Survey

The Irish Wetland Bird Survey (I-WeBS) monitors non-breeding waterbirds in the Republic of Ireland (Crowe 2005). I-WeBS was launched in 1994 as a joint partnership between BirdWatch Ireland, National Parks and Wildlife Service of Dúchas, The Heritage Service of the Department of Environment and Local Government (Ireland), and WWT, with additional funding and support from the Heritage Council and WWF UK (World Wide Fund for Nature). I-WeBS is complementary to and compatible with the UK scheme. The main methodological difference from UK-WeBS is that counts are made only between September and March, inclusive.

Productivity monitoring

Changes in numbers of waterbirds counted in the UK between years are likely to result from a number of factors, including coverage and weather, particularly for European and Russian breeding species which may winter further east or west within Europe according to the severity of the winter. Genuine changes in population size will, however, result from differences in recruitment and mortality between years.

For several species of swans and geese, young of the year can be readily identified in the field and a measure of productivity can be obtained by recording the number of young birds in sampled flocks, expressed as a percentage of the total number of birds aged. Experienced fieldworkers, by observing the behaviour of and relationship between individuals in a flock, can record brood sizes as the number of young birds associating with, usually, two adults.

ANALYSIS AND PRESENTATION

In fulfilment of the WeBS objectives, results are presented in a number of different sections. An outline of the analyses undertaken for each is given here; further details can be provided upon request. A number of limitations of the data or these analytical techniques necessitate caution when interpreting the results presented in this report (see *Interpretation of Waterbird Counts*).

Count accuracy and completeness

Counts at individual sites may be hampered by poor conditions, or parts of the site may not be covered. This may result in counts missing a significant proportion of one or more species.

It is important to flag such counts since using them at face value would under-represent the importance of the site and give misleading results, *e.g.* when used for trend calculations and assessment of site importance.

Counts at sites - and at individual sectors of large sites that are counted using a series of sub-divisions (known as 'complex sites') - are flagged as 'OK' or 'Low' by the counter, where 'Low' indicates that the counter feels a significant proportion of the birds present at the time of the count may have been missed, *e.g.* because all of the site or sector was not visited, or because a large flock of birds flew before counts were complete. Such assessments may be provided for individual species, or for all species present.

Similarly, at complex sites, one or more sectors may be missed in a particular month, again rendering the total count for the site incomplete to a greater or lesser degree for one or more species.

For single sector sites, counts are assessed as incomplete based on the 'OK/Low' information provided by the counter. For complex sites, an algorithm is used to assess whether missed sectors and/or 'Low' counts in some sectors constitute an incomplete count at the site level. The mean count of each sector is calculated based on 'OK' counts from a window extending a month either side of the month of the count in question, and using earlier or subsequent years, such that within this window the 15 nearest counts are used to make the assessment. The total count for the site in any one month is considered incomplete if the sectors for which the count is missing or 'Low' in that month tend to hold, on the basis of their mean values, more than 25% of the sum of all sector means. The assessment is made on a species-by-species basis, recognising the fact that species distribution is not uniform across a site that and a missed sector may be particularly important for some species but not for others.

Completeness assessments are made for all WeBS Core Counts, and for most goose roost counts (which, as single-sector sites, are made on the basis of the 'OK/Low' assessment provided by the counter).

Because the completeness calculation for complex sites is based on a moving window of counts, and the use of different parts of the site by species may change, the addition of new data each year may result in counts flagged in previous *Waterbirds in the UK* (prior to 2004/05 published as *Wildfowl and Wader Counts*) as complete now being considered incomplete, and *vice versa*.

Counts are not flagged as 'Low' if a large number of the birds present is routinely missed, *e.g.* because they are cryptic, secretive, or hide in reeds - such as Snipe, Teal and Water Rail. 'Low' indicates that a significant proportion of the birds that could reasonably be expected to be counted under normal conditions was considered to have been missed. Similarly, many counts of waterbirds on the sea may be undercounts. Indeed, if the distribution of a flock stretches beyond the limits of visibility, the counter - as with birds hidden in reeds - can never know with confidence whether the count included all birds present. Counts flagged as incomplete are treated differently in trend analysis and site importance assessments.

The WeBS Year

Different waterbird species occur in the UK at different times of year. Most occur in largest numbers during winter, some are residents with numbers boosted during winter, while others occur primarily as passage migrants or even just as summer visitors.

Although WeBS counts concentrate primarily on winter months, survey is made year-round. Accordingly, different 12-month periods are used to define a year to report upon different species, in particular, to define the 'annual' maximum and to identify the peak 'annual' count for assessing site importance.

For most species, the year is defined as July to June, inclusive. Thus, for species present in largest numbers during winter, counts during autumn passage and spring passage the following calendar year are logically associated with the intervening winter. For species present as summer visitors - notably terns, Garganey and Little Ringed Plover - the calendar year is used to derive national and site maxima. The different format used for column headings (*e.g.* 04/05 or 2004) in the 'header' and tables in each species account identify whether a 'winter' or calendar year has been used.

Note that national totals (reported in Tables 3 and 4) now present data for the period July 2004 to June 2005, as opposed to April to

March as in previous WeBS reports. This means that for the first time spring data are presented alongside that of the adjoining autumn and winter periods.

National totals and annual maxima

Total numbers of waterbirds recorded by WeBS and other schemes are presented (within Tables 3 and 4 and within individual species accounts). It is very important to appreciate that these national totals are not population estimates, as WeBS does not cover 100% of the population of any species. The totals are presented separately for Great Britain (including the Isle of Man but excluding the Channel Islands) and Northern Ireland in recognition of the different legislation that applies to each. Separate totals for England, Scotland, Wales, and the Channel Islands can be obtained from the BTO upon request. The count nearest the monthly priority date or, alternatively, the count coordinated with nearby sites if there is considered to be significant interchange, is chosen for use in this report if several accurate counts are available for the same month. A count from any date is used if it is the only one available.

Totals from different count methods are mostly not combined to produce national totals because the lack of synchronisation may result in errors, *e.g.* birds counted at roost by one method may be effectively double counted during the WeBS count at a different site in that month. Total counts from several national goose surveys are, however, used instead of WeBS Core Counts where the census total provides a better estimate of the total numbers, as follows:

- Pink-footed and Icelandic Greylag Geese in October and November;
- Greenland White-fronted Goose in December and March;
- Greenland Barnacle Geese in November and March;
- NW Scotland Greylag Geese in August and February;
- Canadian Light-bellied Brent Geese in October.

Additionally, counts of Svalbard Barnacle Geese from North Cumbria and Dumfries & Galloway are replaced by Solway-wide dedicated counts between October and March. Finally, the maximum British totals for both Bewick's and Whooper Swan do include roost counts from the Ouse and Nene Washes and Martin Mere in place of Core Counts at this site, given the particular concentration of these species feeding around and roosting at this site. Counts from other site or regional-based surveys, for example of seaducks, are not included in national totals. Where a census total replace standard Core Count data these are indicated by '*'.

Some of the goose populations are identified according to location (from research into movements of marked birds) as they cannot be separated in the field by appearance alone. In such cases, a standard region of the UK is used each year to assign individual birds to particular populations and thus to derive national totals. For full details please contact BTO but broadly, the breakdown is as follows:

- NW Scotland Greylag Goose - Inner and Outer Hebrides plus Southwest Highland.
- Icelandic Greylag Goose - all other areas of Scotland plus Northumberland and North Cumbria.
- Re-established Greylag Goose - other areas.
- Greenland Barnacle Goose - Scottish west coast plus Shetland and Orkney.
- Svalbard Barnacle Goose - other Scottish regions plus Northumberland and North Cumbria.
- Naturalised Barnacle Goose - other areas.
- Canadian Light-bellied Brent Goose - Northern Ireland, Wales, western and northern Scotland, Cornwall, Devon and Channel Islands.
- Svalbard Light-bellied Brent Goose - other areas.

(*Note that the separate populations overlap to some extent, and some birds are thus likely to be mis-assigned using these areas. This is particularly so in the case of Greylag Goose and future surveys are planned to help rectify this issue*).

Data from counts at all sites are used, irrespective of whether they are considered complete or not. Numbers presented in this report are not rounded. National and site totals calculated as the sum of counts from several sectors or sites may imply a false sense of accuracy if different methods for recording numbers have been used, *e.g.* 1,000 birds estimated on one sector and a count of seven individuals on another is presented as 1,007. It is safe to assume that any large count includes a proportion of estimated birds. Reproducing the submitted counts in this way is, however, deemed the most appropriate means of presentation and avoids the summation of 'rounding error'.

In the accounts of some scarcer species, including many escaped or introduced species, summed site maxima - calculated by summing the highest count at each site, irrespective of the month in which it occurred - have also been quoted. For some species, particularly more numerous ones, this is likely to result in double counting where birds have moved between sites.

Annual indices

Because the same sites are not necessarily covered by WeBS on every month in every year, relative changes in waterbird numbers cannot be determined simply by comparing the total number of birds counted each year (Tables 3 and 4). This issue is addressed by using indexing techniques that have been developed to track relative changes in numbers from incomplete data.

In summary, for occasions when a particular site has not been visited, an expected count for each species is calculated (imputed) based on the pattern of counts across months, years and other sites. This effectively means that a complete set of counts are available for all years and all months for a sample of sites. Only sites that have a good overall level of coverage are used (at least 50% of possible visits undertaken) and the underlying assumption is that the pattern of change in numbers across these sites (the index) is representative of the pattern of change in numbers at the country level (see *Interpretation of Waterbird Counts* below). Annual index values are expressed relative to the most recent year, which takes an arbitrary value of 100.

The 'Underhill index' was specifically developed for waterbird populations (see Underhill 1989, Prŷs-Jones *et al.* 1994, Underhill & Prŷs-Jones 1994 and Kirby *et al.* 1995 for a full explanation of this indexing process and its application for WeBS data). This report uses Generalized Additive Models (GAMs; Hastie & Tibshirani 1990) to fit both index values and a smoothed trend to the WeBS count data (see Maclean *et al.* 2005 for a full explanation of this process and its application for WeBS data) whilst retaining elements from the Underhill method that allows the assessment of whether or not counts flagged as incomplete should be treated as missing data. The generated smoothed trends are less influenced by years of abnormally high or low numbers and sampling 'noise' than are the raw index values. This makes them especially useful when assessing changes through time (*e.g.* WeBS Alerts; Maclean *et al.* 2005). Months used for indexing are assigned in a species-specific manner following established recommendations (Underhill & Prŷs-Jones 1994 and Kirby *et al.* 1995).

Not all species are included in the indexing process. Gulls and terns are excluded because counting of these species is optional. Species that occur substantially on habitat not well monitored by WeBS (*e.g.* Moorhen and Snipe) are excluded as are species that occur at sites sporadically and/or in small numbers (*e.g.* Bean Goose and Smew).

The periods of years for which indices are calculated have been revised slightly in the light of recent analyses. Data for wildfowl continue to be presented for the period 1966/67 to the present. Data from 1974/75 onwards have been used for waders as a high proportion of counts before this winter were imputed. For species added later to the scheme, (*i.e.* Great Crested Grebe and Coot in 1982/83, Little Grebe in 1985/86, Cormorant in 1986/87 and gulls, terns, divers, rare grebes and other species from 1993/94), data from the first two years following their inclusion have been omitted from indices, as initial take-up by counters appears not to have been complete, resulting in apparent sharp increases in numbers during this time. For similar reasons the first two years of data have been excluded from Northern Ireland indices.

Index values, where calculated, are graphed within each account. The underlying trend, where calculated, is shown using a broken line. The actual index values used to produce the graphs in this report can be obtained on request from the British Trust for Ornithology (see *Contacts*).

Monthly indices

The abundance of different waterbird species varies during the winter due to a number of factors, most notably the timing of their movements along the flyway, whilst severe weather, particularly on the continent, may also affect numbers in the UK. However, due to differences in site coverage between months, such patterns cannot be reliably detected using count totals. Consequently, an index is calculated for each month to reflect changes in relative abundance during the season.

The imputing process used to derive missing data for generating annual trends also allows monthly indices to be calculated across the same suite of sites. This reveals patterns of seasonality for the species considered. These are presented as graphs in the species accounts, giving the value for the most recent winter and the average value and range over the five preceding winters. Monthly graphs are not presented for the goose species for which annual indices are based on censuses as data for these are available for a limited number of months only.

Broad differences in the monthly values between species reflect their status in the UK. Resident species, or those with large UK breeding populations, *e.g.* some grebes and Mallard, are present in large numbers early in the winter. Declines through the winter result in part from mortality of first year birds, but also birds returning to remote or small breeding sites that are not covered by WeBS. The majority of UK waterbirds either occur solely as winter visitors, or have small breeding populations that are swelled by winter immigrants, with peak abundance generally occurring in mid winter.

The vast majority of the wintering populations of many wader species are found on estuaries, and, since coverage of this habitat is relatively complete and more or less constant throughout winter, meaningful comparisons of total monthly counts can be made for many species.

Site importance

Criteria for assessing the international importance of wetlands have been agreed by the Contracting Parties to the Ramsar Convention on Wetlands of International Importance (Ramsar Convention Bureau 1988). Under criterion 6, a wetland is considered internationally important if it regularly supports 1% of the individuals in a population of one species or subspecies of waterbird, whilst any site regularly supporting 20,000 or more waterbirds qualifies under criterion 5. Similar criteria have been adopted for identification of SPAs under the EC Birds Directive in the UK legislation. A wetland in Britain is considered nationally important if it regularly holds 1% or more of the estimated British numbers of one species or subspecies of waterbird, and in Northern Ireland, important in an all-Ireland context if it holds 1% or more of the all-Ireland estimate. More detailed information about SPAs and Ramsar sites in the UK can be accessed via the JNCC website at http://www.jncc.gov.uk/page-4. There are currently 251 SPAs and 146 Ramsar sites in the UK.

Population estimates are revised once every three years, in keeping with internationally agreed timetables (Rose & Stroud 1994). International estimates used in this report follow recent revisions of international populations (Wetlands International 2002) and of estimates for Great Britain (Kershaw & Cranswick 2003, Rehfisch *et al.* 2003). The relevant 1% thresholds are given in Appendix 1. and are also listed at the start of each individual species account. (It should be noted that the estimates and thresholds for some species or populations which should be the same at an international and national level because all birds are found in Britain, *e.g.* for Pink-footed Goose, differ slightly because of the rounding conventions applied. In most species accounts, these differences have been rationalised and only one or other of the estimates used).

For some species (*e.g.* Lapwing) no national thresholds are available and arbitrary levels have been used to compile the table of sites, the chosen level being given in the sub-heading of the table. Passage thresholds, applied to counts of some wader species in Great Britain, are also listed.

'National threshold' is used as a generic term to imply the 1% British threshold for sites in Great Britain, and the all-Ireland threshold for sites in Northern Ireland. Similarly, the term 'national importance' implies sites in Great Britain and in Northern Ireland that meet the respective thresholds.

Tables in the Species accounts rank the principal sites for each species according to the mean of annual maxima for the last five years (the five-year peak mean), in line with recommendations of the Ramsar Convention, and identify those meeting national and international qualifying levels (see also *Interpretation of Waterbird Counts*). For each site, the maximum count in each of the five most recent years, the month of occurrence of the peak in the most recent year, and the five-year peak mean are given. Incomplete counts are bracketed.

In accounts for most wildfowl, divers, grebes, Cormorant, herons, gulls, terns and Kingfisher, annual maxima are derived from any month in the appropriate 12-month period (see *The WeBS Year*). Average maxima for sites listed in the wader accounts that are based on a 'winter' year are calculated using data from only the winter period, November to March. Data from other sources, often involving different methods, *e.g.* goose roost censuses, are used where these provide better, *i.e.* larger, counts for individual sites. The source of all counts, if not derived from WeBS Core Counts, is indicated using a superscripted number after the count (a list of sources is given at the beginning of the accounts).

In the first instance, five-year peak means are calculated using only complete counts; incomplete counts are not used if they depress the mean count. Incomplete counts are, however, included in the calculation of the mean if they raise the value of the mean. Where all annual maxima are incomplete, the five-year peak mean is the highest of these individual counts. Averages enclosed by brackets are based solely on incomplete counts.

Sites are selected for presentation using a strict interpretation of the 1% threshold (for convenience, sites in the Channel Islands and Isle of Man are identified using 1% thresholds for Great Britain and included under the Great Britain section of the tables). For some species with very small national populations, and consequently very low 1% thresholds, an arbitrary, higher level has been chosen for the inclusion of sites. Where no thresholds are given, *e.g.* for introduced species, and where no or very few sites in the UK reach the relevant national qualifying levels, an arbitrary threshold has been chosen to select a list of sites for this report. These adopted thresholds are given in the sub-headings of the table. A blank line has been inserted in the table to separate sites that qualify as nationally important from those with five-year peak mean counts of less than 50 birds.

All sites that held numbers exceeding the relevant national threshold (or adopted qualifying level) in the most recent year, but with five-year peak means below this value, are listed separately. This serves to highlight important sites worthy of continued close attention.

For a number of wader species, where different thresholds exist for passage periods, the peak count during this period and month of occurrence are also listed. This list includes all those sites with counts above the relevant threshold, even if already listed in the main part of the table by virtue of the five-year winter peak mean attaining the national threshold.

Where the importance of a site has changed since the previous *Waterbirds in the UK* (prior to 2004/05 published as *Wildfowl and Wader Counts*) as a result of the data collected since then - *i.e.* it has become nationally or internationally important but was not following the previous year, or it has changed from international to national importance or *vice versa* - this is indicated in the table to the right of the five-year peak mean. Sites with elevated status have a black triangle pointing up (▲) to the right of the average, whilst those with lowered status are indicated using a triangle pointing down (▼). Sites for which the average fell below the threshold for national importance following 2003/04 are listed at the end of the table.

It should be noted that a site may appear to have been flagged erroneously as having elevated status if the most recent count was below the relevant threshold. However, a particularly low count six years previously will have depressed the mean in the previous report. The converse may be true for sites with lowered status and thus, in exceptional circumstances, a site may be listed in the relevant sections of the table as both no longer being of national importance yet also with a peak count in the most recent year exceeding the national threshold.

WeBS Alerts

WeBS Alerts have been developed to provide a standardised method of measuring and reporting on changes in wintering waterbird numbers at different temporal and spatial scales using WeBS data. Generalized Additive Models (GAMs) are used to fit smoothed trends to annual population indices (changes in population size calculated using these smoothed values are less susceptible to the effects of short-term fluctuations in population size or to errors when sampling than are results produced using raw data plots). Alerts are triggered for populations that have undergone major declines, and are intended to help identify where research into causes of decline may be needed and inform conservation management.

Proportional changes in the smoothed index value of a population over short- (5-year), medium- (10-year) and long- (25-year) term time frames are categorised according to their magnitude and direction. Population declines of between 25% and 50% trigger Medium Alerts and declines of greater than 50% trigger High Alerts. Increases of 33% and 100% (values chosen to be those necessary to return a population to its former size following declines of 25% and 50% respectively) are also identified, albeit that these are rarely of conservation concern.

National Alerts are generated for species (or specific populations of a species) using data from across the WeBS site network, for Great Britain and the constituent countries of the UK (Maclean *et al.* 2005). Alerts status for Great Britain and Northern Ireland are given in the header information of the species accounts. These Alerts provide some context for understanding finer scale changes in numbers. Alerts are calculated only for native species for which WeBS annual indices are calculated. Alerts are not available for some species over long time periods because there were only relatively recently included in WeBS Core Counts. Full results from the latest Alerts report are available for download from http://www.bto.org/survey/webs/webs-alerts-index.htm.

Principal sites

In addition to the assessment of sites against 1% thresholds in Species accounts, sites are identified for their importance in terms of overall waterbird numbers in the section *Principal Sites*. The peak count at each site is calculated by summing the individual species maxima during the season, irrespective of the month in which they occurred, or whether counts were complete or not. Data from all sources used for site assessment within the species accounts are used here, including wader numbers during passage periods. Non-native introduced or escaped species (*i.e.* those not in BOURC category A) are not included in these totals.

Counts made using methodologies that employ different site definitions to those used by WeBS (*e.g.* seaducks on the Moray Firth) are not incorporated into the calculations. Such sites are, however, listed at the end of the table.

INTERPRETATION OF WATERBIRD COUNTS

Caution is always necessary in the interpretation and application of waterbird counts given the limitations of these data. This is especially true of the summary form, which, by necessity, is used in this report. A primary aim here remains the rapid feedback of key results to the many participants in the WeBS scheme. More detailed information on how to make use of the data for research or site assessment purposes can be obtained from the British Trust for Ornithology (see *Contacts*).

Whilst the manner of presentation is consistent within this report, information collated by WeBS and other surveys can be held or used in a variety of ways. Data may also be summarised and analysed differently depending on the requirements of the user. Consequently, calculations used to interpret data and their presentation may vary between this and other publications, and indeed between organisations or individual users. The terminology used by different organisations may not always highlight these differences. This particularly applies to summary data. Such variations do not detract from the value of each different method, but offer greater choice to users according to the different questions being addressed. This should always be borne in mind when using data presented here.

For ease of reference, the caveats provided below are broadly categorised according to the presentation of results for each of the key objectives of WeBS. Several points, however, are general in nature and apply to a broad range of uses of the data.

National totals

The majority of count data are collected between September and March, when most species of waterbird are present in the UK in highest numbers. Data are collected during other months and have been presented where relevant. Caution is urged, however, regarding their interpretation both due to the relative sparsity of counts from this period and the different count effort for different sites. Data are presented for the months July to June inclusive (see *The WeBS Year*), matching the period for which data are provided *en masse* by counters.

A number of systematic biases of WeBS or other count methodology must be borne in mind when considering the data. Coverage of estuarine habitats and large, standing waters by WeBS is good or excellent. Consequently, counted totals of those species which occur wholly or primarily on these habitats during winter will approach a census. Those species dispersed widely over rivers, non-estuarine coast or small inland waters are, however, likely to be considerably under-represented, as will secretive or cryptic species, such as snipes, or those which occur on non-wetlands, *e.g.* grassland plovers. Species which occur in large numbers during passage are also likely to be under-represented, not only because of poorer coverage at this time, but due to the high turnover of birds in a short period. Furthermore, since counts of gulls and terns are optional, national totals are likely to be considerable underestimates of the number using the WeBS network of sites. Only for a handful of species, primarily geese, can count totals be considered as a census.

One instance of possible over-estimation may occur if using summed site maxima as a guide to the total number of scarcer species. For species with mobile flocks in an area well covered by WeBS, *e.g.* Snow Goose in southeast England, it is likely that a degree of double counting will occur, particularly if birds move between sites at different times of the year.

The publication of records of vagrants in this report does not imply acceptance by the British Birds Rarities Committee (*e.g.* Rogers and the Rarities Committee 2004).

Annual indices

For most species, the long-term trends in index values can be used to assess changes in overall wintering numbers with confidence. However, the comments above concerning the differential coverage of different habitats remain important. For some species, a substantial proportion of wintering birds occur away from those sites monitored by the WeBS Core Count scheme or use these sites at certain times of day that make them unlike to be encountered by WeBS counters. Consequently, this incomplete coverage needs to be borne in mind when interpreting the indices for some species. The proportion of some of these species being monitored by the WeBS Core Count scheme can be quantified and biases understood by comparison to other surveys. For example, from the Non-estuarine Coastal Waterbird Survey (NEWS) it is known that WeBS Core Counts monitor between one quarter and one half of wintering Ringed Plover, Purple Sandpiper, Sanderling and Turnstone and that the indices and trends reported will be biased towards changes occurring on estuaries. Similarly, trends reported for seaduck and grassland plovers will be biased towards changes occurring within estuaries although in these species the proportion of overall numbers monitored by WeBS Core counts is less well understood. In the case of winter swans, although the sites on which they occur are generally well monitored by WeBS Core counts they are mainly used as roost sites by the birds and therefore changes in the birds' daily routine with weather or local feeding opportunities may have considerable influence on whether they are present during the WeBS count and thus affect the reported indices and trends.

Indices and trends for Pink-footed Goose, Greenland White-fronted Goose, Icelandic Greylag Goose and Svalbard Barnacle Goose can be considered to be especially representative of national patterns. The numbers of these species are not well monitored by monthly WeBS Core Counts but rather are preferentially monitored by the annual coordinated censuses that cover the majority of British wintering birds. Indices for strictly or principally estuarine species (*e.g.* Wigeon and Knot) can also be considered especially representative as over 90% of British estuaries, including all major sites, are counted each month between September and March. Similarly, species that occur principally on larger inland waterbodies (*e.g.* Pochard and Goldeneye) are well monitored by WeBS Core Counts although the proportion of the numbers not being monitored is largely unquantified. For these species the indices and trends reported can be considered representative of the national pattern. For more widespread species (*e.g.* Mallard, Tufted Duck and Curlew) a large proportion of birds occur at small inland sites and habitats not well monitored by WeBS Core Counts. The selection of such sites follows no formal sampling pattern and therefore it is unclear as to whether these wetlands are a representative sample of the country as a whole.

Because short-term fluctuations provide a less rigorous indication of population changes, care should be taken in their interpretation. The underlying trend, denoted by the smoothed line in the annual index graphs, will give a better overall impression of trends for species with marked inter-annual variation, although it should be noted that unusually high or low index values in the most recent year will have a disproportionate effect on the trend at that point.

Caution should be used in interpreting figures for species that only occur in small numbers. Thus, numbers tend to fluctuate more widely for many species in Northern Ireland, largely as a result of the smaller numbers of birds involved but also, being at the western most limit of their range, due to variable use being made of Ireland by wintering waterbirds.

It should be borne in mind that the imputed values, used in place of missing and incomplete counts, are calculated anew each year, as in the completeness calculation for 'complex sites' which may cause the same count to change from complete to incomplete or *vice versa* with the addition of a new year's data. Because the index formula uses data from all years, each new year's counts will slightly alter the site, month and year factors. In turn, the assessment of missing counts may differ slightly and, as a result, the index values produced each year are likely to differ from those published in the previous *Waterbirds in the UK* (prior to 2004/05 published as *Wildfowl and Wader Counts*). Additionally, data submitted too late for inclusion are subsequently added to the dataset. The indices published here represent an improvement on previous figures as the additional year's data allow calculation of the site, month and year factors with greater confidence.

Monthly indices

As for annual indices, the reduced numbers of both sites and birds in Northern Ireland result in a greater degree of fluctuation in numbers used in the analyses of data from the province.

Site definition

To compare count data from year to year requires that the individual sites - in terms of the area surveyed - remain the same. The boundary of many wetlands are readily defined by the extent of habitat (*e.g.* for reservoirs and gravel pits), but are less obvious for other sites (*e.g.* some large estuaries) and here count boundaries have often been defined over time by a number of factors to a greater or lesser degree, including the distribution of birds at the time of the count, known movements of birds from roost to feeding areas, the extent of habitat, and even ease of access.

Sites are defined for a variety of purposes, and the precise boundary of sites describing ostensibly the same wetland may differ accordingly. For example, the boundaries used to define a large lake may differ for its definition as a wetland (based on habitat), as a waterbird count area (some birds may use adjacent non-wetland habitat), and as a statutorily designated site for nature conservation (which may be constrained by the need to follow boundaries easily demarcated in planning and legal terms). It should be recognised that the boundary of a site for counting may even differ between different waterbird surveys, particularly where different methodologies are employed, *e.g.* the 'Forth Estuary' comprises one large site for WeBS Core Counts, a slightly different area for Low Tide Counts, and two roost sites for Pink-footed Geese.

Data from different waterbird surveys have been used for assessment of site importance in this report if collected for ostensibly the same site, and are unlikely to cause significant discrepancies in the vast majority of cases (though see *Site importance*).

Particular caution is urged, however, in noting that, owing to possible boundary differences, totals given for WeBS or other sites in this report are not necessarily the same as totals for designated statutory sites (ASSIs/SSSIs, SPAs or Ramsar Sites) having the same or similar names.

It should also be borne in mind that whilst discrete wetlands may represent obvious sites for waterbirds, there is no strict definition of a site as an ecological unit for birds. Thus, some wetlands may provide all needs - feeding, loafing and roosting areas - for some species, but a 'site' for other species may comprise a variety of disparate areas, not all of which are counted for WeBS. Similarly, for some habitats, particularly linear areas such as rivers and rocky coasts, and marine areas, the definition of a site as used by waterbirds is not readily discerned without extensive survey or research that is usually beyond the scope of WeBS or other similar surveys. The definitions of such sites may thus evolve, and therefore change between *Waterbirds in the UK* (prior to 2004/05 published as *Wildfowl and Wader Counts*). Further, the number of birds recorded by WeBS at particular sites should not be taken to indicate the total number of birds in that local area.

In some cases, for example where feeding geese are recorded by daytime WeBS Core Counts over large sites, and again at discrete roosts within or adjacent to that same site, data are presented for both sites in the table of key sites given the very different nature or extent of the sites and often number of birds, even though the same birds will be counted at both.

Site importance

Sites are selected for presentation in this report using a strict interpretation of the 1% threshold. It should be noted, however that where 1% of the national population is less than 50 birds, 50 is normally used as a minimum qualifying threshold for the designation of sites of national importance. It should also be noted that the 'qualifying levels' used for introduced species are used purely as a guide for presentation of sites in this report and do not infer any conservation importance for the species or the sites concerned since protected sites would not be identified for these non-native birds.

It is necessary to bear in mind the distinction between sites that regularly hold wintering populations of national or international importance and those which may happen to exceed the appropriate qualifying levels only in occasional winters. This follows the Ramsar Convention, which states that key sites must be identified on the basis of demonstrated regular use (calculated as the mean winter maxima from the last five seasons for most species in this report), otherwise a large number of sites might qualify as a consequence of irregular visitation by one-off large numbers of waterbirds. However, the Convention also indicates that provisional assessments may be made on the basis of a minimum of three years' data. These rules of thumb are applied to SPAs and national assessments also. Sites with just one or two years' data are also included in the tables if the mean exceeds the relevant threshold for completeness but this does not, as such, imply qualification. This caveat applies also to sites that are counted in more than two years but, because one or more of the peak counts are incomplete, whose means surpass the 1% threshold based on counts from only one or two years.

Nevertheless, sites which irregularly support nationally or internationally important numbers may be extremely important at certain times, *e.g.* when the UK population is high, during the main migratory periods, or during cold weather, when they may act as refuges for birds away from traditionally used sites. For this reason also, the ranking of sites according to the total numbers of birds they support (particularly in *Principal Sites*) should not be taken as a rank order of the conservation importance of these sites, since certain sites, perhaps low down in terms of their total 'average' numbers, may nevertheless be of critical importance to certain species or populations at particular times.

Peak counts derived from a number of visits to a particular site in a given season will reflect more accurately the relative importance of the site for the species than do single visits. It is important to bear this in mind since, despite considerable improvements in coverage, data for a few sites presented in this report derive from single counts in some years. Similarly, in assessing the importance of a site, peak counts from several winters should ideally be used, as the peak count made in any one year may be unreliable due to gaps in coverage and disturbance- or weather-induced effects. The short-term movement of birds between closely adjacent sites may lead to altered assessments of a site's apparent importance for a particular species. More frequent counts than the once-monthly WeBS visits are necessary to assess more accurately the rapid turnover of waterbird populations that occurs during migration or cold weather movements.

It should also be borne in mind that because a count is considered complete for WeBS, it does not imply that it fully represents the importance of the site. A site of importance for a wintering species may have been counted only in autumn or spring, and thus while a valid complete count is available for that year, it under-represents the importance of the site for that species. This problem is overcome to some extent by the selection of counts from a limited winter window for wader species, although this will also tend to underestimation of the mean if it excludes large counts at other times of year. A similar issue arises for counts derived from different survey methods. For example, many sites important as gull roosts are identified on the basis of evening roost counts. Valid and complete counts may have been made by WeBS Core Counts during daytime over the course of a particular winter but, if no roost counts were made, the mean will be depressed by the much lower Core Count in that year. Thus, when counts appear to fluctuate greatly between years at individual sites on the basis of data from different sources - particularly for geese and gulls in the absence of roost counts, and for seaducks in the absence of dedicated survey - the five-year means and apparent trends over time should be viewed with caution.

Caution is also urged regarding the use of Low Tide Count data in site assessment. Whilst this survey serves to highlight the importance of some estuaries for feeding birds that, because they roost on other sites, are missed by Core Counts, the objectives of Low Tide Counts do not require strict synchronisation across the site and this may result in double counting of birds on some occasions. It should also be noted that count completeness assessments are not made for Low Tide Count totals at complex sites, and any undercounts from this scheme are not flagged in the tables, leading to under-estimation of the site's importance.

This list of potential sources of error in counting wetland birds, though not exhaustive, suggests that the net effect tends towards under- rather than over-estimation of numbers and provides justification for the use of maximum counts for the assessment of site importance or the size of a population. Factors causing under-estimation are normally constant at a given site in a given month, so that while under-estimates may occur, comparisons between sites and years remain valid.

It should be recognised that, in presenting sites supporting nationally or internationally important numbers of birds, this report provides just one means of identifying important sites and does not provide a definitive statement on the conservation value of individual sites for waterbirds, let alone other conservation interests. The national thresholds have been chosen to provide a reasonable amount of information in the context of this report only. Thus, for example, many sites of regional importance or those of importance because of the assemblage of species present are not included here. European Directives and conservation Conventions stress the need for a holistic approach to effect successful conservation, and lay great importance on maintaining the distribution and range of species, in addition to the conservation of networks of individual key sites.

For the above reasons of poor coverage, geographically or temporally, outlined above, it should be recognised that lists of sites supporting internationally and nationally important numbers of birds are limited by the availability of WeBS and other survey data. Whilst the counter network is likely to cover the vast majority of important sites, others may be missed and therefore will not be listed in the tables due to lack of appropriate data.

Some counts in this report differ from those presented previously; this results from the submission of late data and corrections, and in some cases, the use of different count seasons or changes to site structures. Additionally, some sites may have been omitted from tables previously due to oversight. It is likely that small changes will continue as definitions of sites are revised, in the light of new information from counters. Most changes are minor, but comment is made in the text where they are significant.

Note that sites listed under 'Sites no longer meeting table qualifying levels' represent those that would have been noted of national importance based on the preceding five years (*i.e.* 1999/2000 to 2003/04) but which, following the 2004/05 counts, no longer met the relevant threshold. It is not an exhaustive list of sites, which at any time in the past have been of national or all-Ireland importance.

COVERAGE

WeBS Core Counts

Coordinated, synchronous counts are advocated to prevent double counting or birds being missed. Consequently, priority dates are recommended nationally. Due to differences in tidal regimes around the country, counts at a few estuaries were made on other dates to match the most suitable conditions. Weather and counter availability also result in some counts being made on alternative dates.

Table 2. WeBS Core Count priority dates in 2004/05

25 April	12 December
23 May	16 January
20 June	13 February
18 July	13 March
22 August	10 April
19 September	08 May
17 October	26 June
14 November	

Standard Core Counts were received from 2,106 sites of all habitats for the period July 2004 to June 2005, comprising 3,336 count sectors (the sub-divisions of large sites for which separate counts are provided).

WeBS and I-WeBS coverage in 2004/05 is shown by in Figure 1. The location of each count sector is shown using only its central grid reference. The region and grid reference of all sites mentioned by name in this report are given in Table A2. in Appendix 2. Principal core sites are shown in Figure A1. in Appendix 2.

As ever, areas with few wetlands (*e.g.* inland Essex/Suffolk) or small human populations (*e.g.* much of Scotland) are apparent on the map as areas with little coverage. Northwest Scotland is usually poorly covered, although in 2004/05 this area was covered by surveys by the RAF Ornithological Society, which are reported upon in this report. Northern Ireland remains relatively uncovered aware from the major sites and further volunteers from here, or indeed anywhere in the UK, are always welcome.

Swan censuses

The International Swan Census, aimed to estimate population sizes and identify important wintering sites for Bewick's and Whooper Swans, took place in January 2005.

Goose censuses

In 2004/05, supplementary counts of Bean Geese were submitted by the Bean Goose Action Group (Slamannan Plateau) (Simpson & Maciver 2005) and the RSPB (Middle Yare Marshes). National surveys of Pink-footed and Icelandic Greylag Geese (the Icelandic-breeding Goose Census) were undertaken at roosts in October and November 2004 (Rowell 2005). A census of the Northwest Scotland Greylag Goose population on the Uists was made in August 2004 and February 2005 (R MacDonald *in litt.*), and counts of this population at other key sites (e.g. Tiree) were timed to coincide with these counts. Censuses of Greenland White-fronted Geese were carried out in autumn 2004 and spring 2005 by the Greenland White-fronted Goose Study (Fox & Francis 2005). Greenland Barnacle Geese were counted regularly by SNH and others on Islay and other key locations (SNH data) whilst the Svalbard Barnacle Geese on the Solway were counted regularly by WWT staff and volunteers (Griffin & Mackley 2004). Data were also provided by the All-Ireland Light-bellied Brent Goose census (Colhoun *pers. comm.*).

Seaduck surveys

Coastal counts of seaduck, divers and grebes were received from several sites. Aerial and/or shore-based counts from Orkney, the Hebrides, Aberdeen coast, Tay Estuary, St Andrews Bay were provided by JNCC (Dean *et al.* 2004). Aerial surveys were undertaken by WWT of extensive areas for DTI, windfarm developers and DEFRA/Country Agencies, in the Irish Sea from Anglesey to the Solway, and in the North Sea from Flamborough Head (Yorkshire) to Kent. Continuing surveys of the Moray Firth were carried out between November 2004 and January 2005 (RSPB Scotland/Talisman Energy (UK) Ltd). Monthly aerial and/or land-based counts of Common

Scoter in Carmarthen Bay were carried out between November 2004 and March 2005 (Banks *et al.* 2005). Continuing counts of key sites around Shetland were provided by SOTEAG (Heubeck and Mellor 2005). Continuing survey of the Eiders of the wider Firth of Clyde area were carried out in August and September 2004 (Waltho *pers. comm.*).

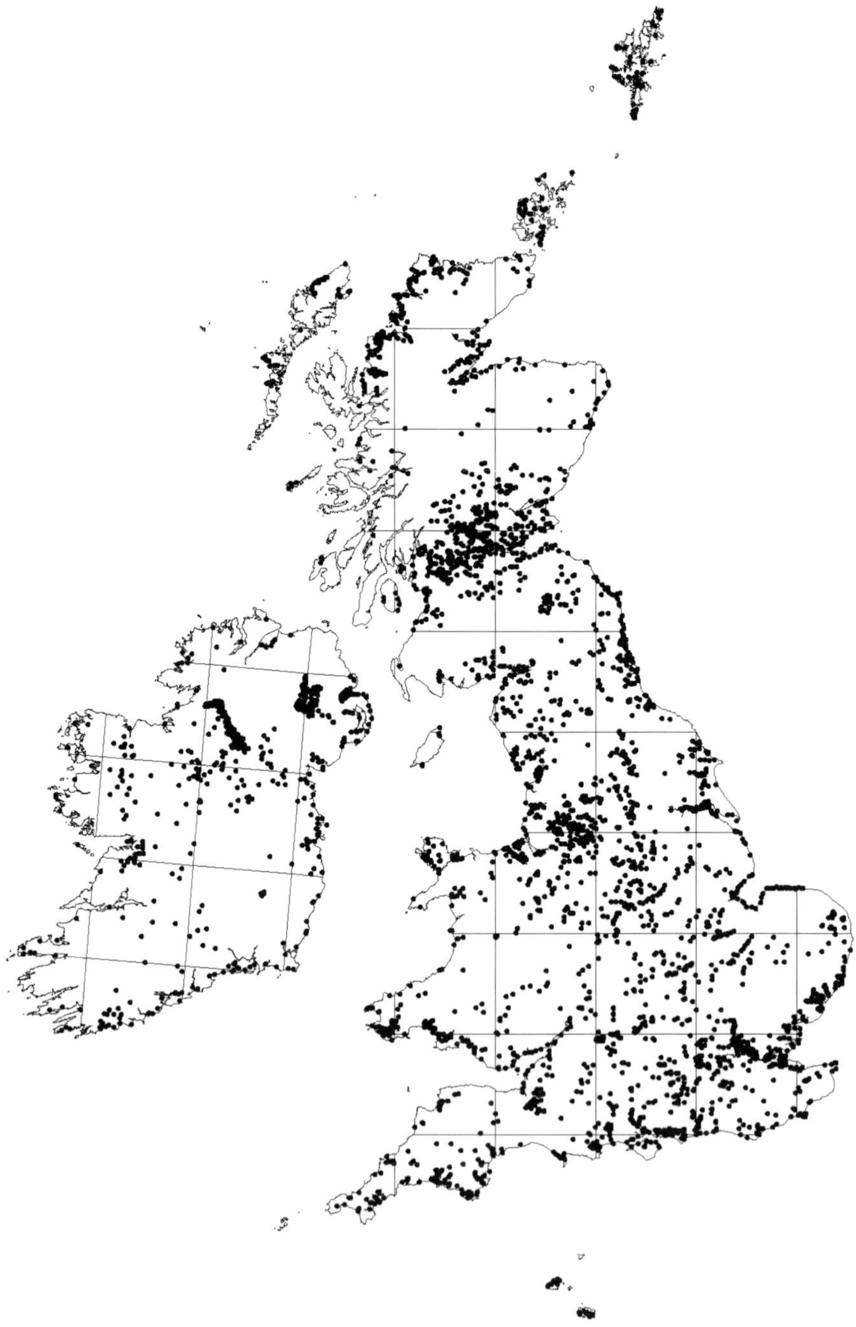

Figure 1. Position of all locations counted for standard WeBS and I-WeBS counts between July 2004 and June 2005.

The total numbers of waterbirds recorded by WeBS in 2004/05 are given in Tables 3 and 4 for Great Britain (including the Isle of Man, but excluding the Channel Islands) and Northern Ireland, respectively. Counts of waterbirds in the Republic of Ireland by I-WeBS are provided in Table 5.

Site coverage for gulls and terns is given separately since recording of these species was optional.

Totals for April to June 2004 are given in Tables A3 and A4 in Appendix 3.

Introduced and escaped waterbirds

Many species of waterbird occur in the UK as a result of introductions, particularly through escapes from collections. Several have become established, such as Canada Goose and Ruddy Duck. The British Ornithologists' Union Records Committee recently established a category 'E' for 'species that have been recorded as introductions, transportees or escapes from captivity, and whose breeding populations (if any) are not thought to be self-sustaining' (BOURC 1999). WeBS records of these species are included in this report both for the sake of completeness and in order to assess their status and monitor any changes in numbers, a key requirement given the need, under the African-Eurasian Waterbird Agreement of the Bonn convention '. . . to prevent the unintentional release of such species . . .' and once introduced, the need '. . . to prevent these species from becoming a threat to indigenous species' (Holmes *et al.* 1998).

Numbers of established populations (*e.g.* Canada Goose and Ruddy Duck, which are placed in category 'C') are excluded from Figure 2 below since the large numbers involved would swamp numbers of other species. Additionally, species that occur in both categories A and E (*e.g.* Pink-footed Goose) are also excluded since separation of escaped from wild birds is not readily possible using WeBS methods.

A total of 21 category E species were recorded in 2004/05 at 171 sites, both somewhat lower than the previous few years. The summed site maxima of 391 birds was the lowest during the past five years. As usual over half the total was attributable to Black Swan and Muscovy Duck, however, numbers of Bar-headed Goose were their lowest during the past five-years.

Although this figure will undoubtedly include some duplication of individual birds recorded at more than one site and occasional records of pinioned birds, this figure probably provides a truer reflection of the numbers of introduced or escaped waterbirds frequenting WeBS sites than the peak monthly total of 167 birds in November 2004.

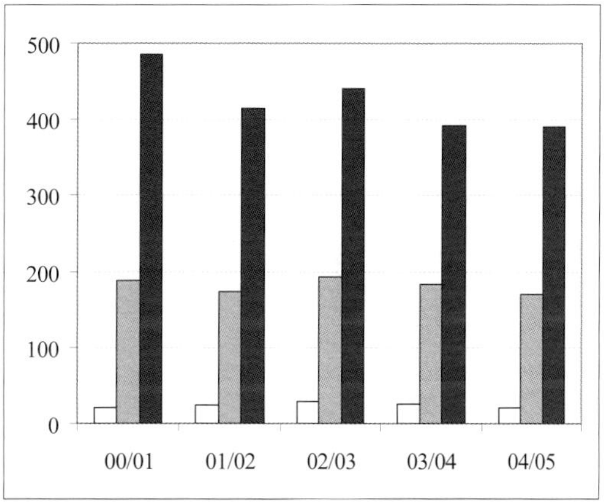

Figure 2. Number of species (white bars), number of sites at which birds were recorded (grey bars) and summed site maxima (black bars) for waterbirds in the BOURC's category E.

Table 3. Total numbers of waterbirds recorded by WeBS Core Counts in Great Britain in 2004/05. Census total are indicated by '*'.

	Species	Jul	Aug	Sep	Oct	Nov
Number of sites visited		*752*	*800*	*1453*	*1589*	*1636*
MS	Mute Swan	10912	11686	17772	20585	20355
AS	Black Swan	30	41	40	42	41
AT	Black-necked Swan	0	0	0	4	0
BS	Bewick's Swan	0	0	2	148	2520
WS	Whooper Swan	25	20	62	2330	6250
HN	Chinese Goose	3	8	9	12	25
BE	Bean Goose	0	0	0	254	493
PG	Pink-footed Goose	24	26	10216	271934	251948
WG	White-fronted Goose	0	0	0	1	0
EW	European White-fronted Goose	2	0	9	41	214
NW	Greenland White-fronted Goose	0	2	1	289	579
LC	Lesser White-fronted Goose	1	0	1	0	0
JI	Icelandic Greylag Goose	0	0	0	35334	85582
JH	NW Scotland Greylag Goose	1548	*8516	1002	394	506
JE	Re-established Greylag Goose	12198	16476	28120	24346	25482
HD	Bar-headed Goose	8	4	20	15	14
SJ	Snow Goose	3	3	15	14	19
RJ	Ross's Goose	2	1	1	1	0
EM	Emperor Goose	6	8	21	21	19
CG	Greater Canada Goose	27707	33963	54324	56609	54379
ZE	Canada Goose x Bar-headed Goose	0	0	0	0	0
ZI	Canada Goose x domesticated Greylag	2	0	0	1	0
LQ	Lesser Canada Goose	0	0	0	1	1
YE	Naturalised Barnacle Goose	141	267	520	844	779
YN	Nearctic Barnacle Goose	5	5	4	72	*40572
YS	Svalbard Barnacle Goose	106	9	310	*24335	*21382
ZH	Barnacle Goose x Canada Goose	0	0	2	1	0
ZG	Barnacle Goose x Greylag Goose	0	0	0	0	0
BG	Brent Goose	0	0	1	0	0
DB	Dark-bellied Brent Goose	24	51	200	33896	62324
QN	Nearctic Light-bellied Brent Goose	0	0	423	46	184
QS	Svalbard Light-bellied Brent Goose	0	0	1576	1947	1830
BB	Black Brant	0	0	0	0	1
EB	Red-breasted Goose	2	2	2	0	3
ZL	Feral/hybrid Goose	156	175	680	749	681
ZM	Hybrid goose	0	0	0	1	1
EG	Egyptian Goose	272	183	442	387	159
UB	Paradise Shelduck	1	1	1	1	1
UE	Cape Shelduck	1	0	1	1	0
UD	Ruddy Shelduck	2	9	10	3	0
SU	Shelduck	32002	26513	41675	57071	50550
ZT	Hybrid shelduck	0	0	0	0	0
MY	Muscovy Duck	24	23	49	49	55
DC	Wood Duck	1	0	2	5	6
MN	Mandarin	195	158	293	373	293
WN	Wigeon	259	1814	43103	250074	346552
AW	American Wigeon	0	0	0	0	2
HL	Chiloe Wigeon	1	2	0	0	0
GA	Gadwall	1810	3846	9500	10325	12739
T.	Teal	1113	14075	69424	126637	117934
TA	Green-winged Teal	0	0	0	0	3
KQ	Speckled Teal	0	0	2	0	2
MA	Mallard	45027	64187	106190	124379	125101
ZF	Feral/hybrid mallard type	294	342	500	446	545
QB	Chestnut Teal	0	0	0	0	1
PT	Pintail	11	91	6186	20336	20930
PN	Bahama Pintail	0	0	1	0	0
GY	Garganey	5	37	53	10	0
TB	Blue-winged Teal	0	0	0	0	0

Table 3. continued

	Dec	Jan	Feb	Mar	Apr	May	Jun
sites	*1664*	*1774*	*1728*	*1685*	*686*	*583*	*566*
MS	20955	19708	17149	16469	7694	6494	7897
AS	37	39	26	39	28	39	22
AT	0	0	0	0	0	0	0
BS	5692	6673	4826	342	3	0	0
WS	8088	7889	7989	3993	405	23	15
HN	17	14	12	7	5	0	1
BE	549	424	438	35	3	0	1
PG	99982	104750	53852	46925	29957	2000	42
WG	13	0	0	0	0	0	0
EW	1176	1913	1983	528	1	0	0
NW	*14079	616	386	*14030	84	0	0
LC	0	0	0	0	0	0	0
JI	23950	20440	20710	19975	1618	0	0
JH	660	1164	*8498	1038	76	51	13
JE	23252	20320	16584	13067	5528	5358	9558
HD	5	12	9	7	6	6	6
SJ	33	22	32	34	17	7	14
RJ	0	0	1	1	0	0	0
EM	15	18	8	13	4	9	0
CG	53428	52549	42602	30709	11806	8446	21741
ZE	0	0	0	0	1	0	0
ZI	5	4	0	0	3	5	11
LQ	1	1	0	0	0	0	0
YE	763	724	521	334	151	67	127
YN	727	476	923	*48966	17	1	0
YS	*28356	*25876	*24852	*20169	16291	50	95
ZH	0	1	1	0	0	0	1
ZG	0	0	0	0	0	2	3
BG	2	0	0	0	0	0	0
DB	65282	85469	69271	53153	14960	9525	26
QN	316	81	147	84	0	0	0
QS	1989	1424	330	609	50	0	1
BB	2	2	3	2	0	0	0
EB	3	4	4	4	2	0	0
ZL	746	807	529	566	155	147	176
ZM	0	0	0	0	1	0	0
EG	154	145	132	107	87	114	192
UB	1	1	1	1	0	0	0
UE	0	0	0	0	0	0	0
UD	1	2	2	0	2	0	0
SU	49876	53776	44561	41442	19691	12444	17801
ZT	0	1	1	0	0	0	0
MY	57	70	57	46	29	27	19
DC	2	3	4	7	4	3	3
MN	271	298	306	275	90	64	34
WN	376743	359920	271479	213932	11354	451	222
AW	2	2	0	0	0	0	1
HL	0	0	0	0	0	0	1
GA	14649	15639	13137	9151	3512	1601	1811
T.	125808	151875	96640	65791	13764	442	719
TA	3	5	7	3	2	0	0
KQ	5	0	2	0	0	0	0
MA	140112	128387	84895	63639	24709	18326	26792
ZF	602	435	447	389	180	128	280
QB	0	0	0	0	0	0	1
PT	24042	24745	18333	9173	1104	62	8
PN	14	0	0	0	2	1	0
GY	0	1	0	0	41	15	3
TB	0	0	0	0	0	1	0

Table 3. continued

	Species	Jul	Aug	Sep	Oct	Nov
Number of sites visited		*752*	*800*	*1453*	*1589*	*1636*
SV	Shoveler	370	2581	8397	10785	9940
IE	Ringed Teal	0	0	0	0	0
RQ	Red-crested Pochard	13	24	152	175	135
PO	Pochard	2418	5627	6704	18315	23116
AZ	Redhead	0	0	0	1	1
NG	Ring-necked Duck	1	1	1	1	4
NZ	New Zealand Scaup	0	0	1	0	1
FD	Ferruginous Duck	0	1	1	0	0
TU	Tufted Duck	19808	28917	41308	56338	51997
SP	Scaup	3	25	48	1062	344
AY	Lesser Scaup	0	0	0	0	0
ZD	Aythya hybrid	1	1	2	2	2
E.	Eider	13544	15646	16026	13282	9948
KE	King Eider	0	0	0	0	0
LN	Long-tailed Duck	0	0	0	510	597
CX	Common Scoter	3036	994	1773	7161	11640
DX	Black Scoter	0	0	0	0	0
FS	Surf Scoter	0	0	0	0	0
VS	Velvet Scoter	107	266	159	640	1153
UX	Unidentified scoter	0	0	0	0	0
VH	Bufflehead	0	0	0	1	1
GN	Goldeneye	96	158	205	2092	7568
SY	Smew	1	0	0	1	11
RM	Red-breasted Merganser	622	746	898	2634	3149
GD	Goosander	897	700	769	1315	1692
RY	Ruddy Duck	634	1193	2469	2922	3636
OI	Argentine Bluebill	0	0	1	1	0
WQ	White-headed Duck	0	0	1	1	0
RH	Red-throated Diver	35	57	155	395	233
BV	Black-throated Diver	2	2	4	9	22
ND	Great Northern Diver	0	0	1	31	58
LG	Little Grebe	1200	2389	4531	4882	4561
GG	Great Crested Grebe	4356	6288	8495	9193	8712
RX	Red-necked Grebe	7	24	30	11	7
SZ	Slavonian Grebe	1	2	13	219	80
BN	Black-necked Grebe	32	36	46	59	49
UV	Unidentified grebe	0	0	0	0	0
CA	Cormorant	7397	10071	16612	17535	17192
SA	Shag	271	791	1223	2238	947
XU	Unidentified Cormorant/Shag	1	0	14	7	13
BI	Bittern	3	1	4	10	15
QH	Squacco Heron	0	0	0	1	0
ET	Little Egret	1149	2197	2780	2377	2031
HW	Great White Egret	0	1	0	2	1
H.	Grey Heron	2031	2453	3331	3767	3550
OR	White Stork	2	4	2	0	2
IS	Sacred Ibis	1	1	1	1	0
NB	Spoonbill	4	2	2	3	12
NA	African Spoonbill	0	0	0	1	1
WA	Water Rail	39	63	94	273	509
AK	Spotted Crake	0	2	2	0	0
MH	Moorhen	4565	5884	10897	12409	13431
CO	Coot	36754	55821	92795	105550	109032
AO	American Coot	0	0	0	0	0
AN	Crane	0	1	2	0	1
KF	Kingfisher	146	186	375	443	456
	TOTAL WILDFOWL	**233470**	**318746**	**613089**	**1340867**	**1491790**

Table 3. continued

	Dec	Jan	Feb	Mar	Apr	May	Jun
sites	*1664*	*1774*	*1728*	*1685*	*686*	*583*	*566*
SV	10442	11131	11141	10099	5170	625	375
IE	0	0	0	0	0	0	1
RQ	77	107	118	101	28	24	16
PO	27211	30553	23971	15298	1203	837	994
AZ	1	0	0	0	0	0	0
NG	6	8	4	4	2	0	0
NZ	0	0	0	0	0	0	0
FD	0	0	0	1	0	0	0
TU	54390	55151	49606	42890	16018	7250	7185
SP	7599	2446	1268	1213	113	45	8
AY	2	2	1	3	0	1	0
ZD	1	4	3	4	1	0	0
E.	18661	11603	12164	14544	9271	12242	8833
KE	1	1	0	1	0	0	0
LN	3586	3205	1322	693	165	322	0
CX	12197	3554	5907	3569	8841	5789	85
DX	0	0	0	1	0	0	0
FS	3	4	0	1	0	0	0
VS	2014	678	1117	1001	268	92	0
UX	0	2570	2500	0	0	0	0
VH	1	1	0	0	0	0	0
GN	11392	12946	12194	11031	1722	68	75
SY	127	170	146	84	3	1	0
RM	3232	3053	2530	3018	1416	382	435
GD	2650	2597	2429	2230	305	227	382
RY	3139	3856	3061	2511	993	507	404
OI	0	0	0	0	1	1	0
WQ	0	1	1	0	0	0	0
RH	345	210	188	303	108	68	6
BV	20	17	67	78	15	9	8
ND	59	77	138	82	17	12	0
LG	4125	3686	3099	3200	1090	694	671
GG	8660	7058	7304	7453	3131	2280	2264
RX	13	19	14	14	3	1	0
SZ	159	163	134	138	15	0	0
BN	80	51	51	52	40	32	30
UV	0	0	0	1	0	0	0
CA	16164	14555	13023	11991	5818	4029	3906
SA	1549	859	1229	435	231	99	104
XU	0	9	0	0	0	0	0
BI	26	16	18	11	6	3	0
QH	0	0	0	0	0	0	0
ET	1577	1397	1112	1162	772	551	692
HW	2	2	0	0	0	1	0
H.	3670	3329	3404	3172	1547	1251	1478
OR	3	2	2	2	0	0	0
IS	1	0	0	0	2	0	0
NB	4	9	7	9	10	9	7
NA	1	1	1	1	1	1	0
WA	502	359	316	341	96	35	41
AK	0	0	0	0	0	0	0
MH	13334	13464	11613	12223	4672	2711	2349
CO	106571	97403	69498	54318	19954	10538	18040
AO	1	0	0	0	0	0	0
AN	0	0	0	0	0	0	0
KF	355	285	219	248	68	56	66
	1365259	**1360124**	**1020782**	**818566**	**246553**	**116702**	**136092**

Table 3. continued

	Species	Jul	Aug	Sep	Oct	Nov
Number of sites visited		*752*	*800*	*1453*	*1589*	*1636*
OC	Oystercatcher	51791	161966	185119	202112	214842
IT	Black-winged Stilt	1	1	1	1	1
AV	Avocet	2326	2417	2451	5379	4281
TN	Stone-curlew	0	1	0	0	0
LP	Little Ringed Plover	203	60	10	2	3
RP	Ringed Plover	1645	18317	12115	10527	7006
KP	Kentish Plover	0	0	0	0	0
DO	Dotterel	0	0	0	1	0
GP	Golden Plover	5471	40538	36383	104189	99299
GV	Grey Plover	850	12056	30474	42690	26789
L.	Lapwing	33981	71964	60760	140494	166446
KN	Knot	19031	104363	156262	209500	265938
SS	Sanderling	1501	11836	11215	9679	7490
LX	Little Stint	0	37	66	50	9
TK	Temminck's Stint	0	0	0	0	0
WU	White-rumped Sandpiper	0	0	1	1	0
BP	Baird's Sandpiper	0	0	0	1	0
PP	Pectoral Sandpiper	4	0	4	1	0
CV	Curlew Sandpiper	12	181	316	39	3
PS	Purple Sandpiper	211	41	103	464	668
DN	Dunlin	31812	93803	86651	238706	307786
RU	Ruff	170	373	530	452	647
JS	Jack Snipe	0	0	5	91	144
SN	Snipe	178	1665	2425	5084	8167
WK	Woodcock	0	2	0	12	41
BW	Black-tailed Godwit	8519	23519	35666	23921	26866
BA	Bar-tailed Godwit	4592	11744	20677	26570	24009
WM	Whimbrel	652	1143	171	79	13
CU	Curlew	42146	64049	67678	65864	55557
DR	Spotted Redshank	49	116	121	129	97
RK	Redshank	24554	64629	78243	93619	79164
GK	Greenshank	592	1630	1621	572	382
LY	Lesser Yellowlegs	0	0	1	0	1
GE	Green Sandpiper	264	478	181	164	131
OD	Wood Sandpiper	2	72	2	0	0
CS	Common Sandpiper	841	1058	237	76	40
PQ	Spotted Sandpiper	0	0	1	0	0
TT	Turnstone	1060	7695	11229	13287	12562
WF	Wilson's Phalarope	0	0	0	0	0
NK	Red-necked Phalarope	1	1	0	0	0
PL	Grey Phalarope	0	0	0	1	0
	TOTAL WADERS	**232459**	**695755**	**800719**	**1193757**	**1308382**

Table 3. continued

	Dec	Jan	Feb	Mar	Apr	May	Jun
sites	*1664*	*1774*	*1728*	*1685*	*686*	*583*	*566*
OC	234868	214561	207041	135276	55267	32789	21273
IT	1	0	0	0	1	0	0
AV	4729	5760	3893	5203	2212	1485	1681
TN	0	0	0	0	0	2	0
LP	0	2	0	6	122	128	154
RP	6631	7662	6322	3793	3173	7545	806
KP	0	0	0	0	2	0	0
DO	0	0	0	0	0	2	0
GP	177717	208610	154960	66621	4836	117	128
GV	28883	32890	27270	26682	21793	21719	879
L.	345413	410308	264206	44588	5294	3452	10799
KN	244583	232627	227793	133041	67872	61088	15753
SS	6933	6320	6730	6554	9746	8765	303
LX	8	12	4	9	13	4	2
TK	0	0	0	0	0	2	1
WU	0	0	0	0	0	0	0
BP	0	0	0	0	0	0	0
PP	0	0	0	0	0	0	0
CV	0	0	0	5	2	6	1
PS	1042	1298	950	597	450	451	1
DN	332051	334093	315440	186708	56694	87951	3158
RU	632	884	874	695	205	40	52
JS	183	147	141	179	15	0	0
SN	7558	7732	6788	5837	996	89	59
WK	28	28	16	21	3	2	3
BW	21528	23879	21953	21224	19058	6022	3620
BA	30582	37497	31315	29524	4519	1830	340
WM	10	7	9	14	132	1790	52
CU	61524	75140	67378	53442	23804	2809	12228
DR	66	56	59	65	61	13	54
RK	75722	77058	72366	58002	32824	3734	4842
GK	267	317	247	257	134	81	54
LY	1	1	1	0	1	0	0
GE	96	98	94	109	41	5	62
OD	0	0	0	0	0	3	3
CS	43	35	26	30	87	239	193
PQ	0	0	0	0	0	0	0
TT	13007	12495	12120	10501	5329	2598	1007
WF	0	1	0	0	0	0	0
NK	0	0	0	0	0	0	1
PL	0	2	0	0	0	0	0
	1594106	**1689520**	**1428010**	**788983**	**314686**	**244761**	**77509**

Table 3. continued

	Species	Jul	Aug	Sep	Oct	Nov
Number of sites visited		*612*	*668*	*1142*	*1286*	*1308*
MU	Mediterranean Gull	97	136	107	131	64
LU	Little Gull	127	358	7038	37	1
ON	Bonaparte's Gull	0	0	1	0	0
BH	Black-headed Gull	58407	101690	144576	148348	167090
IN	Ring-billed Gull	0	1	2	2	2
CM	Common Gull	3419	13717	24307	52972	31305
LB	Lesser Black-backed Gull	45841	22990	16243	16735	10848
YM	Western Yellow-legged Gull	109	164	132	27	9
YC	Caspian Gull	0	1	1	1	2
YG	Unidentified Yellow-legged Gull	26	0	25	105	23
HG	Herring Gull	24592	36407	39729	33698	46665
IG	Iceland Gull	0	0	0	0	0
GZ	Glaucous Gull	0	0	0	0	1
GB	Great Black-backed Gull	2057	3454	4784	7539	7116
KI	Kittiwake	1100	514	1117	682	89
UU	Unidentified gull	5588	10594	3943	150	0
VU	Unidentified large gull	0	0	0	0	0
	TOTAL GULLS	**141363**	**190026**	**242005**	**260427**	**263215**

	Species	Jul	Aug	Sep	Oct	Nov
Number of sites visited		*628*	*668*	*1100*	*1206*	*1190*
AF	Little Tern	760	329	8	2	0
BJ	Black Tern	1	86	38	5	0
TE	Sandwich Tern	11877	5546	2353	71	2
CN	Common Tern	3430	5222	703	58	0
RS	Roseate Tern	1	39	0	0	0
AE	Arctic Tern	922	275	43	8	0
UI	Common/Arctic Tern	2	0	0	0	0
UT	Unidentified tern	2	0	0	0	0
	TOTAL TERNS	**16995**	**11497**	**3145**	**144**	**2**

Table 3. continued

	Dec	Jan	Feb	Mar	Apr	May	Jun
sites	*1322*	*1411*	*1321*	*1320*	*582*	*466*	*472*
MU	73	58	93	101	73	11	30
LU	1	12	7	3	24	34	24
ON	0	1	0	1	0	0	0
BH	165133	208015	187238	136117	38108	26507	33747
IN	2	4	0	3	0	0	1
CM	27207	44475	57817	44099	4952	1698	4366
LB	11462	11058	10270	10662	32523	30777	31798
YM	11	3	2	5	10	0	4
YC	3	2	3	0	0	0	0
YG	20	13	6	6	1	0	52
HG	53964	44282	88754	43084	31421	24756	24646
IG	0	12	25	12	2	1	0
GZ	0	9	21	5	1	0	0
GB	7007	7085	5441	3199	1831	1377	1611
KI	270	147	245	56	647	272	740
UU	157	45	50	0	620	262	417
VU	0	0	0	0	0	0	160
	265310	**315221**	**349972**	**237353**	**110213**	**85695**	**97596**

	Dec	Jan	Feb	Mar	Apr	May	Jun
sites	*1192*	*1257*	*1192*	*1163*	*540*	*447*	*435*
AF	0	0	0	0	1	320	642
BJ	0	0	0	0	0	0	0
TE	1	2	0	4	1155	2477	6535
CN	0	0	0	0	116	1861	3409
RS	0	0	0	0	0	0	3
AE	0	0	0	0	0	143	579
UI	0	0	0	0	0	6	5
UT	0	0	0	0	0	0	10
	1	**2**	**0**	**4**	**1272**	**4807**	**11183**

Table 4. Total numbers of waterbirds recorded by WeBS Core Counts in Northern Ireland in 2004/05. . Census total are indicated by '*'.

	Species	Jul	Aug	Sep	Oct	Nov
Number of sites visited		*3*	*3*	*11*	*18*	*23*
MS	Mute Swan	53	12	196	1048	1275
BS	Bewick's Swan	0	0	0	39	0
WS	Whooper Swan	1	0	20	1145	2002
PG	Pink-footed Goose	0	0	0	4	0
NW	Greenland White-fronted Goose	0	0	0	0	0
JE	Re-established Greylag Goose	12	0	77	107	182
CG	Greater Canada Goose	0	0	230	227	0
YE	Naturalised Barnacle Goose	0	0	0	0	0
QN	Nearctic Light-bellied Brent Goose	0	0	1022	*27988	1645
SU	Shelduck	15	28	652	1509	1639
WN	Wigeon	0	0	2071	9561	5834
GA	Gadwall	0	1	48	87	164
T.	Teal	0	0	1917	2212	5180
TA	Green-winged Teal	0	0	0	0	2
MA	Mallard	383	150	2803	6625	6694
PT	Pintail	0	0	113	134	364
SV	Shoveler	0	0	6	101	150
PO	Pochard	0	0	1	2346	1938
TU	Tufted Duck	0	0	182	2284	5200
SP	Scaup	0	0	0	239	2181
AY	Lesser Scaup	0	0	0	0	0
E.	Eider	431	3	1117	1502	846
LN	Long-tailed Duck	0	0	0	0	11
CX	Common Scoter	0	0	0	0	2
GN	Goldeneye	0	0	9	83	2869
SY	Smew	0	0	0	1	0
RM	Red-breasted Merganser	2	154	286	514	434
GD	Goosander	0	0	1	1	1
RY	Ruddy Duck	0	0	0	16	1
RH	Red-throated Diver	0	0	5	14	11
BV	Black-throated Diver	0	0	0	0	0
ND	Great Northern Diver	0	0	0	1	1
UL	Unidentified diver	0	0	0	1	0
LG	Little Grebe	7	0	127	619	686
GG	Great Crested Grebe	0	49	1546	1633	963
SZ	Slavonian Grebe	0	0	0	0	0
BN	Black-necked Grebe	0	0	0	0	1
CA	Cormorant	105	302	1024	2798	1703
SA	Shag	0	1	245	493	334
XU	Unidentified Cormorant/Shag	0	0	198	300	350
ET	Little Egret	0	5	6	0	3
H.	Grey Heron	60	28	207	405	316
WA	Water Rail	0	0	1	1	3
MH	Moorhen	8	0	38	197	146
CO	Coot	0	0	263	1896	2213
KF	Kingfisher	0	0	0	5	1
	TOTAL WILDFOWL	**1077**	**733**	**14411**	**39666**	**45345**

Table 4. continued

	Dec	Jan	Feb	Mar	Apr	May	Jun
sites	*23*	*26*	*21*	*22*	*2*	*1*	*1*
MS	1922	1868	1692	987	2	18	3
BS	0	10	0	0	0	0	0
WS	2311	2652	2706	2188	0	2	0
PG	1	2	0	5	0	0	0
NW	67	102	115	39	0	0	0
JE	208	530	375	1402	0	0	0
CG	191	622	513	18	0	0	0
YE	0	1	0	0	0	0	0
QN	1325	1021	670	1893	538	0	0
SU	5010	5620	3328	3350	239	92	56
WN	6889	4463	3863	4789	8	0	0
GA	182	78	105	125	0	0	0
T.	3622	5847	4281	2452	290	0	0
TA	0	0	0	1	0	0	0
MA	6318	4996	3640	2346	61	10	60
PT	76	351	330	117	0	0	0
SV	95	156	83	19	0	0	0
PO	3016	7108	7110	2886	0	0	0
TU	7799	8774	11536	7623	0	0	0
SP	3473	3036	5806	4605	0	0	0
AY	0	0	1	0	0	0	0
E.	1712	1026	1285	581	26	0	0
LN	8	15	12	14	0	0	0
CX	26	0	0	0	0	0	0
GN	2653	4386	5667	6151	4	0	0
SY	0	1	4	4	0	0	0
RM	304	285	298	478	33	0	0
GD	2	0	0	1	0	0	0
RY	18	9	33	14	0	0	0
RH	7	20	3	48	2	0	0
BV	0	1	0	0	0	0	0
ND	1	6	2	8	0	0	0
UL	0	0	0	0	0	0	0
LG	673	419	442	243	0	0	0
GG	804	2065	841	844	20	0	0
SZ	0	4	2	10	0	0	0
BN	0	0	0	0	0	0	0
CA	1977	1267	1465	1300	43	9	10
SA	229	400	67	107	6	0	0
XU	136	0	127	0	40	0	0
ET	2	2	2	3	3	0	0
H.	319	258	241	130	11	1	3
WA	7	3	3	2	0	0	0
MH	171	239	236	190	1	0	0
CO	3758	2675	2861	1329	0	0	0
KF	3	2	1	2	0	0	0
	55315	**60320**	**59746**	**46304**	**1327**	**132**	**132**

Table 4. continued

	Species	Jul	Aug	Sep	Oct	Nov
Number of sites visited		*3*	*3*	*11*	*18*	*23*
OC	Oystercatcher	926	1468	12937	20366	14266
RP	Ringed Plover	53	197	420	747	597
KL	Killdeer	0	0	0	0	0
GP	Golden Plover	0	0	1796	6876	11637
GV	Grey Plover	0	0	37	87	977
L.	Lapwing	27	110	744	4023	13159
KN	Knot	1	7	669	508	2405
SS	Sanderling	4	20	11	10	11
CV	Curlew Sandpiper	0	0	1	1	0
PS	Purple Sandpiper	0	0	3	18	1
DN	Dunlin	176	693	1663	3126	8970
RU	Ruff	0	0	2	24	4
JS	Jack Snipe	0	0	1	0	0
SN	Snipe	2	5	15	91	135
BW	Black-tailed Godwit	2	116	1715	1462	782
BA	Bar-tailed Godwit	2	35	91	670	1189
WM	Whimbrel	20	4	2	4	1
CU	Curlew	1159	925	3881	5929	4612
RK	Redshank	132	1181	7640	10498	6959
GK	Greenshank	15	31	112	181	320
GE	Green Sandpiper	1	0	0	0	0
CS	Common Sandpiper	0	4	0	0	0
TT	Turnstone	0	112	731	1940	835
	TOTAL WADERS	**2520**	**4908**	**32471**	**56561**	**66860**

	Species	Jul	Aug	Sep	Oct	Nov
Number of sites visited		*3*	*2*	*10*	*12*	*14*
MU	Mediterranean Gull	0	0	3	3	3
LU	Little Gull	0	0	0	0	0
BH	Black-headed Gull	1199	551	5873	9241	11937
IN	Ring-billed Gull	0	0	0	0	2
CM	Common Gull	1005	254	2926	3210	4217
LB	Lesser Black-backed Gull	49	53	369	600	313
HG	Herring Gull	13	68	3236	4759	8499
IG	Iceland Gull	0	0	0	0	0
GZ	Glaucous Gull	0	0	0	0	0
GB	Great Black-backed Gull	11	14	342	519	477
KI	Kittiwake	0	0	7	0	0
	TOTAL GULLS	**2277**	**940**	**12756**	**18332**	**25448**

	Species	Jul	Aug	Sep	Oct	Nov
Number of sites visited		*3*	*3*	*9*	*10*	*8*
TE	Sandwich Tern	113	220	289	15	0
CN	Common Tern	0	0	10	0	0
	TOTAL TERNS	**113**	**220**	**299**	**15**	**0**

Table 4. continued

	Dec	Jan	Feb	Mar	Apr	May	Jun
sites	*23*	*26*	*21*	*22*	*2*	*1*	*1*
OC	14631	15772	12333	10849	423	34	40
RP	481	381	380	157	81	20	0
KL	0	0	1	0	0	0	0
GP	15296	12537	16252	13208	1500	0	0
GV	149	163	122	98	2	0	0
L.	21055	16487	11426	1158	0	0	0
KN	3122	1808	5919	1705	0	1	0
SS	56	16	12	9	282	6	0
CV	0	0	0	0	0	0	0
PS	0	38	4	86	0	0	0
DN	8266	6782	9185	5501	52	290	25
RU	2	5	0	5	0	0	0
JS	1	0	0	1	0	0	0
SN	103	57	47	91	0	0	0
BW	484	596	460	703	26	1	0
BA	1028	277	1494	894	6	0	0
WM	0	0	2	0	0	0	0
CU	4187	7653	5664	4535	124	1	51
RK	6484	6880	5844	7903	1035	1	4
GK	140	94	116	78	1	0	0
GE	0	0	0	0	0	0	0
CS	0	0	0	0	0	0	0
TT	759	1404	706	1882	368	0	0
	76244	**70950**	**69967**	**48863**	**3900**	**354**	**120**

	Dec	Jan	Feb	Mar	Apr	May	Jun
sites	*16*	*17*	*13*	*13*	*1*	*1*	*1*
MU	0	1	0	0	0	0	0
LU	0	1	0	0	0	0	0
BH	11822	11063	14218	10759	40	16	150
IN	1	0	1	0	0	0	0
CM	4853	3193	3623	3015	14	14	40
LB	186	92	347	375	0	3	50
HG	6990	8012	4057	3347	11	8	12
IG	2	11	8	4	0	0	0
GZ	4	1	15	2	0	0	0
GB	882	1250	930	1348	6	3	12
KI	0	0	0	0	0	0	0
	24740	**23624**	**23199**	**18850**	**71**	**44**	**264**

	Dec	Jan	Feb	Mar	Apr	May	Jun
sites	*9*	*10*	*8*	*9*	*1*	*1*	*1*
TE	0	7	0	1	6	30	20
CN	0	0	0	0	0	0	1
	0	**7**	**0**	**1**	**6**	**30**	**21**

Table 5. Total numbers of waterbirds recorded by I-WeBS in the Republic of Ireland in 2004/05.

Species	Sep	Oct	Nov	Dec	Jan	Feb	Mar
Number of sites visited	*98*	*114*	*161*	*138*	*233*	*159*	*119*
Mute Swan	943	1576	3565	2164	3659	1799	1153
Black Swan	0	0	1	1	0	0	0
Bewick's Swan	0	8	13	28	220	156	0
Whooper Swan	27	1431	5330	2924	5468	2331	2591
Pink-footed Goose	0	0	4	11	9	3	2
Greater Canada Goose	24	19	98	36	203	7	7
Greenland White-fronted Goose	8	401	7603	9909	10062	769	384
Greylag Goose	190	392	5766	2657	1732	681	2191
Snow Goose	0	0	0	1	1	0	0
Barnacle Goose	0	76	856	884	3508	727	1
Dark-bellied Brent Goose	0	0	0	0	0	1	0
Light-bellied Brent Goose	1008	3846	17346	15752	11542	13539	9313
Black Brant	0	0	1	0	0	0	0
Feral/hybrid Goose	26	20	0	18	17	25	24
Shelduck	246	2186	3760	5834	6618	5557	3031
Wigeon	2978	18221	37558	31801	47440	20047	7239
American Wigeon	1	1	1	0	0	1	0
Gadwall	92	168	358	244	616	207	95
Teal	2401	5751	14854	13748	23565	9936	5399
Green-winged Teal	0	0	2	0	0	1	0
Mallard	7277	7179	9459	7448	9559	4273	2547
Pintail	85	73	401	899	1024	694	171
Garganey	0	0	0	0	0	1	0
Shoveler	56	339	1368	1685	1444	1154	324
Pochard	5	296	9531	2201	6854	3683	414
Ring-necked Duck	0	0	0	2	1	0	1
Ferruginous Duck	0	0	0	0	0	1	0
Tufted Duck	404	1016	8007	3213	10256	2900	1294
Scaup	0	612	184	862	582	403	4
Eider	14	90	5	3	18	20	19
Long-tailed Duck	0	3	33	22	31	45	26
Common Scoter	2622	786	2991	1275	3000	2227	2236
Surf Scoter	0	1	3	0	0	3	0
Velvet Scoter	0	0	6	2	0	0	0
Goldeneye	1	78	803	805	1436	616	313
Smew	0	0		1	1	0	0
Red-breasted Merganser	240	439	804	436	761	452	576
Goosander	0	0	0	3	0	3	0
Ruddy Duck	0	2	0	1	2	2	0
Unidentified Duck	0	0	0	0	50	0	0
Red-throated Diver	128	27	195	266	196	103	112
Black-throated Diver	0	0	9	2	14	4	21
Great Northern Diver	3	42	226	82	443	178	389
Unidentified Diver	0	0	0	33	0	0	0
Little Grebe	344	419	609	533	688	333	171
Great Crested Grebe	186	531	771	500	987	474	160
Red-necked Grebe	0	0	0	0	1	0	0
Slavonian Grebe	0	1	0	6	4	3	10
Black-necked Grebe	0	0	1	1	0	0	0
Cormorant	2245	1876	2831	1532	2911	1546	1019
Unidentified Cormorant/Shag	198	300	350	136	0	127	0
Little Egret	334	169	202	157	206	192	124
Grey Heron	534	508	723	440	629	289	245
Water Rail	13	6	20	13	20	18	13
Moorhen	357	401	344	365	558	380	302
Coot	1541	3566	23144	8757	12073	2421	526
Kingfisher	10	15	14	8	10	9	9
TOTAL WILDFOWL	**24541**	**52871**	**160150**	**117701**	**168419**	**78341**	**42456**

Table 5. continued

Species	Sep	Oct	Nov	Dec	Jan	Feb	Mar
Oystercatcher	18264	23273	27293	19866	30158	22062	13365
Avocet	0	0	0	3	1	1	0
Ringed Plover	1743	2678	5071	2672	2493	1972	549
Golden Plover	2655	35837	86540	84160	111756	47875	25512
Grey Plover	171	445	1113	788	1547	1693	392
Lapwing	1095	8271	59046	53615	118985	25892	909
Knot	1864	4675	4870	12296	11463	6483	3073
Sanderling	102	493	1524	1917	1330	1877	794
Little Stint	0	0	1	0	0	0	0
Baird's Sandpiper	1	0	0	0	0	0	0
Pectoral Sandpiper	0	1	0	0	0	0	0
Curlew Sandpiper	31	4	0	0	2	0	0
Purple Sandpiper	0	2	71	13	61	24	66
Dunlin	4692	11258	38357	27977	46050	29379	6508
Ruff	13	20	3	686	456	3	2
Jack Snipe	0	1	17	23	24	19	23
Snipe	104	271	450	564	625	507	223
Woodcock	0	0	3	2	1	1	4
Black-tailed Godwit	8247	5270	11738	8745	9149	8678	5091
Bar-tailed Godwit	1152	4517	5831	5829	6480	5552	4846
Whimbrel	16	61	0	2	1	1	1
Curlew	8942	9570	16044	13615	23912	15669	6883
Spotted Redshank	2	4	3	4	4	2	0
Redshank	10594	11455	16795	10632	14836	10208	9396
Greenshank	395	383	505	378	496	402	274
Green Sandpiper	2	2	1	2	5	7	0
Wood Sandpiper	0	0	0	0	0	0	0
Common Sandpiper	6	3	3	2	2	1	0
Turnstone	1080	1536	2883	1698	2056	1805	1883
Grey Phalarope	1	30	0	0	0	0	0
TOTAL WADERS	**61172**	**120060**	**278162**	**245489**	**381893**	**180113**	**79794**

Species	Sep	Oct	Nov	Dec	Jan	Feb	Mar
Mediterranean Gull	32	16	16	27	15	22	12
Little Gull	2	1	5	0	0	0	2
Black-headed Gull	14771	13342	14906	14325	22796	15008	12207
Ring-billed Gull	1	1	3	3	2	3	1
Common Gull	4844	6445	7401	6339	9200	3463	7267
Lesser Black-backed Gull	2028	1167	873	1340	1812	366	811
Herring Gull	1549	1386	1945	1226	2785	1465	1117
Iceland Gull	0	0	0	2	10	14	4
Glaucous Gull	0	0	0	0	17	16	13
Great Black-backed Gull	1244	1354	1341	1197	1523	1176	740
Unidentified gull	0	0	0	69	0	0	45
Yellow-legged Gull	0	0	0	0	0	1	0
TOTAL GULLS	**24471**	**23712**	**26490**	**24528**	**38160**	**21534**	**22219**

Species	Sep	Oct	Nov	Dec	Jan	Feb	Mar
Sandwich Tern	2103	3	0	2	0	0	5
Common Tern	100	0	0	1	1	0	0
Roseate Tern	117	0	0	0	0	0	0
Arctic Tern	78	1	0	0	0	0	0
Unidentified Tern	248	0	0	0	0	0	0
TOTAL TERNS	**2646**	**4**	**0**	**3**	**1**	**0**	**5**

SPECIES ACCOUNTS

Key to symbols commonly used in the species accounts.

As footnotes to thresholds:

?	population size not accurately known
+	population too small for meaningful threshold
*	where 1% of the national population is fewer than 50 birds, 50 is normally used as a minimum threshold for national importance
**	a site regularly holding more than 20,000 waterbirds (excluding non-native species) qualifies as internationally important by virtue of absolute numbers
†	denotes that a qualifying level different to the national threshold has been used for the purposes of presenting sites in this report

In tables of important sites:

-	no data available
( )	incomplete count
†	same meaning as used for thresholds
▲	site was of a higher importance status in the previous five-year period
▼	site was of a lower importance status in the previous five-year period
1,2	count obtained using different survey methodology from WeBS Core Counts (see table below)

To denote WeBS alerts status in headers:

++	>100% increase
+	33-100% increase
o	stable trend
-	25-50% decrease
--	>50% decrease

S	short term (5 years)
M	medium term (10 years)
L	long term (25 years or maximum period if <25 years)

Sources of additional information used in compiling tables of important sites are listed below. Non-WeBS counts are identified in the tables by the relevant number below given in superscript following the count.

1 RSPB/Talisman Energy studies, *e.g.* Stenning (1998)
2 M. Howe (*in litt.*)
3 WWT studies, *e.g.* Rees *et al.* (2000)
4 Bean Goose Working Group, *e.g.* Smith *et al.* (1994)
5 RSPB *pers comm.*
6 Lancashire Goose Report, *e.g.* Forshaw (1998)
7 SNH 'adopted' counts
8 WWT data
9 Greenland White-fronted Goose Study, *e.g.* Fox and Francis (2004)
10 SOTEAG reports, *e.g.* Heubeck and Mellor (2005)
11 WeBS Low Tide Counts
12 Roost counts
13 Supplementary daytime counts
14 WWT/JNCC National Grey Goose Census
15 Firth of Clyde Eider counts, *e.g.* Waltho (2004)
16 R. Godfrey (*in litt.*)
17 Argyll Bird Report
18 SNH Greenland Goose Census
19 R. MacDonald (*in litt.*)
20 Little Egret Roost counts
21 D Carrington (*in litt.*)
22 P Reay (*in litt.*)
23 C Hartley (*in litt.*)
24 WWT unpublished data
25 Woolmer *et al.* 2001, Carmarthen Bay Common Scoter count
26 Dorset Bird Report
27 Judith Smith, Gr. Manchester County recorder
28 BTO/ Lucy Smith
29 Paul Daw, County recorder for Argyll
30 Steve Percival's counts of Lindisfarne – Svalbard Light-bellied Brents
31 JNCC report of aerial surveys for seaducks, divers
32 WWT report to DTI. Aerial survey of Greater Thames strategic area
33 WWT report to DTI. Aerial survey of Greater Wash strategic area
34 All Wales Common Scoter Survey. WWT reports to CCW
35 All-Ireland Light-bellied Brent Goose Census
36 Cormorant Roost Survey 2003
37 Scottish Bird Report records
38 Worden *et al.* 2004
39 RSPB data
40 SNH data
41 WWT UK-breeding Greylag Goose Survey
42 Frank Mawby (*in litt.*)
43 Shetland coordinated swan count
44 CSL supplementary Ruddy Duck counts
45 Winter Gull Roost Survey
46 BTO/CCW Carmarthen Bay surveys
47 KOS Great Crested Grebe records
48 B McMillan (*in litt.*)
49 C Langton (*in litt.*)
50 B Yates (*in litt.*)
51 Tiree non-estuarine counts, per J Bowler
52 A Stevenson (*in litt.*)
53 D Tate (*in litt.*)
54 Uist Greylag Goose Management Committee
55 Uists SPA wader survey (Ecology UK Ltd 2005)
56 P Wilson / Lancs Bird Report
57 W Aspin (*in litt.*)
58 Winter Swan Census
59 JNCC shore-based count

Mute Swan

Cygnus olor

International threshold (British population):	380
International threshold (Irish population):	100
Great Britain threshold:	375
All-Ireland threshold:	100

GB max:	20,955	Dec
NI max:	1,922	Dec

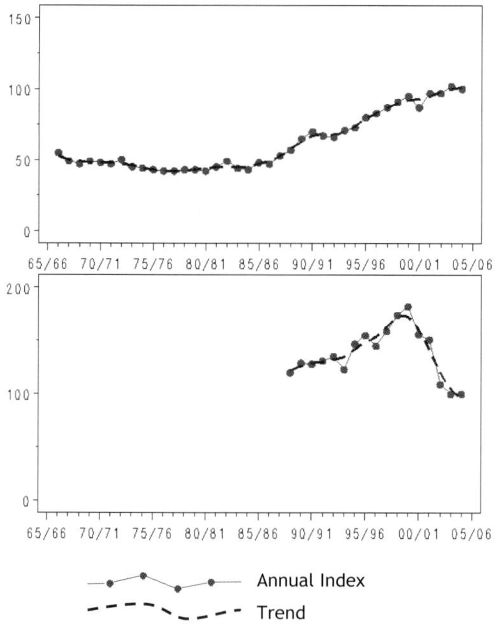

Figure 3.a, Annual indices & trend for Mute Swan for GB (above) & NI (below).

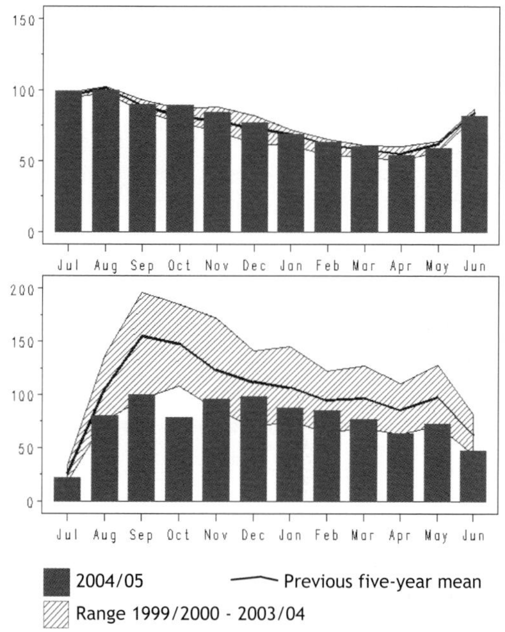

Figure 3.b, Monthly indices for Mute Swan for GB (above) & NI (below).

The Mute Swan is, for the general public, one of our most familiar waterbirds. Although easily detected during WeBS counts, this is a species that is widely dispersed throughout lowland wetland habitats and, consequently, many birds on smaller lakes and rivers go uncounted. The peak British count during 2004/05 was the highest ever recorded by WeBS but still represents just over half of the latest population estimate. The British annual index indicates that Mute Swans remain at a similar, high level to 2003/04 and about twice that seen up to the mid-1980s before the recent steady rise in numbers. In Northern Ireland, there were also similar numbers to 2003/04, but in this case these represent a large decline since the late 1990s. The decline in Northern Ireland is largely due to falling numbers at Loughs Neagh and Beg where peak counts have halved over five winters. I-WeBS data from the Republic of Ireland suggests this decline is not representative of Ireland as a whole (O. Crowe *pers. comm.*) Although British and Irish Mute Swans seem relatively secure, this species is one that featured in outbreaks of highly pathogenic avian influenza H5N1 in Western Europe in early 2006. Whether or not the species has a particular vulnerability to the disease remains to be seen.

Monthly indices for Britain peak in the late summer then drop steadily through the winter, the expected pattern for a resident species where there is no significant immigration. A similar pattern is seen in Northern Ireland, although the pattern is less clear here, due mostly to the smaller number of sites involved but perhaps also due to interchange with sites in the Republic of Ireland.

At most sites, peak counts remained broadly similar to those in previous years. The most noteworthy increase was seen at the Tweed Estuary where the peak was highest since 1998. As well as the aforementioned Loughs Neagh and Beg, relatively low counts were also noted from Loch of Harray, Loch Leven and Strangford Lough. It should be noted that Broad Water Canal, listed as a key site in previous reports, is now treated as part of Loughs Neagh and Beg.

A census of the British Mute Swan breeding population was undertaken in spring 2002, revealing an estimated 6,150 breeding pairs and a further 19,400 non-breeding individuals, totalling 31,700 birds at the start of the breeding season (Ward *et al.* in prep). This represents an increase of 23% over the period 1990-2002. The sites with the largest breeding season concentrations were Abbotsbury on the Fleet and at the Lochs of Harray and Stenness in Orkney.

	00/01	01/02	02/03	03/04	04/05	Mon	Mean
Sites of international importance in the UK							
Loughs Neagh and Beg	1,950	1,423	1,510	920	949	Nov	1,350
Fleet and Wey	1,150	1,228	1,368	1,092	1,118	Oct	1,191
Somerset Levels	1,110	1,084	(1,098)	(883)	1,076	Feb	1,092
Ouse Washes	726 [13]	1,110 [13]	782 [13]	606	693	Nov	783
Rutland Water	547	590	594	542	593	Aug	573
Loch of Harray	597	597	672	522	467	Jan	571
Tweed Estuary	575	464 [13]	414	582	614	Jul	530
Loch Leven	496	506	550	526	202	Sep	456
Hornsea Mere	346	217	486	527 [13]	455	Jul	406
Upper Lough Erne	445	306	323	272	449	Jan	359
Lower Lough Erne			199	(286)	(300)	Dec	262
Strangford Lough	174	183	180	193	94	Nov	165
Other sites surpassing table qualifying levels in WeBS-Year 2004/2005 in Great Britain							
Tring Reservoirs	306	282	447	322	404	Nov	352
Loch Bee (South Uist)	343	200	297	407	394	Dec	328
Severn Estuary	337	339	284	(318)	390	Dec	338
Other sites surpassing table qualifying levels in WeBS-Year 2004/2005 in Northern Ireland							
Upper Quoile River	32	117	(71)	108	108	Jan	91

Black Swan

Cygnus atratus

Escape†
Native Range: Australia

GB max:	**42**	**Oct**
NI max:	**0**	

The British maximum of 42 was unremarkable, being slightly lower than in recent years. Peak counts from single sites reached seven at both Arnot Park Lake and Fleet & Wey. The peak at the 2003/04 top site, Woburn Park Lakes, fell to just two birds. In total the species was noted at 85 sites, slightly more than in past years, whilst the summed site maxima rose again to 146.

Sites with three or more birds in 2004/05†		
Arnot Park Lake	7	Mar
Fleet and Wey	7	Oct, Mar-May
R. Kennet: R'bury-Chilton Foliat	6	May
Abberton Reservoir	4	Sep
Belvide Reservoir	4	Apr
Chilton Foliat	4	Nov
Lee Valley Gravel Pits	4	Dec
Walthamstow Reservoirs	4	Jul
Blatherwyke Lake	3	Mar
Fort Henry Ponds & Exton Park Lake	3	Feb
Pugneys Country Park Lakes	3	Jul
Thames Estuary	3	Aug

† *as no British or All-Ireland thresholds have been set a qualifying level of three has been chosen to select sites for presentation in this report*

Black-necked Swan

Cygnus melanocoryphus

Escape
Native Range: South America

GB max:	**4**	**Oct**
NI max:	**0**	

The only record of Black-necked Swans was of four at Ramsbury Lake in Wiltshire; this was the first ever record of this species during a WeBS count.

Bewick's Swan

Cygnus columbianus

International threshold (*bewickii*):	290	
Great Britain threshold:	81	
All-Ireland threshold:	25*	

GB max:	6,673	Jan
NI max:	39	Oct

	S	M	L
GB change	o	o	+
NI change	(--)	--	--

**50 is normally used as a minimum threshold*

% Young	12.3
Brood size	2.0

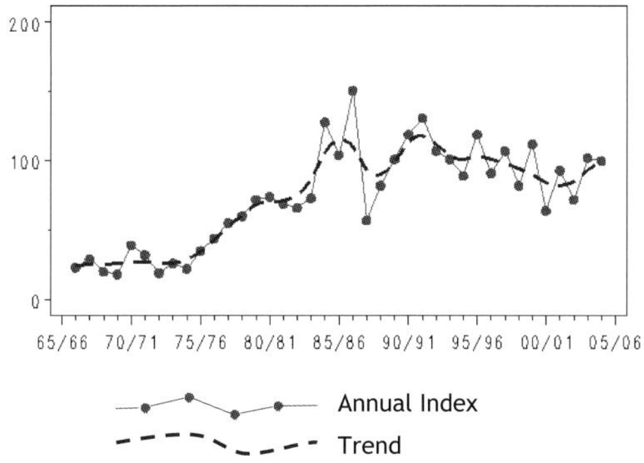

Figure 4.a, Annual indices & trend for Bewick's Swan for GB.

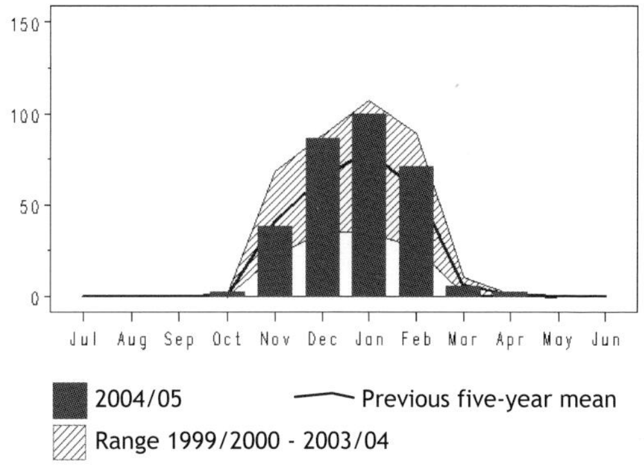

Figure 4.b, Monthly indices for Bewick's Swan for GB.

The British index remained similar to 2003/04, suggesting a stabilisation of numbers since the slow decline seen over the previous decade. Conversely, numbers of Bewick's Swans in Northern Ireland remain very low. It should be noted that the index plots published in Wildfowl and Wader Counts for 2001/02-02/03 and 2003/04 were erroneous as they included an incomplete set of roost counts from the Ouse Washes, Nene Washes and Martin Mere. The index plot now takes the roost counts from these sites into account and is deemed more representative of the true picture.

The peak roost count of 7,491 on the Ouse Washes on 30th January 2005 was the highest yet recorded at the site being over 1,000 higher than the previous peak, counted in January 2004. Indeed, this supplementary count was higher than the British total of 6,673 made around the recommended count date of 16th January 2005. This means that the Ouse Washes is by far the most important single site in northwest Europe for this species. Numbers on the Nene Washes were relatively low in 2004/05, perhaps involving local redistribution to the Ouse Washes. Peak counts at the Severn Estuary, Martin Mere and Ribble Estuary and the Somerset Levels, all in the west, were also low, whereas other sites in eastern England retained more normal numbers. A flock of 120 birds at Cantley in the Middle Yare Marshes was an interesting development at a site that generally holds only single figures, if any, but which is close to a number of other established flocks. The peak total in Northern Ireland was the highest since 1998/99 but was still only a fraction of that seen up to the mid-1990s. The majority of birds were at Loughs Neagh and Beg, with one record from Lough Foyle.

The international Bewick's Swan Census, conducted in mid January 2005, counted 7,216 individuals in Britain and Ireland combined (Worden *et al.* accepted). Analyses of data from previous censuses confirm the trends derived from WeBS annual indices. Although numbers have increased by 16% on those recorded during the first international Bewick's Swan census in 1984 (Beekman *et al.* 1985), numbers recorded in January 2005 were 4.6% lower than those found during the 1995 census.

Numbers of Bewick's Swans recorded during the international census in both Northern Ireland and the Republic of Ireland reflect the low numbers recorded by WeBS. Combined totals have decreased dramatically from 1,244 birds recorded in 1984, to just 224 birds in January 2005. This downturn in numbers appears to be part of a long term trend. Declines have also occurred in many regions of England with notable decreases in many counties. Britain and Ireland are located along the western edge of this population's wintering range and declines in the western-most regions such as Ireland, Southwest England and Northwest England suggest a contraction of the wintering range. Although annual fluctuations at the Ouse Washes occur, census results show a rising trend with a 7% increase between January 1995 and January 2005. Southeast England has also seen an increase in Bewick's Swan records, supporting the notion that an eastward retraction of range may be occurring.

Annual monitoring in the Netherlands show stable numbers in the north of the country but a downward trend in the west Netherlands since 1995 (van Roomen *et al.* 2004). Index values generated from monitoring data from 1992/93 to 2002/03 suggest a downward trend for the country as a whole (van Roomen *et al.* 2004). Analysis of data from across Europe is needed to determine whether declining numbers in parts of Britain and Ireland reflect a shift in winter distribution or a decrease in total population size for Bewick's Swans in Northwest Europe.

During winter 2004/05, British flocks of Bewick's Swans were assessed as containing 12.3% young, with an average brood size of 2.0 cygnets per successful pair. Overall, this suggests breeding success had been a little below average, with the proportion of young at the Ouse Washes (8.8%) particularly low.

	00/01	01/02	02/03	03/04	04/05	Mon	Mean
Sites of international importance in the UK							
Ouse Washes	4,693 [12]	5,735 [8]	5,177 [8]	6,330 [12]	7,491 [12]	Jan	5,885
Nene Washes	1,100	347 [12]	1,068	790	262 [12]	Dec	713
Sites of national importance in Great Britain							
Severn Estuary	272 [8]	310 [8]	345 [8]	230	223 [8]	Feb	276
Martin Mere and Ribble Estuary	322 [8]	296 [8]	315	221	175	Mar	266
St Benet`s Levels	(206)	147	287	280			238
Breydon Wtr & Berney Marshes	186	85	240	220	237	Jan	194
Old Romney				184 [13]			184
Hickling Broad	0				282 [58]	Jan	141 ▲
Confidential SE England Site	10	180	220	148	140	Jan	140
Dee Estuary (England & Wales)	(118)	(78)	(70)	(92)	(101)	Jan	(118)
Somerset Levels	(146)	108	(69)	(112)	21	Jan	91
Other sites surpassing table qualifying levels in WeBS-Year 2004/2005 in Great Britain							
Middle Yare Marshes	(6)	0	0	1	(120)	Feb	25
Other sites surpassing table qualifying levels in WeBS-Year 2004/2005 in Northern Ireland							
Loughs Neagh and Beg	15	19	6	5	(39)	Oct	17

Whooper Swan

Cygnus cygnus

International threshold:	210	
Great Britain threshold:	57	
All-Ireland threshold:	100	

GB max:	8,088	Dec
NI max:	2,706	Feb

% Young	16.9
Brood size	2.3

	S	M	L
GB change	+	+	+
NI change	o	+	o

The British Whooper Swan index dropped a little from the peak seen in 2003/04 but was still at its second-highest value to date, with counted numbers reaching a new peak. The annual index in Northern Ireland reached its highest value, albeit just slightly above that seen in 2002/03. As with Bewick's Swan, it should be noted that the index plots published in Wildfowl and Wader Counts for 2001/02-02/03 and 2003/04 were erroneous as they included an incomplete set of roost counts from the Ouse Washes, Nene Washes and Martin Mere. The index plots now take the roost counts from these sites into account.

The Ouse Washes remain the key site for this species and numbers here reached a new peak. In fact, the count of 4,397 at roost on 30th January 2005 was the highest ever recorded at a single UK site, and was nearly 800 more than the previous peak, at the same site a year earlier. Counts at Martin Mere and Ribble Estuary also reached a new peak, topping 2,000 birds for the first time. This was in contrast to Bewick's Swan, which is declining in Lancashire. Similarly, unlike the situation for Bewick's Swan, numbers of Whooper Swans at Loughs Neagh and Beg remain healthy. In general, numbers at other Northern Ireland sites were about average, although there were record counts at both Strangford Lough and Lough McNean Lower. In Britain, further high site counts were recorded at Loch a` Phuill, Monk Myre and Lawhill Oxbows, but numbers were low at Black Cart Water and birds were absent at Warkworth Lane Ponds and Mildam & Balfour Main Pools.

Many new sites in the table this year result from the international Whooper Swan census carried out in January 2005. The census recorded 15,062 swans in Britain and Northern Ireland – almost 30% higher than the WeBS peak annual count, demonstrating the importance of a full census to ensure that numbers at all sites are monitored. The census total (26,366 across the range) was the highest recorded to date. The numbers of birds recorded in Britain and Northern Ireland by WeBS annually mirror international census results, which suggest a sustained period of growth since the mid 1990s, with a 32% increase between 1995 and 2000 (Cranswick *et al.* 1997), and 26% between 2000 and 2005 (Worden *et al.* submitted).

Census results indicate a 55% increase in numbers in Britain between 2000 and 2005, compared with only 11% in Ireland. This increase in Britain was not accompanied by a significant increase in the number of flocks recorded, and the distribution remains concentrated at relatively few sites. In fact, the continuing increase in size of flocks using the Ouse Washes in Norfolk accounted for 82% of the rise in total numbers found in England.

Major increases in the Northwest European population of Whooper Swans (which breeds in Fenno-Scandia to northwest Russia and winters in continental Europe) since the 1980s (Laubek *et al.* 1999) have led to the suggestion that the increasing numbers of Whooper Swans in Britain, particularly in East Central England, may be inflated by an increasing proportion of birds from the former population wintering in Britain. A greater understanding of the extent of interchange between these populations, and use of British and Irish wintering grounds by the Northwest European population, is necessary to ascertain the true size of these growing populations.

In 2004/05, Whooper Swans in Britain and Ireland contained 16.9% young with successful pairs being accompanied by an average of 2.3 young per pair, although productivity varied between regions with lower proportions of young in eastern England and southwest Scotland. Overall productivity was fairly typical.

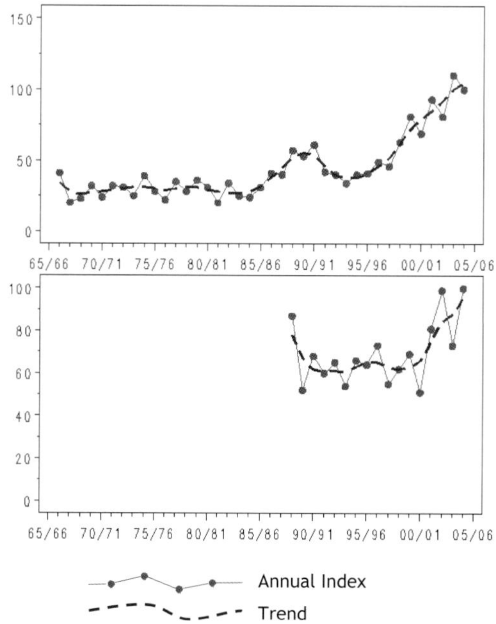

Figure 5.a, Annual indices & trend for Whooper Swan for GB (above) & NI (below).

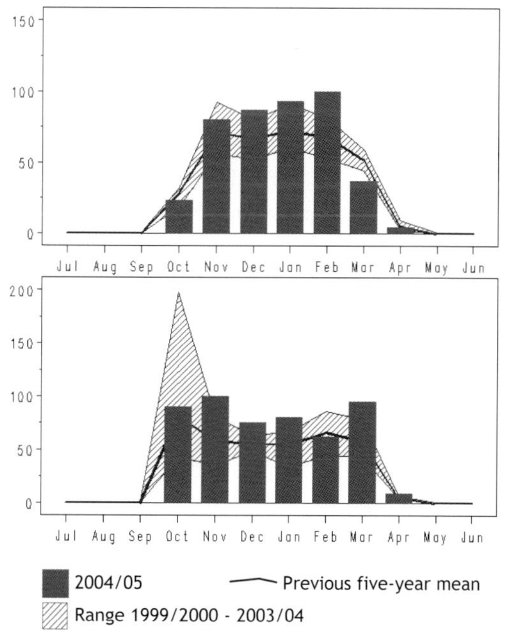

Figure 5.b, Monthly indices for Whooper Swan for GB (above) & NI (below).

	00/01	01/02	02/03	03/04	04/05	Mon	Mean
Sites of international importance in the UK							
Ouse Washes	1,797 [8]	2,894 [8]	2,745 [12]	3,624 [12]	4,397 [12]	Jan	3,091
Martin Mere and Ribble Estuary	1,650	1,762 [8]	1,770 [8]	1,597	2,081 [58]	Jan	1,772
Loughs Neagh and Beg	(735)	(1,532)	1,514	(867)	1,543	Mar	1,530
Lough Foyle	434	548	3,284	680	950 [58]	Jan	1,179
Upper Lough Erne	1,010	1,228 [13]	658	855	1,123	Feb	975
Loch of Strathbeg	424	(223)	(67)	794	355	Nov	524
Solway Estuary	466 [8]	(309)	340 [8]	(250)	508 [58]	Jan	438
Loans of Tullich					253 [58]	Jan	253 ▲
Loch Eye and Cromarty Firth	(39)	230	141	322	275	Oct	242 ▲
Loch of Wester	(86)	341			128	Mar	235 ▲
Bridge of Crathies					(220) [58]	Jan	(220) ▲
Sites of national importance in Great Britain							
Wigtown Bay	(110)	156	(135)	255	205	Mar	205
Black Cart Water	299	238 [3]	176 [3]	151 [3]	112	Dec	195 ▼
Norham West Mains					184 [58]	Jan	184 ▲
Lower Teviot Valley	(179)	(50)	(29)		(58)	Nov	(179)
R Clyde: Carstairs to Thankerton	142	242	(101)	91	110	Feb	146 ▼
R. Nith: Keltonbank to Nunholm	131	125	(108)	165	(104)	Feb	140
Killimster Loch		135					135
Leven Cut					125 [58]	Jan	125
Loch a` Phuill (Tiree)	36	83	168	118	194	Nov	120
Tarbat Ness			0	44	306	Dec	117 ▲
Nene Washes	111	110 [12]	143	111	104 [12]	Feb	116
Ravenstruther	347		0	48	56	Jan	113
Strathearn South Kinkell					111 [58]	Jan	111 ▲
Loch Insh and Spey Marshes	96	92	91	110	124	Feb	103
Inner Moray and Inverness Firth	1	(173)	60	165	(27)	Dec	100
Dornoch Firth	307	53 [11]	23	94	18	Dec	99
Caistron Quarry	71	71	67	164	96	Feb	94
Vasa Loch Shapinsay			68	96	119	Nov	94
East Fenton Farm Reservoir					89	Jan	89 ▲
Loch of Lintrathen	96	10	166	93	69	Nov	87
Kinnordy Loch		116	82	35	96	Feb	82
R. Tweed: Kelso to Coldstream	47	60	116	109	75	Jan	81
Broubster Leans					75 [58]	Jan	75 ▲
Loch of Spiggie	73	47	86	89	69	Nov	73
Lower Derwent Ings			91	52			72
Lindisfarne	45 [11]	15	(90)	(139)	71	Jan	72
Milldam and Balfour Mains Pools	112	98	41	86	0		67
Farmland near Monymusk					65 [58]	Jan	65 ▲
Loch Tuamister (Lewis)					63 [58]	Jan	63 ▲
Farmland near Whitekirk					61 [58]	Jan	61 ▲
Forth Estuary	(95)	(20)	(24)	62 [11]	19	Nov	59
St Benet`s Levels	(26)	6	58	108			57 ▲
Sites of all-Ireland importance in Northern Ireland							
Strangford Lough	220 [11]	212	191	150	244	Dec	203
Sites no longer meeting table qualifying levels in WeBS-Year 2004/2005							
Loch Leven	144	0	13	19	66	Oct	48
Clatto Reservoir	108	62	10	58	10	Dec	50
Barons Folly	107	3	4	126	5	Mar	49
Warkworth Lane Ponds	62	25	128	47	0		52
Threave Estate	74	21					48
Other sites surpassing table qualifying levels in WeBS-Year 2004/2005 in Great Britain							
Monk Myre	0	1	3	1	130	Dec	27
River Earn - Lawhill Oxbows	33	12	0	0	113	Nov	32
Loch Moraig	15	12	20	121	87	Dec	51
Loch Leven	144	0	13	19	66	Oct	48
Morecambe Bay	7	125	6	(20)	63	Dec	50
Loch Heilen				24	60	Oct	42
Other sites surpassing table qualifying levels in WeBS-Year 2004/2005 in Northern Ireland							
Lough McNean Lower	39	76 [13]			124	Feb	80

Whooper Swan (Mark Collier)

Chinese Goose

Anser cygnoides

Escape
Native Range: E Asia

GB max:	25	Nov
NI max:	0	

The British peak was double that of the previous year. The highest count at a single site was of nine at Harrow Lodge Park, where birds were present during September, November and December. The numbers at Diss Mere varied between two and seven and the species was present in eight months between July and March. Other sites at which the species was present in more than one month included Arran, Kirk Loch Lochmaben and River Derwent: Cromford to Matlock Bath.

Bean Goose

Anser fabalis

International threshold (*fabalis*):	1,000
Great Britain threshold:	4*
All-Ireland threshold:	+

GB max:	549	Dec
NI max:	0	

% Young	15.4
Brood size	2.0

**50 is normally used as a minimum threshold*

Whilst the two regular wintering sites for nominate (or Taiga) Bean Geese supported the largest numbers as usual, there was a substantial arrival of *rossicus* (or Tundra) Bean Geese during winter 2004/05. The highest numbers were present between December and February and birds were noted from over 30 different sites. The peak of 87 (*rossicus)* at the Ouse Washes was the highest away from the two key sites since 1995, with double-figure counts also from North Warren, Somerset Levels, Whitemoor Haye, Beaulieu Estuary, the North Norfolk Coast and a further site in southeast England (for where the land-owner has requested the name be with-held), a wide geographic spread. Still, the numbers are insignificant in the context of the estimated 600,000 *rossicus* wintering in continental Europe.

At the regular wintering sites, peak numbers remain low at the Middle Yare Marshes but increased to their highest on record at the Slamannan Plateau. Simpson & Maciver (2005) describe how the first birds arrived at the latter site on 2nd October, with most having arrived by late October. The geese remained mostly as a single flock into December after which time they split into different groups. The last to depart were seen on 15th February, an earlier departure than in recent years.

Productivity was assessed in November at Slamannan Plateau (15.6% young, 2.0 young per successful pair) and in December at the Middle Yare Marshes (15.2% young). However, ageing first-year geese can become increasingly difficult as the winter progresses and the late arrival of Bean Geese in Britain means that productivity measures may be underestimates.

	00/01	01/02	02/03	03/04	04/05	Mon	Mean	
Sites of national importance in Great Britain								
Slamannan Area	183 [4]	192 [4]	231 [4]	235 [4]	262 [4]	Nov	221	
Middle Yare Marshes	276	272 [5]	183 [5]	140	156 [39]	Dec	205	
Upper River Avon - Wester Jaw	187	0	0				62	
Ouse Washes	4 [12]	4 [13]	8 [13]	4	87 [12]	Jan	21	
Confidential SE England Site	0	0	0	0	86	Feb	17	▲
N. Warren & Thorpeness Mere		0	0	3 [13]	38	Jan	10	▲
Fleet and Wey	(0)	(26)	6	0	0		8	
Balnakeil Bay				5 [13]			5	
Somerset Levels	0	0	(0)	0	14	Dec	4	▲
Abberton Reservoir	0	22	0	0	0		4	
Whitemoor Haye	0	3	0	0	17	Jan	4	▲
Hule Moss	4 [13]						4	
Sites no longer meeting table qualifying levels in WeBS-Year 2004/2005								
Lower Derwent Ings				1			1	
Other sites surpassing table qualifying levels in WeBS-Year 2004/2005 in Great Britain								
Beaulieu Estuary	0	0	0	0	15	Dec	3	
North Norfolk Coast	0	0	1	0	10	Feb	2	
Nene Washes	0	0	0	0	8	Dec	2	
Walmore Common	0	0	0	0	8	Dec	2	
Swale Estuary	0	(0)	0	0	7	Dec	2	
The Wash	0	1	0	0	4	Nov	1	

Pink-footed Goose

Anser brachyrhynchus

International threshold: 2,400
Great Britain threshold: 2,400
All-Ireland threshold: +

GB max: 271,934 Oct
NI max: 5 Mar

% Young 19.4
Brood size 2.1

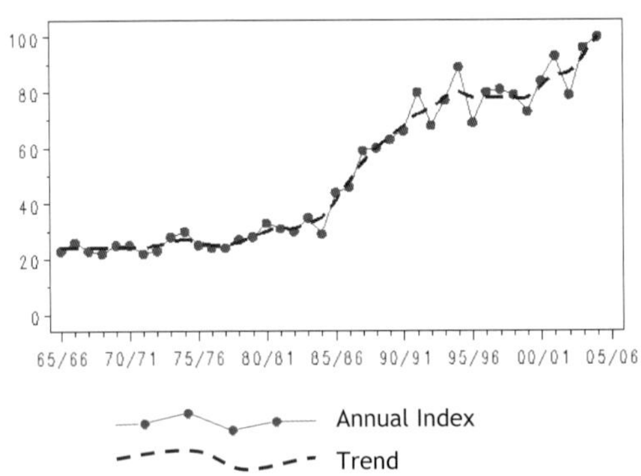

Figure 6.a, Annual indices & trend for Pink-footed Goose for GB.

The peak number of Pink-footed Geese counted by the Icelandic-breeding Goose Census was slightly less than in autumn 2003. However, by adjusting for the few sites where counts were missed (notably West Water Reservoir) the true population was estimated at 292,154 Pink-footed Geese in October 2004, which was 4% higher than the previous year. This increase followed a reasonably successful breeding season, with 19.4% young and an average of 2.1 young per successful pair (Rowell 2005).

The list of sites is again headed by Scolt Head in North Norfolk, where the count was down somewhat on the massive peak in 2003/04. The peak count at Holkham in Norfolk was made a week later than the Scolt peak and there is doubtless much interchange of birds, but simultaneous counts at the two roosts, as well as at Snettisham and Breydon Water on 13th December 2004 revealed a new Norfolk record count of 147,250 (J Scott *pers. comm.*). In Lancashire, the peak of nearly 44,000 birds was the highest to date. At the Loch of Strathbeg, numbers again peaked in September as in 2003; the only key site to see a peak in this month. Elsewhere in Scotland, higher than average counts were noted from Montrose Basin and, especially, Loch Spynie where the peak was more than twice the previous highest count at the site.

Given the national increase of the species, it is not surprising that there are only a few sites with apparently declining numbers, although there is an indication of a slow reduction in the numbers using Carsebreck & Rhynd Lochs, and the roost at Dupplin Lochs was absent on the night of the census. Conversely, looking at the sites at which high numbers were recorded in 2004/05, but which do not yet qualify as important on the basis of five-year means, both South Medwin Pools and East Chevington Pools appear to be showing patterns of sustained increase and could well qualify soon if such increases continue. The roost at Rossie Bog in November was also particularly notable. As usual, Pink-footed Geese were extremely scarce in Northern Ireland, with single figure counts from Loughs Neagh and Beg, Strangford Lough, Belfast Lough and Lough Foyle.

	00/01	01/02	02/03	03/04	04/05	Mon	Mean
Sites of international importance in the UK							
Scolt Head	41,000 [14]	33,900 [14]	62,500 [14]	80,000 [12]	66,000 [13]	Dec	56,680
Loch of Strathbeg	42,615 [14]	46,898 [14]	39,900	66,000 [14]	65,000 [14]	Sep	52,083
Holkham Bay	33,750 [14]	45,000 [14]	33,800 [14]	47,750 [14]	58,000 [13]	Dec	43,660
West Water Reservoir	26,500 [14]	23,270 [14]	(40,000) [14]	34,210 [14]			30,995
Snettisham	18,250 [14]	35,000 [14]	37,050 [14]	27,350 [14]	35,360 [13]	Dec	30,602
Southwest Lancashire	16,885 [14]	33,180 [14]	31,645 [14]	27,025 [14]	43,950 [6]	Jan	30,537
Montrose Basin	29,922 [14]	38,669 [14]	11,500 [14]	10,149 [14]	31,896 [14]	Oct	24,427
Ythan Estuary and Slains Lochs	23,500 [14]	13,900 [14]	19,600 [14]	19,200 [14]	16,200	Oct	18,480
Aberlady Bay	16,750 [14]	13,740 [14]	22,200 [14]	15,040 [14]	18,430 [14]	Oct	17,232
Morecambe Bay	7,143	14,100 [14]	14,600 [6]	17,050 [6]	26,910 [6]	Jan	15,961
Findhorn Bay	5,500 [14]	14,000 [14]		25,000 [14]	18,000 [14]	Oct	15,625
Loch Leven	14,700 [14]	16,200 [14]	(12,874) [14]	15,120 [14]	14,750	Oct	15,193
Loch Spynie	8,000 [14]	9,100 [14]	11,700 [14]	11,100 [14]	27,000 [14]	Nov	13,380
Loch of Skene	13,550 [12]	13,175 [14]	(8,420) [14]	(8,500) [14]	12,000 [14]	Nov	12,908
Carsebreck and Rhynd Lochs	16,500 [14]	14,500 [14]	10,320 [14]	11,450 [14]	8,770 [14]	Oct	12,308
Dupplin Lochs	15,530 [14]	17,500 [14]	9,500 [14]	14,100 [14]	2 [14]	Oct	11,326
Hule Moss	14,700 [14]	8,600 [14]	5,850 [14]	14,200 [13]	7,950 [14]	Oct	10,260
Breydon Wtr & Berney Marshes	5,500 [14]	4,380	7,100	17,100	12,784	Jan	9,373
Cameron Reservoir	5,000	15,823	3,000	8,900 [14]	2,692 [14]	Nov	7,083
Loch of Lintrathen	2,220 [14]	5,920 [14]	(6,440) [14]	11,100 [14]	8,921 [14]	Oct	7,040
R Clyde: Carstairs to Thankerton	(4,850)	11,000	3,350	5,300	(3,050)	Oct	6,550
Tay Estuary	8,930 [14]	11,385 [14]	2,700 [14]	2,425 [14]	4,560 [14]	Nov	6,000
Horsey Mere	3,620 [14]	(5,000) [14]	4,000 [14]	8,200 [14]	7,231 [13]	Jan	5,763
Humber Estuary	(2,700) [14]	4,300 [14]	4,620 [14]	6,562	(5,638)	Oct	5,280
Loch Tullybelton	4,050 [14]				6,500 [14]	Oct	5,275
Wigtown Bay	50	5,316 [14]	(4,747)	8,662 [14]	(7,219)	Mar	5,199
Solway Estuary	2,541 [14]	(5,550)	(4,075)	(10,243)	2,612 [14]	Nov	5,004
Holbeach St Matthew	(5,000) [14]						(5,000)
Lindisfarne	5,881	6,450 [14]	(3,679)	1,496	5,300 [14]	Oct	4,782
Fala Flow	4,910 [14]	7,500 [14]	2,790 [14]	5,450 [14]	741 [14]	Oct	4,278
Strathearn (West)			4,100 [14]				4,100
River Tay - Haughs of Kercock					4,000 [14]	Nov	4,000 ▲
R. Nith: Keltonbank to Nunholm	(1,850)	(1,200)	(470)	(3,710)	(950)	Mar	(3,710)
Lake of Menteith	(4,500) [14]	3	4,515 [14]	4,026 [14]	5,357 [14]	Oct	3,680
Holburn Moss	(1,500) [14]	20	4,250 [14]	6,500 [14]	2,300 [14]	Oct	3,268
Loch Eye and Cromarty Firth	126 [14]	367 [14]	14,050 [14]	546	900	Feb	3,198
Tay and Isla Valley	(2,000) [14]	2,133 [14]	2,497 [14]	4,134 [14]	4,000	Nov	3,191
Thornham	5,180 [14]			950 [14]			3,065
Skinflats	2,750 [14]	3,800 [14]	1,900 [14]	3,250 [14]	2,530 [14]	Oct	2,846
Heigham Holmes		2,500 [14]					2,500
Sites no longer meeting table qualifying levels in WeBS-Year 2004/2005							
Orchardton & Auchencairn Bays	(990)		2,000	3,100	300	Dec	1,800
Gladhouse Reservoir	1,520 [14]	3,200	700	4,570 [14]	0		1,998
Loch Mullion	(660) [14]	900 [14]	1,600 [14]				1,250
Other sites surpassing table qualifying levels in WeBS-Year 2004/2005 in Great Britain							
Rossie Bog	250 [14]	50 [14]			6,290 [14]	Nov	2,197
Loch Flemington	0	400 [14]	0	370	4,700 [14]	Nov	1,094
Duns Dish	1,292 [14]	440 [14]	2,300 [14]	700 [14]	3,000 [14]	Oct	1,546
South Medwin Pools	430	270	1,500	1,000	3,000	Feb	1,240
East Chevington Pools	88	800	1,200	3,032 [14]	2,540 [14]	Nov	1,532
Corby Loch		0	0	11	2,500	Nov	628

European White-fronted Goose

Anser albifrons albifrons

International threshold: 10,000
Great Britain threshold: 58
All-Ireland threshold: +

GB max:	1,983	Feb
NI max:	0	

	S	M	L
GB change	(-)	-	--

% Young	27.5
Brood size	2.42

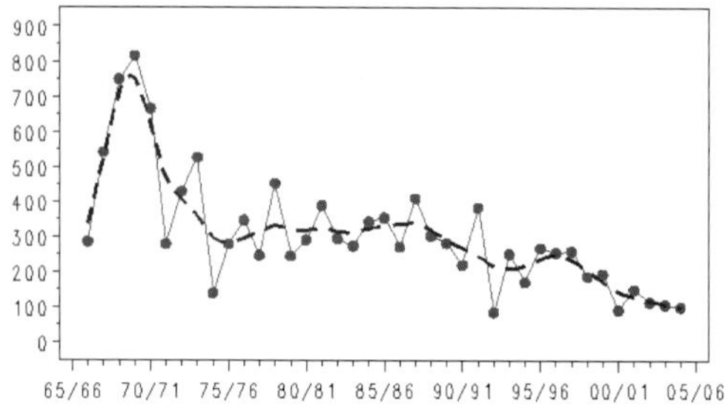

Figure 7.a, Annual indices & trend for European White-fronted Goose for GB.

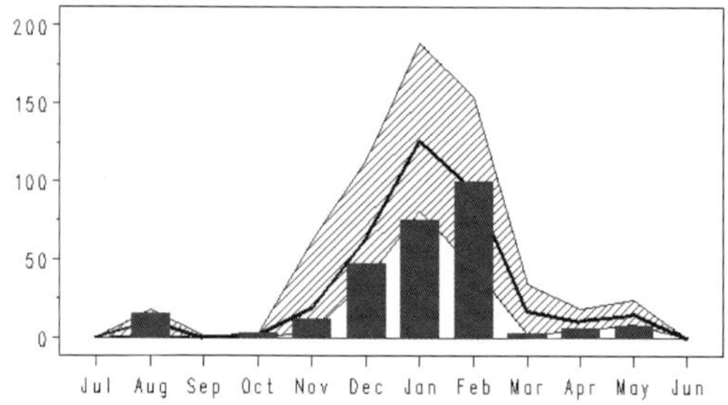

Figure 7.b, Monthly indices for European White-fronted Goose for GB.

There were no signs of a recovery for European White-fronted Geese in Britain in 2004/05, the annual index a little lower than the previous year with the underlying trend suggesting that wintering numbers are now at their lowest since records began. The peak counted total of less than 2,000 birds was the lowest since records began. The monthly indices show that European Whitefronts now visit almost exclusively between November and February. However, the population as a whole continues to do extremely well, with about a million wintering in continental Europe.

The peaks at most individual key sites were below their recent five-year means, with a continuing decline at Slimbridge on the Severn Estuary, and particularly low counts also from Dungeness Gravel Pits, Alde Complex and Minsmere Levels. Only at North Warren and Breydon Water were the peaks above the recent averages, albeit not by a large margin. Away from the listed sites, European White-fronted Geese were noted from 38 more sites, although in double-figures only from nine of these, mostly coastal sites in southeast England.

Breeding success was assessed at ten sites in Britain during January 2005. At Slimbridge on the Severn Estuary, the only site with a comparable long run of data, the percentage of young was the second highest of the past ten years, but the mean brood size was the lowest since 1993.

	00/01	01/02	02/03	03/04	04/05	Mon	Mean
Sites of national importance in Great Britain							
Severn Estuary	996	1,250 [13]	990 [11]	780 [13]	745 [8]	Feb	952
Swale Estuary	432	360	655	327	(398)	Jan	444
North Norfolk Coast	240	380	347	540	340	Feb	369
Dungeness Gravel Pits	234 [13]	355	460	205 [13]	110	Feb	273
N. Warren & Thorpeness Mere		250	310 [13]	190 [13]	302	Jan	263
Breydon Wtr & Berney Marshes	112	110	181	455	267	Feb	225
Confidential SE England Site	26	450	300	140	137	Dec	211
Middle Yare Marshes	298	74	89	120	109	Jan	138
Alde Complex	(0)	5 [11]	385	54	25	Feb	117
Minsmere	200 [13]	120	1	175	9	Jan	101

Greenland White-fronted Goose

Anser albifrons flavirostris

International threshold: 300
Great Britain threshold: 209
All-Ireland threshold: 140

GB max: 14,079 Dec
NI max: 115 Feb

% Young 7.8
Brood size 3.3

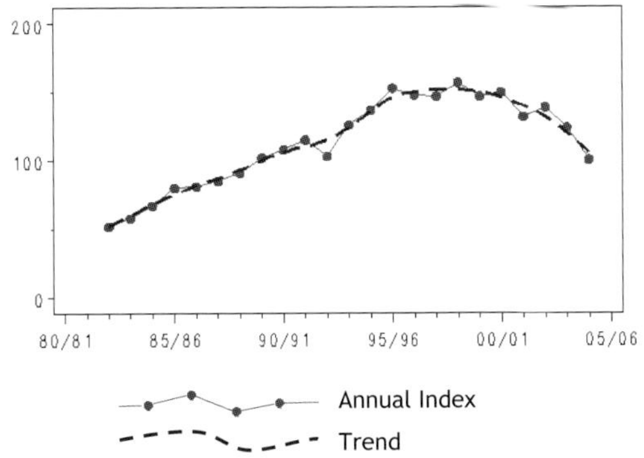

Figure 8.a, Annual indices & trend for Greenland White-fronted Goose for GB.

In autumn 2004, Greenland White-fronted Geese arrived in Britain in their lowest numbers since the late 1980s. The annual census peak fell by 19% since 2003/04, indicating that the recent decline in numbers not only continues, but also does so at an increasingly steep rate. Most of the decline can be attributed to the counts from Islay.

It appears that the entire population has declined by about 10,000 birds since 1999, to less than 24,000 in spring 2005. The decline is linked to poor breeding productivity. During autumn 2004, British flocks comprised only 7.8% young, well below average, although a mean of 3.3 young per successful pair was not especially low. It has been determined that the breeding density of this species in west Greenland has declined three-fold, whilst there has been a concurrent seven-fold increase in non-breeding Canada Geese summering in the same area (although the breeding density of Canada Geese has remained the same). Further studies are required to discover whether the trends in the two species are linked, and if so, by what precise mechanism (Fox *et al.* 2006). However, there has been recent good news from Iceland, where hunting of Greenland White-fronted Geese has been banned from autumn 2006 onwards.

At Islay, the peak count fell below 10,000 birds, meaning that some of the secondary sites take on an increased relative importance. Numbers appeared to be holding up at Machrihanish, Tiree and Coll, whilst the count of over 1,700 from Colonsay was a welcome development and means that the site now supports internationally important numbers. Conversely, counts were low from Rhunahaorine, Keills & Danna, Stranraer Lochs and Loch Ken. Lough McNean Lower was the main site in Northern Ireland, with 102 present in January. At the Dyfi Estuary, peak winter numbers were 92 compared to 116 the previous year.

	00/01	01/02	02/03	03/04	04/05	Mon	Mean	
Sites of international importance in the UK								
Island of Islay	13,281 [9]	12,261 [18]	12,254 [7]	11,272 [7]	8,350 [9]	Nov	11,484	
Machrihanish	1,386 [9]	1,448 [9]	1,501 [9]	1,377 [7]	1,407 [9]	Mar	1,424	
Rhunahaorine	1,551 [9]	1,594 [9]	1,450 [9]	1,156 [7]	894 [9]	Mar	1,329	
Tiree	1,221 [9]	1,076 [9]	1,093 [9]	1,093 [18]	1,133 [39]	Mar	1,123	
Isle of Coll	721 [9]	705 [9]	611 [9]	495 [9]	814 [9]	Feb	669	
Isle of Colonsay		104 [9]	87 [9]	79 [7]	1,718 [7]	Nov	497	▲
Keills Peninsula / Isle of Danna	443 [9]	403 [9]	411 [9]	377 [7]	338 [9]	Dec	394	
Stranraer Lochs	550 [9]	500 [9]	365 [9]	281 [9]	257 [9]	Jan	391	
Sites of national importance in Great Britain								
Isle of Lismore	275 [9]	295 [9]	310 [9]	290 [9]	310 [9]	Oct	296	
Loch Lomond	200 [9]	294 [9]	450 [9]	260 [9]	240 [9]	Dec	289	
Loch Ken	325 [9]	326 [9]	275 [9]	300 [9]	215 [9]	Nov	288	▼
Clachan and Whitehouse	366 [9]	100 [9]	250 [9]	215 [7]	209 [9]	Nov	228	
Loch of Mey	232 [9]	260 [9]	208 [9]	196 [9]	193 [9]	Nov	218	

Lesser White-fronted Goose

Anser erythropus

Vagrant and escape
Native Range: SE Europe, Asia

GB max:	1	Jul
NI max:	0	

There were just two records of Lesser White-fronted Goose during 2004/05, involving single birds at Llyn Traffwll in July and Theale Gravel Pits in September, both of which are likely to be escapes.

Greylag Goose

Anser anser

Icelandic Population

International threshold:	1,000
Great Britain threshold:	819
All-Ireland threshold:	40*

GB max:	85,582	Nov
NI max:	0	

% Young	28.2
Brood size	2.8

**50 is normally used as a minimum threshold*

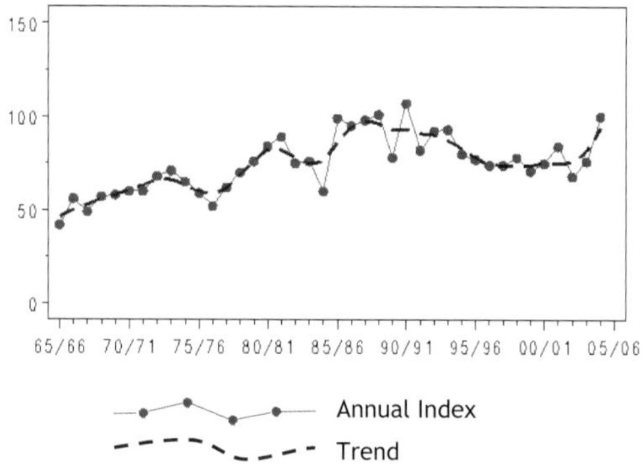

Figure 9.a, Annual indices & trend for Icelandic Greylag Goose for GB.

The Icelandic-breeding Goose Census (IGC) in autumn 2004 estimated a total population of 107,207 Icelandic Greylag Geese in November, the vast majority in Scotland. Whilst this is over 30% higher than the estimate made in autumn 2003, much of the increase was due to an estimate of approximately 20,000 geese remaining in Iceland this autumn. Without this figure, there appears to have been an increase of 7.5% since 2003. Although the counts made in Iceland were not complete, and based to some extent on anecdotal information, there would appear to be an increasing tendency for birds to remain there longer into the autumn than was the case in the past (Rowell 2005). Trials of carrying out the IGC in December are in progress, if deemed a more appropriate month this will be extended throughout the flyway. Assessment of breeding success suggested that 2004 was a very good breeding season, with young birds forming 28.2% of those aged (the highest proportion since 1973), and an average of 2.8 young per successful pair.

The British index shows the wintering population to be at its highest level since 1990/91. Following the rapid increase over the last few years, numbers on Orkney stabilised but still represented by far the most important area in Britain for wintering Greylag Geese. Within Orkney, about half of the total was found on West Mainland, with the remainder spread more widely across the other islands. Following several years of declining numbers, the peaks counted at both Caithness Lochs and Loch Eye & Cromarty Firth were the highest since 1998/99. However, at most other sites numbers were either about average, or lower than usual, the latter being particularly the case at Loch Spynie (in contrast to the record numbers of Pink-footed Geese at this site), Lochs Davan & Kinord, Munlochy Bay and the Upper Tay, whilst six sites no longer qualify as nationally important for the species. As has been mentioned in previous reports, assigning some flocks of Greylag Geese as belonging to the Icelandic population is not straightforward, given the expansion of both the northwest Scotland and re-established birds.

	00/01	01/02	02/03	03/04	04/05	Mon	Mean	
Sites of international importance in the UK								
Orkney	15,914 [14]	22,665 [14]	26,505 [14]	43,097 [14]	42,697 [14]	Nov	30,176	
Loch Eye and Cromarty Firth	(6,192)	5,680 [14]	(7,028) [14]	6,523 [14]	8,313 [14]	Nov	6,886	
Caithness Lochs	8,326 [14]	(7,854)	2,792 [14]	2,971 [14]	11,755 [14]	Nov	6,740	
Loch of Skene	9,660 [14]	2,100 [14]	(1,021) [14]	(2,600) [14]	4,500 [14]	Nov	5,420	
Lower Teviot Valley	3,500 [13]	(598)	(1,800)		(833)	Jan	3,500	
Loch Spynie	5,500 [14]	5,300 [14]	3,200 [14]	2,200 [14]	1,000 [14]	Nov	3,440	
Lochs Davan and Kinord	4,560 [14]	5,277 [14]	2,700 [14]	920 [14]	135	Dec	2,718	
Dornoch Firth	3,339	2,386 [14]	2,916	2,259	1,720	Feb	2,524	
Tay and Isla Valley	2,490 [14]	2,092 [14]	1,700	2,425 [14]	1,930	Oct	2,127	
Loch Garten	2,700 [14]	2,800 [14]	1,000 [14]	1,000 [14]	2,100 [14]	Nov	1,920	
R. Eden: Warcop-Little Salkeld	1,900						1,900	
Bute	1,530 [14]	2,300 [14]	1,380 [14]	2,000 [14]	1,780 [14]	Dec	1,798	
Loch Fleet Complex	1,700 [14]	4,210 [14]	817 [14]	905 [14]	990 [14]	Oct	1,724	
Munlochy Bay	424 [14]	3,500 [14]	3,130 [14]	110 [14]	20 [14]	Oct	1,437	
Kilconquhar Loch	1,096	1,380 [14]	1,552	1,620	1,200 [14]	Nov	1,370	
Beauly Firth	2,980 [14]	840 [14]	2,010 [14]	280 [14]	600 [14]	Oct	1,342	
Loch Ken	(971)	(1,368)	(1,106)	(1,280)	1,023	Feb	1,194	
Inner Firth of Tay	1,116 [14]	1,900 [14]		754 [14]	842 [14]	Nov	1,153	
East Mains Flood					1,131 [14]	Nov	1,131	▲
Strathearn (West)			1,050 [14]	1,050 [14]			1,050	
Haddo House Lakes	1,100 [14]	980	975	1,100 [14]			1,039	▲
Sites of national importance in Great Britain								
Findhorn Bay	620 [14]	1,950 [14]		190 [14]	1,100 [14]	Nov	965	▼
Whitrig Bog			1,000 [14]		850	Jan	925	
Gadloch	1,550 [14]	685	994	650	650	Oct	906	
Upper Tay	1,189 [14]	1,022 [14]	943 [14]	1,197 [14]	181 [14]	Nov	906	
Marlee Loch					880 [14]	Nov	880	▲
Forth Estuary	321	826	1,564	792	802	Oct	861	
Caistron Quarry	850	1,100	1,000	800	513	Feb	853	▲
Cochrage Loch		850 [14]					850	
Loch of Strathbeg	993	1,744 [14]	415 [14]	295 [14]	801 [14]	Dec	850	▲
Sites no longer meeting table qualifying levels in WeBS-Year 2004/2005								
Loch of Lintrathen	(905) [14]	1,330 [14]	400 [14]	616 [14]	540 [14]	Nov	758	
River Earn - Lawhill Oxbows	2,316	1,138	0	0	400	Mar	771	
East Chevington Pools	598	700	1,500	650	400	Nov	770	
R. Eamont/Eden: H'pot-Edenhall	(920)		(1,400)	100	351	Feb	693	
Carsebreck and Rhynd Lochs	1,160 [14]	953 [14]	610 [14]	494 [14]	301 [14]	Feb	704	
Threipmuir / Harlaw Reservoirs	1,390	530 [14]	350 [14]	447	(228)	Jan	679	
Other sites surpassing table qualifying levels in WeBS-Year 2004/2005 in Great Britain								
Birns Farm Gravel Pit		4	400	420	1,500	Dec	581	
Endrick Water				208	1,000	Jan	604	

Northwest Scottish Population

International threshold: 90
Great Britain threshold: 90

GB max: 8,516 Aug
NI max: 0

% Young 27.7
Brood size 2.6

Although the maximum British count was up substantially on that reported for the last three years, this is due in large part to those three August peaks not including counts for Tiree. Nonetheless, comparison of counts of similar areas on the Uists between August 2003 and August 2004 suggested a 5.8% increase, whilst comparison between February 2004 and February 2005 suggested a 16.8% increase. Productivity, recorded on Tiree, was at 27.7% young in August, with a mean of 2.6 young per successful pair, which was slightly below average.

The peak counts for both Tiree and North Uist in 2004/05 were both record counts, whilst the steady increase in the number of birds at Kentra Moss and Lower Loch Shiel has reached the point where the site is now considered to support nationally important numbers. However, the issue of whether some birds should be assigned to this population, or to either the Icelandic wintering birds, or the re-established birds remains difficult to resolve.

	00/01	01/02	02/03	03/04	04/05	Mon	Mean
Sites of international importance in the UK							
Tiree	3,535 [5]	3,674 [39]	3,516 [39]	3,563 [39]	4,005 [39]	Feb	3,659
North Uist	2,877 [19]	2,076 [19]	2,261 [19]	2,642 [19]	2,970 [54]	Aug	2,565
South Uist	1,862 [19]	2,303 [19]	2,095 [19]	2,102 [19]	2,111 [54]	Aug	2,095
Isle of Coll	679 [18]		675 [18]	411 [7]			588
Benbecula	431 [19]	376 [19]	488 [19]	319 [19]	414 [54]	Feb	406
Melbost / Tong / Broad Bay	(394)	197		(4)	(86)	Feb	296
Loch Broom	(197)						(197)
Isle of Colonsay	112 [7]			116 [7]			114
Branahuie Saltings		101					101
Kentra Moss / Lower Loch Shiel	79	90	93	102	136	Jan	100 ▲
Loch Urrahag	(167)	(27)			31	Dec	99
Sites no longer meeting table qualifying levels in WeBS-Year 2004/2005							
Isle Of Jura			90 [7]		52 [7]	Nov	71
Other sites surpassing table qualifying levels in WeBS-Year 2004/2005 in Great Britain							
Lochs Eaval & Hosta (North Uist)	0	1	4	28	150	Dec	37

Re-established Population

Naturalised re-establishment†

GB max: 25,482 Nov
NI max: 1,402 Mar

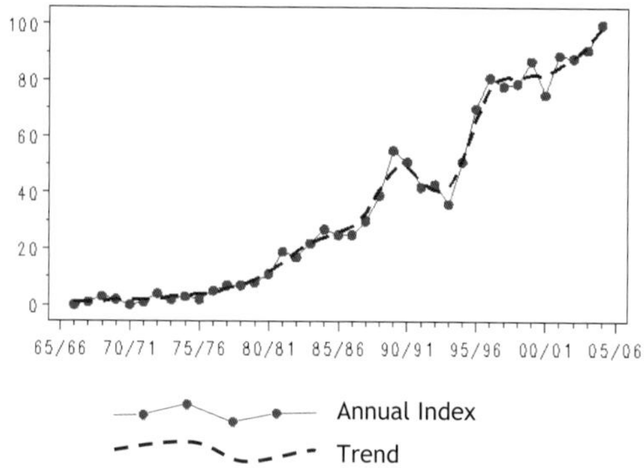

Figure 10.a, Annual indices & trend for Re-established Greylag Goose for GB.

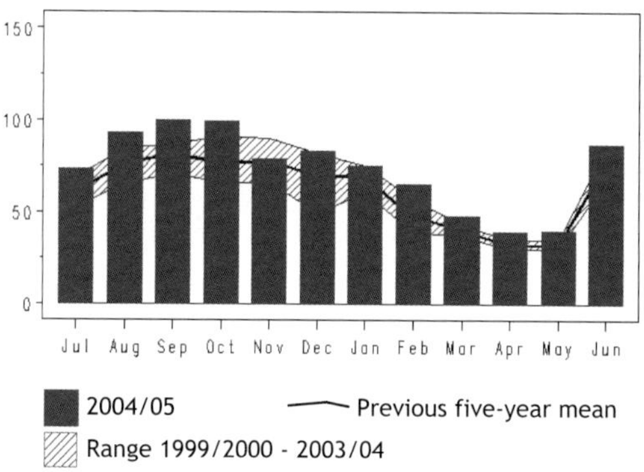

Figure 10.b, Monthly indices for Re-established Greylag Goose for GB.

Although the peak total counted by WeBS in Britain fell slightly, the British annual index rose by 10% in 2004/05 to a new record high, continuing the long-term spread of re-established Greylag Geese across the country. This agreed with the findings of the Breeding Bird Survey that showed a 179% increase between 1994 and 2004 (Raven *et al.* 2005). The number of British sites with mean peak counts in excess of 400 increased again to 46, hence the table displayed below now only lists sites with mean peak counts in excess of 500 birds. The biggest rises in numbers have been at Breydon Water and Livermere whilst steady increases have continued at the Tees Estuary and Martin Mere. Conversely, there has been a sustained decline on the North Norfolk Coast and at the Swale Estuary, whilst numbers at Bough Beech were extremely low. In fact, examination of the counts shows that numbers fluctuate greatly at any given site, presumably reflecting the fact that flocks can be quite nomadic within the wider countryside.

The Northern Ireland peak total was only about half of that recorded by 2003/04, although the peak at Loughs Neagh and Beg was only slightly down on the previous year, whilst there was a high count at Lough Foyle. As in previous years, the overlap between re-established and Icelandic Greylag Geese in the province remains difficult to disentangle. Moreover, this problem is becoming increasingly apparent in Scotland. Whilst WeBS analyses have traditionally not assigned Scottish Greylag Geese to the re-established population, however, the fact that over 300 birds were counted from sites in eastern Scotland in May 2005 demonstrates that re-established birds do occur here, as indeed they have done for the last 75 years. Whilst a forthcoming survey may help to resolve some of these issues, it is possible to foresee that within a relatively short time the distinction between native and re-established birds may become impossible to maintain.

	00/01	01/02	02/03	03/04	04/05	Mon	Mean
Sites with mean peak counts of 500 or more birds in Great Britain†							
North Norfolk Coast	3,431	1,850	1,657	(1,767)	1,371	Dec	2,077
Nosterfield Gravel Pits	678	1,084	1,746	1,338	1,389	Sep	1,247
Lower Derwent Ings			1,219	1,047			1,133
Tophill Low Reservoirs	1,126	1,183	828	683	867	Sep	937
Sutton and Lound Gravel Pits		1,057	1,176	407	950	Nov	898
The Wash	563	967	895	1,011	1,038	Sep	895
Eccup Reservoir	742	760	1,000	1,084	750	Nov	867
Bolton-on-Swale Gravel Pits	1,110	699	1,060	710	729	Sep	862
Ouse Washes	964	958	691 [13]	883 [13]	782	Oct	856
Humber Estuary	(590)	648	1,053	(769)	821	Mar	841
Kirkby-on-Bain Gravel Pits	562	635	900	1,072	925	Dec	819
Baston and Langtoft Gravel Pits	(152)	(330)	(600)	(803)			(803)
Swale Estuary	(907)	830	760	718	625	Oct	768
Lavan Sands	903	609	1,037	623	406	Nov	716
Abberton Reservoir	469	(2,500)	80	77	278	Jan	681
Morecambe Bay	327	867	741	629	786	Aug	670
Breydon Wtr & Berney Marshes	335	340	723	720	1,148	Nov	653
Hornsea Mere	625	745	465	642	785	Jul	652
Llyn Traffwll	450	700	769	891	341	Jul	630
Scorton Quarry				460	800	Aug	630
Livermere	249	490	806	280	1,176	Oct	600
Thames Estuary	(465)	(329)	(378)	(456)	593	Dec	593
Medway Estuary	(146)	(311)	(135)	(146)	589 [11]	Jan	589
Tattershall Pits	403	400	730	1,015	372	Jul	584
Dungeness Gravel Pits	(472)	554	(502)	667	529	Oct	583
Orwell Estuary	449	604 [11]	587 [11]	677 [11]	(543)	Jan	579
Little Paxton Gravel Pits	457	467	746	652	518	Nov	568
Besthorpe & Girton Gravel Pits	(181)	(236)	(4)	(331)	(539)	Dec	(539)
Alton Water	624	490	577	571	419	Jan	536
WWT Martin Mere	440	438	580	600	620	Sep	536
Sites with mean peak counts of 50 or more birds in Northern Ireland†							
Loughs Neagh and Beg	785	915	1,179	1,270	1,005	Feb	1,031
Lough Foyle	0	786	1,207	518	1,291	Mar	760
Strangford Lough	166	405	577	373	544	Mar	413
Belfast Lough	242	188	144	132	125	Nov	166
Lower Lough Erne			(71)	(54)	137	Feb	137
Ballysaggart Lough			70	66			68
Other sites surpassing table qualifying levels in WeBS-Year 2004/2005 in Great Britain†							
Ripon Race Course Gravel Pit				333	640	Sep	487
Tees Estuary	222	(422)	301	518	(623)	Oct	417
Messingham Sand Quarries			38	185	600	Jan	274
Bellflask				21	567	Aug	294
Blackwater Estuary	275	174	212	(124)	566	Oct	307
Llyn Maelog	341	187	227	435	565	Oct	351
Confidential SE England Site	123	160	120	140	550	Jan	219
Point of Ayre Gravel Pit		140	185	402	550	Aug	319
Hamford Water	215	(392)	376	258	539	Sep	356
Hardley Flood	307		515	487	515	Jan	456
Other sites surpassing table qualifying levels in WeBS-Year 2004/2005 in Northern Ireland†							
Upper Lough Erne	10	41	18	15	52	Feb	27

† as no British or All-Ireland thresholds have been set qualifying levels of 500 and 50 have been chosen to select sites, in Great Britain and Northern Ireland respectively, for presentation in this report

Bar-headed Goose

Anser indicus

Escape
Native Range: S Asia

GB max:	20	Sep
NI max:	0	

Birds were recorded from 33 sites across Britain with none reported from Northern Ireland. The British peak was again reached during September but at 20 was half that of the previous year. The highest counts were both from the Chichester area (and conceivably involved the same birds), nine in October at Chichester Harbour and eight in November at Chichester Gravel Pits. Most records were of single birds, but more than one were noted at 12 sites. The summed site maxima came to 68 birds, down a little on 2003/04.

Snow Goose

Anser caerulescens

Vagrant and escape
Native Range: N America

GB max:	34	Mar
NI max:	0	

Snow Geese were recorded in every month during 2004/05. However, the recent decline of British maxima continued, reaching only 34 during March. Likewise, the summed site maxima during the year declined to 66. The number of sites at which Snow Geese were recorded increased slightly to 26 (compared to 21 during 2003/04). The highest site totals were of 15 at Blenheim Park Lake in December and 14 at the Lower Windrush Valley Gravel Pits in April. Whilst most of the records doubtless refer to birds escaped from captivity, those at Wigtown Bay, WWT Caerlaverock, Morecambe Bay and the North Norfolk Coast, and perhaps others, were more likely wild birds accompanying flocks of Pink-footed Geese.

Ross's Goose

Anser rossii

Escape and possible vagrant
Native Range: N America

GB max:	2	Jul
NI max:	0	

Ross's Geese were recorded from five sites during 2004/05. Two were on the Tees Estuary in July and one at Tyninghame Estuary in August. One was present between Ardrossan and West Kilbride in September and October (where one was also noted during 2003/04), whilst records at the Orwell Estuary in February and nearby Hamford Water in March could well refer to the same individual.

Emperor Goose

Anser canagicus

Escape
Native Range: Alaska, NE Siberia

GB max:	21	Sep
NI max:	0	

Birds were recorded from just three sites, compared to nine during the previous year. As during 2003/04 the peak count was of 21 at South Walney Island in Morecambe Bay. Seven were recorded at Rutland Water in March; the first recorded during WeBS at this site, and a single bird was present at Reedham Water during September.

Greater Canada Goose

Branta canadensis

Naturalised introduction†
Native Range: N America

GB max:	**56,609**	**Oct**
NI max:	**622**	**Jan**

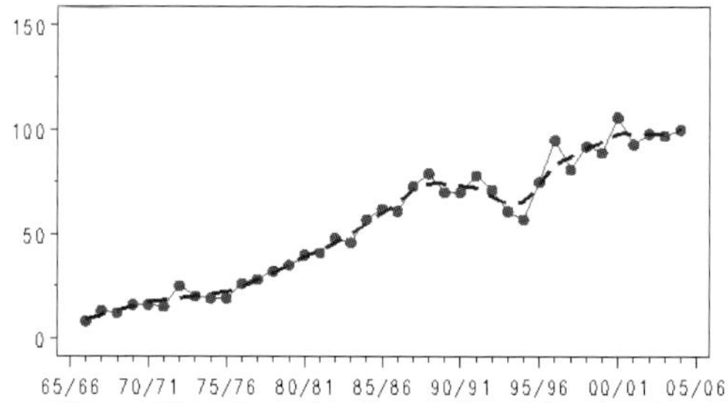

Figure 11.a, Annual indices & trend for Greater Canada Goose for GB.

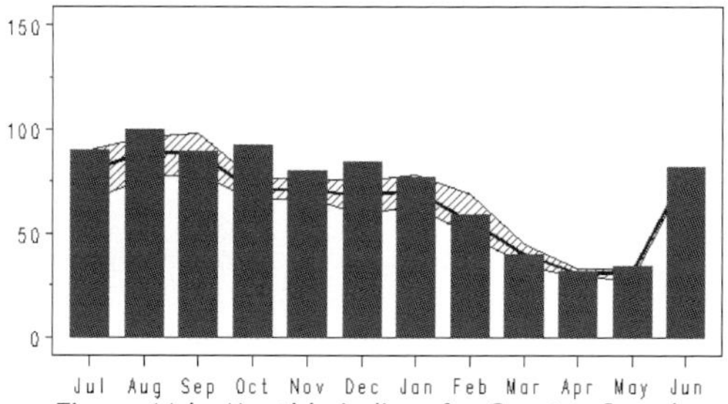

Figure 11.b, Monthly indices for Greater Canada Goose for GB.

Following the recent split of Canada Goose (Sangster *et al.* 2005) into Greater and Lesser Canada Goose, records of this species are now listed separately. Both the peak counted total and the annual index in Great Britain rose slightly compared to 2003/04. In Northern Ireland, counted numbers were down slightly but the annual index (not shown here) rose. Most birds in the province are found at Upper and Lower Loughs Erne and Strangford Lough. Greater Canada Goose displays a typical pattern of monthly indices for a resident species, peaking in the late summer and early autumn then declining gradually throughout the winter. In 2004/05, autumn monthly indices were above their recent average values, but from February onwards were more similar to normal, suggesting a good breeding season not being carried forward throughout the winter.

The number of sites with counts in excess of 1,000 Greater Canada Geese has risen from five in 2003/04 to nine this year. Particularly notable increases have occurred at Ellesmere Lakes, Tring Reservoirs, Alde Complex, Llangorse Lake, Doddington Pool and Blenheim Park Lake. However, at most sites numbers fluctuate a fair amount between years (as is also seen with re-established Greylag Geese). Recent declines have been noted from the Cleddau Estuary and Pitsford Reservoir, whilst the peak at the Dyfi Estuary is about a third less than that of two years previously.

Whilst a large number of Greater Canada Geese have been at large in the UK for many years now, numbers have traditionally remained very low in the Netherlands. However, in recent years, and especially since the late 1990s, numbers there have increased extremely rapidly (van Roomen *et al.* 2005).

	00/01	01/02	02/03	03/04	04/05	Mon	Mean
Sites with mean peak counts of 600 or more birds in Great Britain†							
Dee Estuary (England & Wales)	(1,664)	(2,268)	(2,568)	(1,529)	(2,316)	Oct	(2,568)
Dyfi Estuary	2,180	2,156	3,029	2,437	2,170	Jul	2,394
Mersey Estuary	1,738	737	1,437	1,177	1,496	Jul	1,317
Rutland Water	1,539	1,120	1,276	1,369	1,244	Jul	1,310
Arun Valley	1,139	1,550	(1,754)	860	1,235	Oct	1,308
Colliford Reservoir	946	894	1,884	1,284	1,095	Jul	1,221
Abberton Reservoir	1,217	(2,000)	270	639	(616)	Sep	1,032
Ellesmere Lakes	912	906	751	812	1,348	Oct	946
Tring Reservoirs	626	893	962	560	1,550	Oct	918
Middle Tame Valley Gravel Pits	889	(456)	(402)	(334)	(171)	Jan	889
Bewl Water	1,078	500	885	960	986	Aug	882
Stour Estuary	485	713	983	1,135	978	Dec	859
Fairburn Ings	950	709	823	893			844
Cleddau Estuary	1,080	1,000	765	655	622	Sep	824
Taw-Torridge Estuary	591	888	1,179	526	(912)	Jan	819
Walthamstow Reservoirs	781	662	945	837	784	Jul	802
Southampton Water	735	(1,084)	609	777	(548)	Nov	801
Chew Valley Lake	720	810	830	785	720	Jul	773
King`s Bromley Gravel Pits	850	669	712	776	721	Jul	746

	00/01	01/02	02/03	03/04	04/05	Mon	Mean
Harewood Lake	750	700	700	686	870	Feb	741
Alde Complex	303	510 [11]	(514)	(896)	1,246	Oct	739
Holme Pierrepont Gravel Pits	(209)			(364)	(714)	Sep	(714)
Lee Valley Gravel Pits	591	955	678	699	577	Sep	700
Somerset Levels	(704)	399	1,378	555	432	Dec	694
Doxey Marshes SSSI	349	638	(637)	881	893	Oct	690
Pitsford Reservoir	516	722	967	727	441	Aug	675
Thames Estuary	527	598	(706)	(329)	786	Oct	654
Llangorse Lake	570	600	700	415	936	Nov	644
Old Moor	670	768	(730)	380			637
Carsington Water		511	848	680	500	Jan	635
Arlington Reservoir	329	823	523	703	750	Sep	626
Exe Estuary	753	445	510	617	772	Aug	619
Lower Derwent Ings			642	594			618
Sites with mean peak counts of 50 or more birds in Northern Ireland†							
Upper Lough Erne	289	347	293	263	384	Jan	315
Strangford Lough	310	238	323	307	229	Sep	281
Lower Lough Erne			(110)	(343)	217	Feb	280
Drumgay Lough	70						70
Other sites surpassing table qualifying levels in WeBS-Year 2004/2005 in Great Britain†							
College Lake Reserve	64	476	444	773	919	Nov	535
Humber Estuary	346	425	(456)	525	868	Aug	541
Doddington Pool	236	88	574	249	829	Nov	395
Dorchester Gravel Pits	468	81	405	610	820	Sep	477
Roadford Reservoir		507	501	611	763	Dec	596
Barton Pits	572	209	579	477	732	Aug	514
Blenheim Park Lake	164	246	160	248	724	Oct	308
Grimley New Workings				75	682	Feb	379
Hallington Reservoir	(0)	468	683	287	663	Sep	525
Ouse Washes	654	380	388	429 [13]	649	Nov	500
Watermead Gravel Pits	370	574	610	632	616	Aug	560
Blithfield Reservoir	570	321	386	756	604	Oct	527
Bittell Reservoirs	420	228	303	518	600	Sep	414

† *as no British or All-Ireland thresholds have been set qualifying levels of 600 and 50 have been chosen to select sites, in Great Britain and Northern Ireland respectively, for presentation in this report*

Lesser Canada Goose

Branta hutchinsii

Vagrant and escape
Native Range: N America

GB max: 1 Oct
NI max: 0

Following the recent split of Canada Goose (Sangster *et al.* 2005) into Greater and Lesser Canada Goose, records of this species are now listed separately. Whilst vagrant Lesser Canada Geese can be expected to occur in the winter, particularly in the northwest, the only report during 2004/05 was of a single bird at Chichester Gravel Pits that remained from October until January and which undoubtedly refers to an escape.

Barnacle Goose

Branta leucopsis

Greenland Population

International threshold: 540
Great Britain threshold: 450
All-Ireland threshold: 75

GB max: 48,966 Mar
NI max: 0

% Young 16.1
Brood size 2.4

The dedicated survey of Greenland Barnacle Geese found totals of 40,572 in November and 48,966 in March. The peak was 2.3% higher than the count in the previous spring. This followed a very good breeding season in Greenland, where there were relatively mild conditions and the spring arrived early. The productivity statistics presented here reflect this, with 16.1% a high percentage of young overall although there was variation between different flocks. The vast majority of the birds were found on Islay, where the spring count of 44,186 was the highest there to date. However, numbers on Tiree also continue to increase with a new record count there too, whilst sustained high numbers wintering on Keills Peninsula & Isle of Danna, and on Colonsay/Oronsay, means that these sites also support internationally important numbers now. Extensive survey of key sites on North Uist found similar numbers to 2002/03. Another new site with internationally important numbers, albeit currently based only on a single visit, is Rispond Bay at the mouth of Loch Eriboll on the north coast of Scotland, where an expedition by members of the RAF Ornithological Society found a flock of 550 birds. At the Dyfi Estuary in west Wales, the wintering flock of Barnacle Geese peaked at 140 birds, the lowest total since 2000/01. It has not yet been conclusively established that these birds are Greenland-breeding birds, but they are tentatively assigned to this population on the basis of the Greenland White-fronted Geese also wintering at the site; observations of colour-marked birds at this site would be very helpful.

	00/01	01/02	02/03	03/04	04/05	Mon	Mean	
Sites of international importance in the UK								
Island of Islay	38,022 [40]	35,213 [18]	36,478 [40]	40,018 [7]	44,186 [7]	Mar	38,783	
North Uist	1,957 [19]	3,326 [19]	2,732 [19]		2,836 [54]	Feb	2,713	
Tiree	1,442 [39]	2,132 [39]	2,786 [39]	2,796 [39]	3,273 [39]	Jan	2,486	
South Walls (Hoy)		2,600 [40]	1,800 [40]				2,200	
Sound of Barra (Barra)	1,326 [37]						1,326	
Isle of Coll	718 [39]	933 [39]	1,010 [39]	792 [7]			863	
Sound of Harris (NW) (Harris)			706 [38]				706	
North Sutherland			669 [38]				669	
Colonsay/Oronsay	244 [7]		510 [39]	793 [7]	1,000 [7]	Mar	637	▲
Rispond Bay					550	Feb	550	▲
Sites of national importance in Great Britain								
Keills Peninsula / Isle of Danna	280 [7]	420 [7]	400 [7]	640 [7]	708 [7]	Mar	490	▲
Sites no longer meeting table qualifying levels in WeBS-Year 2004/2005								
Balnakeil Bay				826 [13]	0		413	
Islands South Of Barra	500 [37]		271 [38]				386	

Svalbard Population

International threshold: 230
Great Britain threshold: 220

GB max: 28,356 Dec
NI max: 0

% Young 2.1
Brood size 1.5

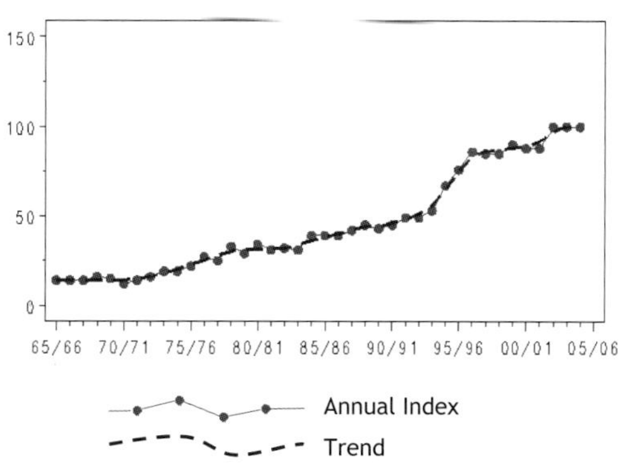

Figure 12.a, Annual indices & trend for Svalbard Barnacle Goose for GB.

The number of Svalbard-breeding Barnacle Geese has remained stable for the past three years. As usual, the vast majority of birds were recorded around the Solway Firth, where the first major arrival of about 6,000 birds occurred on 4th October 2004; in the spring, there were still 4,500 remaining on 4th May 2005. Productivity in summer 2004 was poor, with 2.1% young being the lowest percentage since 1958. In fact, four out of the last five years have now produced less than 4% young and at this rate, overall numbers may start to decline. Within the Solway, geese were widely distributed with four-figure flocks recorded from Calvo Marsh around to Auchencairn Bay, but Caerlaverock and Mersehead were the areas used most frequently and by the largest numbers of geese.

Away from the Solway, there was a large flock at the Loch of Strathbeg in October but then only small numbers throughout the winter until a spring count of 450 in mid-May; this site appears to act as a first and last landfall for some of the population as they make their way between Svalbard and the Solway. Further south, increasing numbers of Barnacle Geese are present at Lindisfarne, with the peak this winter falling in January. Other sites at which double-figure counts of potential Svalbard birds were recorded during winter 2004/05 were the Cromarty Firth, Montrose Basin, East Chevington Pools and Blyth Estuary (Northumberland).

	00/01	01/02	02/03	03/04	04/05	Mon	Mean
Sites of international importance in the UK							
Solway Estuary	23,783 [8]	23,524 [8]	28,447 [8]	27,510 [8]	28,270 [8]	Dec	26,307
WWT Caerlaverock (Inland)	3,350	7,380	14,500	11,030	10,850	Jan	9,422
Mersehead RSPB Reserve	7,490	6,400	8,500	5,700	9,200	Oct	7,458
R. Nith: Keltonbank to Nunholm	(4,485)	(4,730)	(5,030)	(4,850)	(3,000)	Jan	(5,030)
Loch of Strathbeg	3,700 [13]	10,390 [39]	138	95	1,100 [39]	Oct	3,085
Orchardton & Auchencairn Bays	(50)		2,250	3,100	1,100	Nov	2,150
Lindisfarne	93	(400)	(140)	786	160	Jan	360
Wigtown Bay	0	109	(395)	690	(412)	Mar	321

Naturalised Population

Naturalised establishment†

GB max: 739 Oct
NI max: 338 Jan

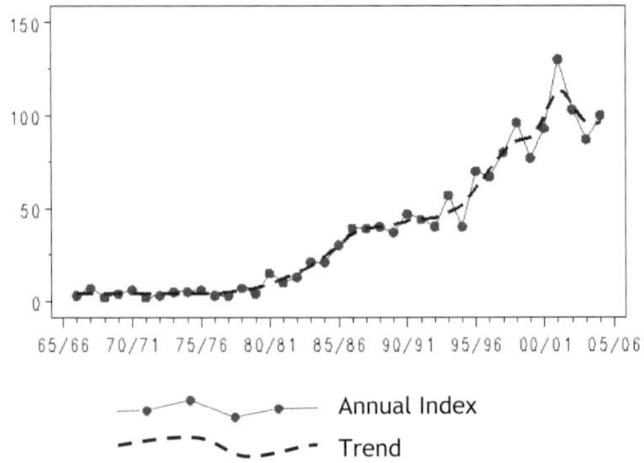

Figure 13.a, Annual indices & trend for Naturalised Barnacle Goose for GB.

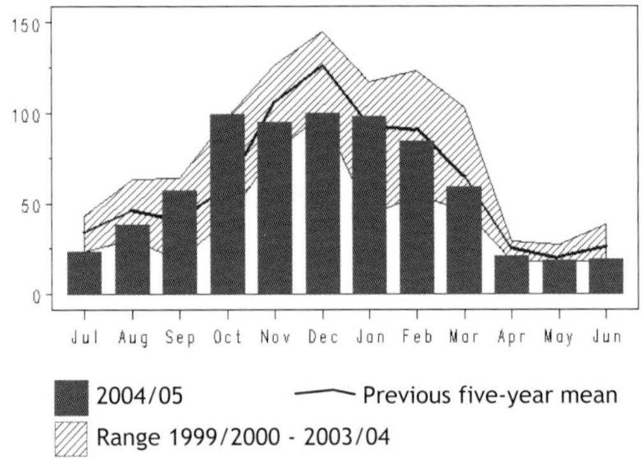

Figure 13.b, Monthly indices for Naturalised Barnacle Goose for GB.

The annual index shows that naturalised Barnacle Geese remain at a high level in Britain. The peak counted number of naturalised Barnacle Geese in Britain was substantially lower than in 2003/04, largely due to the lack of a count from Willington in 2004/05. At most of the sites supporting the largest numbers of naturalised Barnacle Geese, numbers appeared average or low in 2004/05, with peaks from Eversley Cross & Yateley Gravel Pits, Hornsea Mere and Derwent Water well down on recent averages. The sites at which numbers have increased most strongly are Breydon Water and Strangford Lough, the latter site supporting almost all the naturalised Barnacle Geese in Northern Ireland.

Counts of Barnacle Geese are assigned as naturalised birds purely on the basis of geographical location, as there is no way to separate the different populations in the field on the basis of external characteristics. As a result, some extralimital birds from the Svalbard and Greenland populations (as well as those wintering in the Netherlands) may well be incorrectly assigned.

	00/01	01/02	02/03	03/04	04/05	Mon	Mean
Sites with mean peak counts of 50 or more birds in Great Britain†							
Eversley Cross & Yateley GPs	183	236	219	158	107	Sep	181
Thwaite Flat / Roanhead Ponds	152 [13]						152
Duddon Estuary	150	(150)	(1)	(65)	(0)		150
Hornsea Mere	241	202	132	96	73	Jan	149
Willington	47	115	84	298			136
Roxton Gravel Pits	18	105	107	262	107	Aug	120
Benacre Broad	26	42	120	250	130	Sep	114
Humber Estuary	42	(53)	(74)	80	(200)	Oct	99
Middle Yare Marshes	20	141	104	72	82	Dec	84
Severn Estuary	53	73	96	(94)	101	Nov	83
Frampton Pools	37	(75)	79	98	52	Mar	68
Barcombe Mills Reservoir	60	76	64	73	52	Nov	65
Derwent Water	40	61	90	82	34	Jul	61
Ullswater	60	0	2	135	110	Feb	61
Sites with mean peak counts of 50 or more birds in Northern Ireland†							
Strangford Lough	158	214	223	232	337	Jan	233
Other sites surpassing table qualifying levels in WeBS-Year 2004/2005 in Great Britain†							
Breydon Wtr & Berney Marshes	26	2	12	61	70	Dec	34

† *as no British or All-Ireland thresholds have been set a qualifying level of 50 has been chosen to select sites for presentation in this report*

Dark-bellied Brent Goose

Branta bernicla bernicla

International threshold:	2,200
Great Britain threshold:	981
All-Ireland threshold:	+

GB max: 85,469 Jan
NI max: 0

	S	M	L
GB change	o	-	o

% Young 11.9
Brood size 2.3

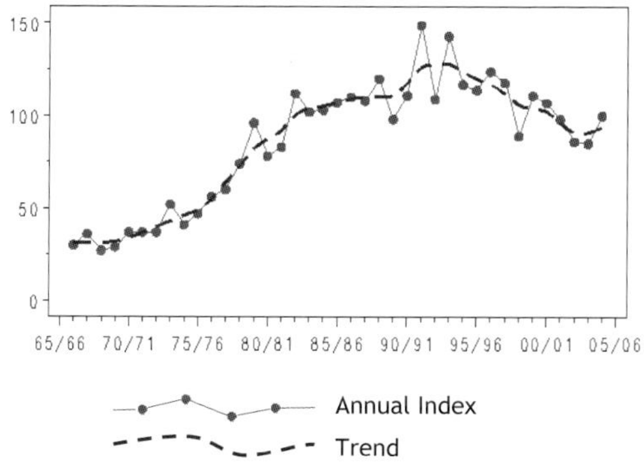

Figure 14.a, Annual indices & trend for Dark-bellied Brent Goose for GB.

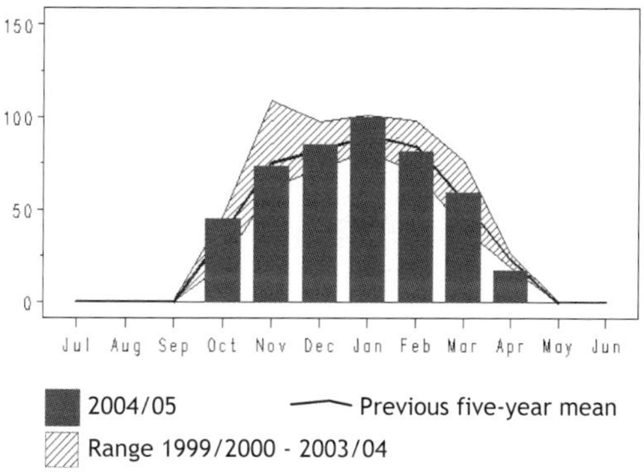

Figure 14.b, Monthly indices for Dark-bellied Brent Goose for GB.

After a long period of declining numbers, starting in the early 1990s, the British annual index increased by 18% compared to 2003/04. This initially gives hope that the decline might be levelling out although given it was another year when productivity was insufficient to replace annual losses it is likely that the rise in numbers reflects a greater proportion of the population wintering in the UK. The monthly indices were mostly similar to those in other recent years, although the January peak was above the recent average, perhaps suggesting a short-lived exodus from the continent. The assessment of productivity suggested that Dark-bellied Brent Geese had their best breeding season in 2004 for five years, although the percentage of young was still only half of that seen in 1999, and 2004 was the fifth consecutive year where the production of young was below the average mortality level of 15%.

No British sites changed in international or national importance status. Higher peak numbers, compared to recent years, were noted from the Wash, Hamford Water, Humber Estuary, Fleet/Wey and Medway Estuary. On the North Norfolk Coast, numbers increased over those recorded in 2003/04 but still remained relatively low. Particularly low peaks were noted at the Deben Estuary and Burry Inlet.

	00/01	01/02	02/03	03/04	04/05	Mon	Mean
Sites of international importance in the UK							
The Wash	19,518	17,924	20,314	18,734	21,969	Jan	19,692
Thames Estuary	7,371	12,157	(8,908)	(6,741)	9,455	Jan	9,661
North Norfolk Coast	10,201	8,033	9,180	5,722	6,607	Jan	7,949
Chichester Harbour	7,412	7,470	7,358	8,290	7,436	Dec	7,593
Blackwater Estuary	(9,860)	(7,195)	6,100	4,892	7,178	Jan	7,045
Langstone Harbour	5,080	4,813	4,686	5,804 [11]	5,069	Jan	5,090
Hamford Water	4,047	4,331	3,567	3,336	5,890	Jan	4,234
Crouch-Roach Estuary	4,446	3,471	3,083	2,914	4,635 [11]	Jan	3,710
Colne Estuary	3,310	2,572	(409)	(1,959)	(2,538)	Dec	2,941
Pagham Harbour	2,520	3,178	2,252	1,210	2,654	Feb	2,363
Sites of national importance in Great Britain							
Humber Estuary	(1,649)	1,432	(2,351)	2,118 [11]	(2,667)	Dec	2,142
Portsmouth Harbour	1,827	1,682	2,185	2,293	1,725	Feb	1,942
Deben Estuary	2,890	2,218	1,251	2,234	984	Jan	1,915
North West Solent	1,616	2,350	1,500	1,790	(2,208)	Jan	1,893
Newtown Estuary	1,800	1,660	1,779	(1,235)	(1,444)	Jan	1,746
Dengie Flats	2,455	1,798	1,160	1,507	(1,538)	Dec	1,730
Stour Estuary	1,716 [11]	1,412	1,753	1,914	1,782	Mar	1,715
Poole Harbour	1,708	(599)	(740)	(868)	(772)	Jan	1,708
Swale Estuary	2,149	1,690 [11]	1,278	1,210	2,111	Mar	1,688
Fleet and Wey	1,813	2,188	398	1,337	2,625	Nov	1,672
Southampton Water	1,742 [11]	(1,455)	(1,326)	(1,274)	1,386	Mar	1,564
Medway Estuary	(1,041)	(1,725)	(1,179)	836	1,834 [11]	Jan	1,465
Exe Estuary	1,345	1,183	1,714	1,368	1,645	Oct	1,451
Beaulieu Estuary	1,334	2,015	1,512	835	1,498	Jan	1,439
Orwell Estuary	1,228 [11]	1,215 [11]	1,525 [11]	1,396 [11]	976	Feb	1,268
Burry Inlet	1,158	1,174	917	(1,255)	811	Jan	1,063

Light-bellied Brent Goose

Branta bernicla hrota

East Canadian High Arctic Population

International threshold: 200
Great Britain threshold: +†
All-Ireland threshold: 200

GB max: 423 Sep
NI max: 27,988 Oct

% Young 21.8
Brood size 2.9

	S	M	L
NI change	o	o	o

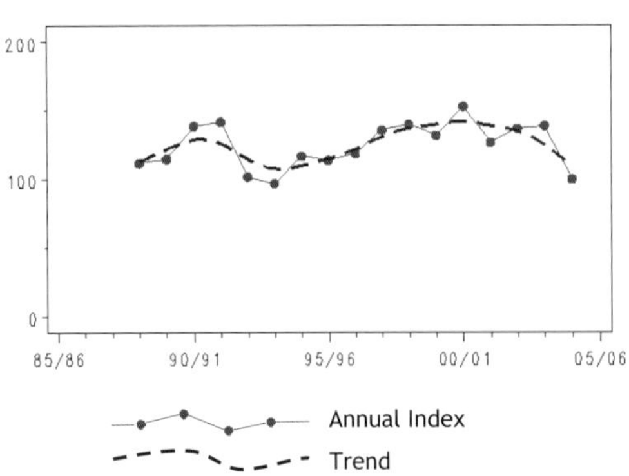

Figure 15.a, Annual indices & trend for Nearctic Light-bellied Brent Goose for NI.

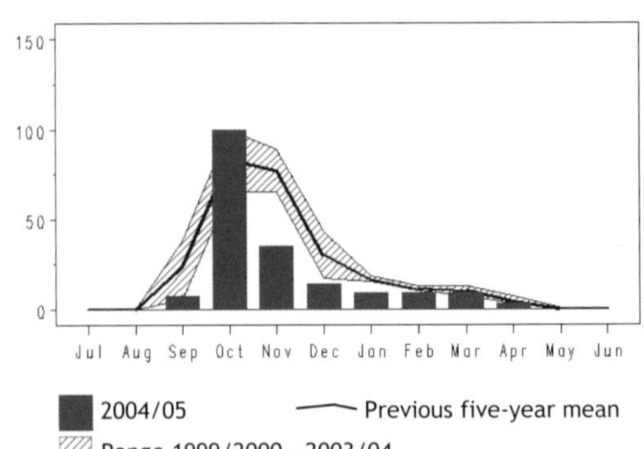

Figure 15.b, Monthly indices for Nearctic Light-bellied Brent Goose for NI.

The annual census, conducted over the weekend of 9th-10th October 2004, found a record total of 33,042 Light-bellied Brent Geese of this population. Whilst the majority of these birds were in Ireland (principally Strangford Lough, Lough Foyle and Tralee Bay in the Republic), about 2,000 were still present on their staging grounds in Iceland. Breeding success during summer 2004 was high, with the highest percentage of young since 2000.

The peak count at Strangford Lough was the highest ever total here, and represented 79% of the census total. As usual, however, numbers dropped at Strangford as the winter progressed and the geese dispersed more widely through Ireland and into western Britain. In fact, although the Strangford peak was at its highest, numbers remaining through the winter were low; the January total of 1,400 was the lowest on record. This explains the decline in the Northern Ireland annual index which represents the overall winter occupancy rather than the autumn peak. Elsewhere in Northern Ireland, site peaks were similar to those in other recent years.

There was an interesting arrival of birds on the Western Isles in September, with 284 at Loch Gruinart, 65 at Ardivachar Point, 31 at North Bay, 22 at Broadford Bay and 17 at Loch Bee. Whilst it is not possible to be completely certain that these refer to Nearctic breeders, rather than Svalbard birds, the northwesterly position of the sites suggests that this is more likely. At none of these sites were birds seen later in the winter. The September peak in Britain was the highest ever recorded. British wintering birds attributed to this population were found as usual at scattered sites around the eastern shores of the Irish Sea. Traeth Melynog and Foryd Bay are close to each other, on opposite sites of the western end of the Menai Straits, and thus the record counts at both of these sites may have involved some of the same birds. The peaks at Loch Ryan and the Dee Estuary were also site records.

	00/01	01/02	02/03	03/04	04/05	Mon	Mean
Sites of international importance in the UK							
Strangford Lough	16,162	19,583 [35]	17,520 [35]	21,500 [35]	26,250 [35]	Oct	20,203
Lough Foyle	3,469	1,841	1,563 [35]	3,277 [35]	1,603 [35]	Oct	2,351
Outer Ards Shoreline	120	210 [35]	700	642	762	Mar	487
Carlingford Lough	(498)	(259)	319	(570)	470	Dec	464
Killough Harbour		489 [11]	472	383	434	Mar	445
Dundrum Bay	(205)	(320)	(242)	(188)	(302)	Dec	(320)
Larne Lough	266	235	139	235	254	Jan	226
Sites with mean peak counts of 25 or more birds in Great Britain†							
Traeth Melynog				117	146	Dec	132
Jersey Shore	86	127					107
Inland Sea / Alaw Estuary	95	80	76	95			87
Loch Gruinart	60		2	0	284	Sep	87
Dee Estuary (England & Wales)	45	32 [11]	25	66	121	Jan	58
Foryd Bay	8	43	9	96	115	Feb	54
Loch Ryan	24	28 [13]	25 [13]	45	67	Mar	38
Morecambo Bay	(3)	7	62 [11]	53	31	Jan	38
Cleddau Estuary	12	(4)	3	106	3	Feb	31
Other sites surpassing table qualifying levels in WeBS-Year 2004/2005 in Great Britain†							
Ardivachar Point (South Uist)	0		0	0	65	Sep	16
North Bay (South Uist)			0	0	31	Sep	10

† *as no British threshold has been set a qualifying level of 25 has been chosen to select sites for presentation in this report*

Svalbard Population

International threshold: 50
Great Britain threshold: 30*

GB max: 1,989 Dec
NI max: 0

	S	M	L
GB change	+	+	++

**50 is normally used as a minimum threshold*

The Light-bellied Brent Geese breeding on Svalbard, northeast Greenland and Franz Josef Land winter mostly in Denmark and at Lindisfarne, Northumberland. The numbers at Lindisfarne in 2004/05 (and hence the British total) were low compared to recent years, although back to a level more typically seen during the 1990s. The geese arrived here in September increasing in October, were present then in similar numbers through to December then declined with no evidence of a mid-winter influx.

The second highest count of birds attributed to this population was a flock of 39 in October at Benacre Broad, Suffolk, but these did not remain throughout the winter. At the Moray Firth, where birds have regularly occurred in recent winters, counts were low and as a result the site no longer supports mean numbers in excess of the international importance threshold. The only other site at which a double-figure count was recorded was Seahouses to Budle Point where 23 were counted in February.

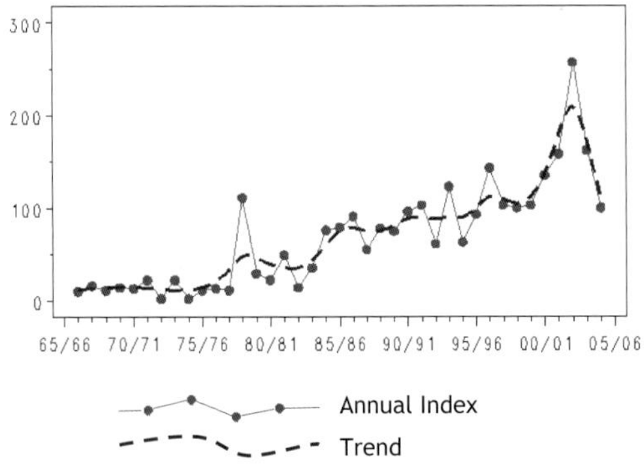

Figure 16.a, Annual indices & trend for Svalbard Light-bellied Brent Goose for GB.

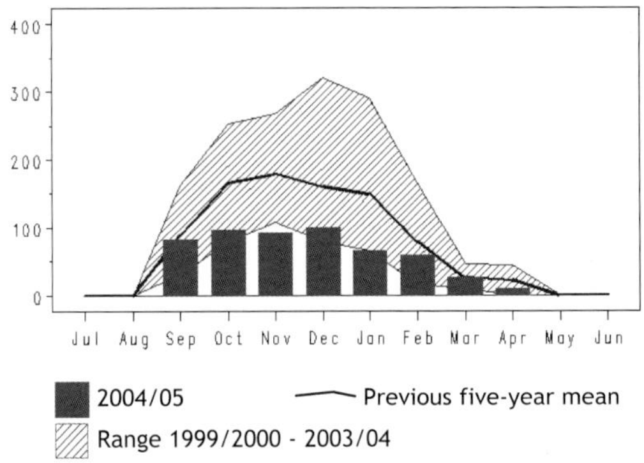

Figure 16.b, Monthly indices for Svalbard Light-bellied Brent Goose for GB.

	00/01	01/02	02/03	03/04	04/05	Mon	Mean
Sites of international importance in the UK							
Lindisfarne	(3,584)	(4,845)	(3,150)	3,716	2,505 [11]	Dec	3,560
Sites of national importance in Great Britain							
Inner Moray and Inverness Firth	10	41	100	55	18	Feb	45 ▼
Other sites surpassing table qualifying levels in WeBS-Year 2004/2005 in Great Britain							
Benacre Broad	0	0	0	0	39	Oct	8

Black Brant

Vagrant

Branta bernicla nigricans

Native Range: N America and E Asia

GB max: 3 Feb
NI max: 0

Black Brants were recorded at six sites, all in the south and east of England. Singles were present on the North Norfolk Coast during November and December, then in the Wash in January. The Thames Estuary held singles in January and February, while the Blackwater and Orwell Estuaries held singles in January and March respectively. The peak count was of two at the Swale Estuary in February, where one bird remained into March.

Red-breasted Goose

Vagrant and escape

Branta ruficollis

Native Range: SE Europe, Asia

GB max: 4 Jan
NI max: 0

Birds were only recorded at five sites, compared to 12 during the previous year. Two birds were present at Diss Mere throughout the year, while a single bird was present at Swarkestone Gravel Pits from November until March. Other sites holding single birds include Blagdon Lake, Chew Valley Lake and Aqualate Mere. None of these appeared to be especially good candidates for wild origin.

Egyptian Goose

Alopochen aegyptiaca

Naturalised introduction†
Native Range: Africa

GB max: 442 Sep
NI max: 0

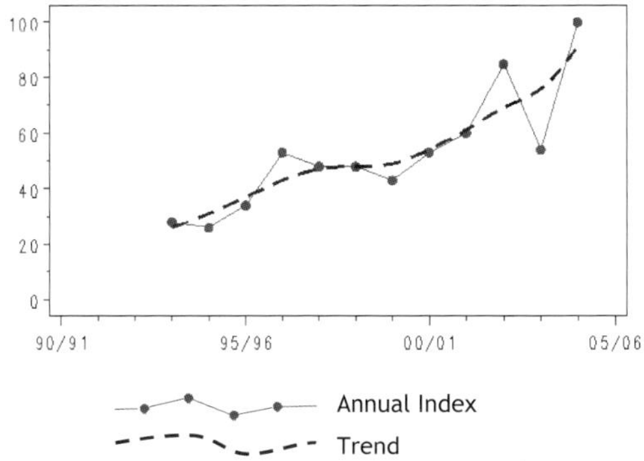

Figure 17.a, Annual indices & trend for Egyptian Goose for GB.

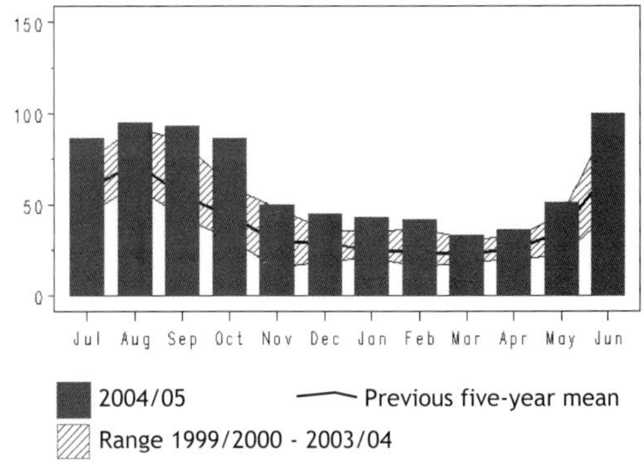

Figure 17.b, Monthly indices for Egyptian Goose for GB.

The annual index for Egyptian Geese in Britain, illustrated this year for the first time, reached its highest level to date and depicts a steady rise in the population of this naturalised introduction. The peak monthly count of 442 was over a third higher than the previous year, although a little below that from 2002/03. Egyptian Geese were recorded from 73 sites this year, just slightly higher than 2003/04, and again most site peaks were in single figures. The top site remains the North Norfolk Coast, even though recent counts from Holkham Park Lake have not been available. Elsewhere, there were new site peaks at Breydon Water, Weybread Pits, Spade Oak Gravel Pits, the Wash and Whitlingham Country Park. The rise in numbers at Rutland Water, however, was checked in 2004/05. Geographically, the main outposts away from Norfolk and Suffolk remain Rutland Water and Spade Oak Gravel Pits, although the increase at the Stour Estuary on the Suffolk/Essex border suggests continued range expansion.

3

	00/01	01/02	02/03	03/04	04/05	Mon	Mean
Sites with mean peak counts of 10 or more birds in Great Britain†							
North Norfolk Coast	218	318	233	(126)	(144)	Sep	256
Sennowe Park Lake Guist			98	85			92
Rutland Water	54	60	58	70	46	Sep	58
Breydon Wtr & Berney Marshes	25	48	63	65	82	Oct	57
St Benet`s Levels	(51)	52	88	23			54
Middle Yare Marshes	(45)	45	72	24	(47)	Sep	47
Nunnery Lakes	20	19	21	51	36	Oct	29
Cranwich Gravel Pits	23	19	34				25
Snetterton Gravel Pits	24						24
Weybread Pits	0	18	31	30	41	Jul	24
Spade Oak Gravel Pit	18	23	33	6	37	Sep	23
Whitlingham Country Park	6	10	7	18	59	Oct	20
Trinity Broads	11	18	20	10	22	Jul	16
Ampton Water			2	25			14
Barton Broad	13	15	18	14	6	Sep	13
Stanford Training Area	(7)	13					13
Lynford Gravel Pit	11						11
Colney Gravel Pits		10					10
Lackford Gravel Pits		10	3	(17)	10	Dec	10
Ranworth and Cockshoot Broads	12[13]	12	14	8	4	Jul	10
Other sites surpassing table qualifying levels in WeBS-Year 2004/2005 in Great Britain†							
The Wash	2	2	12	6	21	Feb	9
Hardley Flood	4		4	10	14	Aug	8
Salhouse Broad	6	8	9	6	11	Aug	8
Stour Estuary	2	6	2	7	10	Dec	5

† *as no British or All-Ireland thresholds have been set a qualifying level of 10 has been chosen to select sites for presentation in this report*

Paradise Shelduck

Tadorna variegata

Escape
Native Range: New Zealand

GB max:	1	Jul
NI max:	0	

One has been present at Par Sands Pools and St Andrews Road, Cornwall, since summer 2003 and this bird was again reported regularly throughout 2004/05.

Cape Shelduck

Tadorna cana

Escape
Native Range: S Africa

GB max:	1	Jul
NI max:	0	

In Cheshire, single birds were reported from Doddington Pool and the Dee Estuary during July and September respectively. This species was recorded from the Mersey Estuary and Ellesmere Lakes in 2003/04, thus perhaps involving the same individuals wandering in the Cheshire area. Another single bird was present on the Colne Estuary, Essex, in October 2004.

Ruddy Shelduck

Tadorna ferruginea

Escape and possible vagrant
Native Range: Asia, N Africa, S Europe

GB max:	10	Sep
NI max:	0	

Ruddy Shelduck were noted from 13 sites this year, records coming from eight months and with a peak of ten in September, the majority of this total was made up of eight birds on the Thames Estuary. The peaks during August and September perhaps suggest the dispersal of birds from the continent. At all other sites single numbers were recorded, except for three at Blithfield Reservoir and Walthamstow Reservoirs and two at Aqualate Mere and Young's Park (Goodrington). The summed site maxima of 26 was lower than in the two previous years.

Shelduck

Tadorna tadorna

International threshold:	3,000
Great Britain threshold:	782
All-Ireland threshold:	70

GB max:	57,071	Oct
NI max:	5,620	Jan

	S	M	L
GB change	o	o	o
NI change	o	+	+

The Northern Irish index rose by 5% to reach its highest ever level in 2004/05. At the same time the British index saw a decline of over 7%, taking it to its lowest ever level. Within Britain the largest declines were evident in England and Scotland, whilst Welsh numbers actually rose slightly. In Britain, monthly indices were above average from July until October after which time they fell below normal. The monthly indices for Northern Ireland show that very few birds were present until December, and the peak was short-lived, with a large decline between January and February. The decline was in line with findings from the Breeding Bird Survey that UK breeding numbers have declined by 38% between 1994 and 2004 (Raven *et al.* 2005).

Peak numbers at the top site, the Mersey Estuary, were lower than those in the previous year yet were similar to the recent average for the site. As seen for several years, numbers peaked on the Mersey in the late summer, representing the moult flock that has been established here since the 1990s (peak numbers have occurred here in July or August since 1996). Proportionally high peak counts were recorded at the Dee Estuary, Solway Estuary and Martin Mere, but counts at the most other listed sites were comparable to recent years. Numbers on the Wash and Chichester Harbour remained low, and the peak from the Forth Estuary was also much lower than normal. In Northern Ireland, a large proportion of the birds are to be found in Strangford Lough, where numbers were roughly average compared to recent years. However, at the other sites in Northern Ireland, many had above-average peaks, particularly Dundrum Bay and Loughs Neagh and Beg.

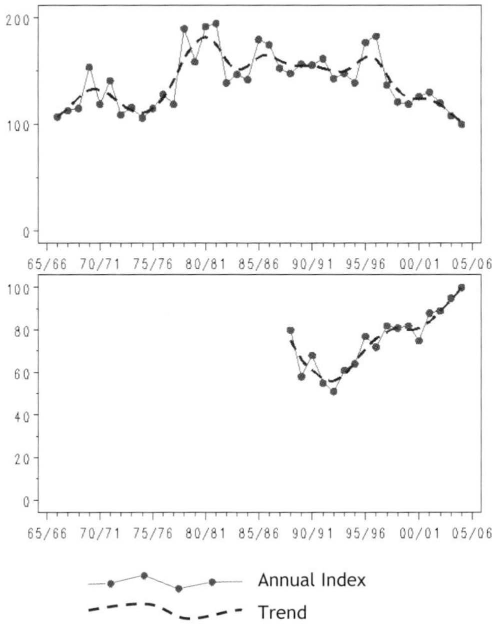

Figure 18.a, Annual indices & trend for Shelduck for GB (above) & NI (below).

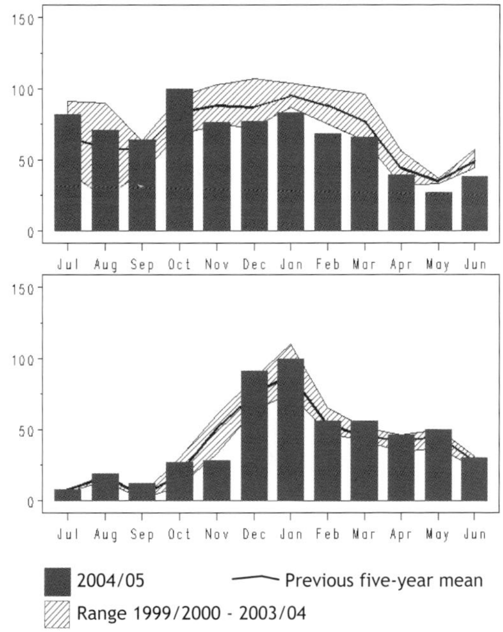

Figure 18.b, Monthly indices for Shelduck for GB (above) & NI (below).

	00/01	01/02	02/03	03/04	04/05	Mon	Mean
Sites of international importance in the UK							
Mersey Estuary	10,084	5,740	19,810	17,823	13,420	Jul	13,375
Dee Estuary (England & Wales)	11,563	10,200	10,533	12,630	13,334	Oct	11,652
The Wash	10,074	11,783	7,834	7,341	7,451	Nov	8,897
Morecambe Bay	(6,707)	6,137	7,164	8,228	7,728	Sep	7,314
Humber Estuary	6,918	3,655	(4,819)	6,426 [11]	(4,188)	Sep	5,666
Solway Estuary	4,626	(2,213)	(4,324)	3,131	5,359	Oct	4,372
Strangford Lough	3,067 [11]	4,162	4,199 [11]	4,475	3,801	Dec	3,941
Ribble Estuary	2,536	3,190	(3,063)	3,829	3,850	Nov	3,351
Severn Estuary	(2,912)	3,776	3,495 [11]	2,579	3,460	Oct	3,328
Forth Estuary	3,009	(2,920)	3,531	3,452	2,284	Sep	3,069
Sites of national importance in Great Britain							
Thames Estuary	(3,385)	2,940	3,285	1,584	(2,318)	Jan	2,799
Medway Estuary	(1,912)	(2,045)	(1,257)	(2,177)	2,360 [11]	Jan	2,360
Blackwater Estuary	(2,873)	(1,808)	(2,572)	1,904	2,073	Mar	2,356
Swale Estuary	2,047	2,342	2,290	1,818	2,207	Jan	2,141
Poole Harbour	1,748	(2,221)	2,385	(2,072)	1,547	Jan	1,995
Hamford Water	2,003	1,737 [11]	1,903	(1,657)	1,951	Jan	1,899
Stour Estuary	2,164 [11]	1,441 [11]	1,916	1,569 [11]	(2,149)	Jan	1,848
Crouch-Roach Estuary	(483)	(478)	(385)	(342)	1,661 [11]	Feb	1,661
Lindisfarne	1,751	1,546 [11]	1,826	1,323 [11]	1,773 [11]	Dec	1,644
North Norfolk Coast	938	2,012 [11]	(1,182)	1,112	1,110	Dec	1,293
WWT Martin Mere	743	950	1,435	1,150	1,510 [12]	Feb	1,158
Alde Complex	1,328	881	945	1,124	1,025	Mar	1,061
Montrose Basin	907	776	1,191	(1,240)	690	Sep	961
Chichester Harbour	990	1,014 [11]	1,019	810	825	Feb	932
Burry Inlet	1,233	963	570	(847)	804	Feb	893
Colne Estuary	773	920	(263)	(804)	(701)	Dec	847
Deben Estuary	772	676	864	802	883	Jan	799
Sites of all-Ireland importance in Northern Ireland							
Larne Lough	710	776	637	633	808	Jan	713
Carlingford Lough	326	(365)	493	423	452	Jan	424
Belfast Lough	319 [11]	437	199 [11]	494 [11]	489 [11]	Dec	388
Lough Foyle	278	536	232	(315)	250	Nov	324
Loughs Neagh and Beg	74	102	146	205	260	Mar	157
Dundrum Bay	79	93	99	138	330	Nov	148
Bann Estuary	50	138	87	104	52	Feb	86

Muscovy Duck

Cairina moschata

Escape†
Native Range: S America

GB max:	**70**	**Jan**
NI max:	**0**	

Muscovy Duck were recorded on 35 sites across England and Scotland, a similar number to the previous year. Nationally, numbers were greatly up on the previous year with the January peak of 70 being half as high again as that of 2003/04. However, at only two sites, Fort Henry Ponds & Exton Park Lake and Derwent Water, were double figures recorded. The summed site maxima was 104, higher than in the two previous seasons but lower than during the whole of the period 1994/95 to 2001/02.

	00/01	01/02	02/03	03/04	04/05	Mon	Mean
Sites with mean peak counts of 5 or more birds in Great Britain†							
Buxton Pavilion Gardens	25						25
Derwent Water	21	10	6	6	11	Oct	11
Ft Henry Ponds / Exton Pk Lake	15	14	0	0	14	Sep	9
Par Sands Pools	4	20	6	8	6	Nov	9
Wilderness Pond	6	10	12	7			9
High Batts Recording Area				8	8	Jul	8
Nafferton Mere	12	7	1	4	6	Nov	6
River Devon: Kersiepow Ponds	13	16	0	0	0		6
Stanley Park Lakes			5				5
Other sites surpassing table qualifying levels in WeBS-Year 2004/2005 in Great Britain†							
Yeadon Tarn	0	0	0	1	5	Feb	1

† *as no British or All-Ireland thresholds have been set a qualifying level of 5 has been chosen to select sites for presentation in this report*

Wood Duck

Aix sponsa

Escape
Native Range: N America

GB max:	**7**	**Mar**
NI max:	**0**	

Wood Duck were recorded from ten sites this year, although the single site peak of four at Stanton Lake was lower than that of the past few years. Three were present at Mere Sands Wood Nature Reserve (February) and two at Arun Valley (November) and River Avon - Salisbury (November and December). Singles were at Castle Park Lochan, Vyne Floods, Cotswold Water Park (East), Lackford Gravel Pits, Darwell Reservoir and Knepp Castle Lake.

Mandarin

Aix galericulata

Naturalised introduction†
Native Range: E Asia

GB max:	**373**	**Oct**
NI max:	**0**	

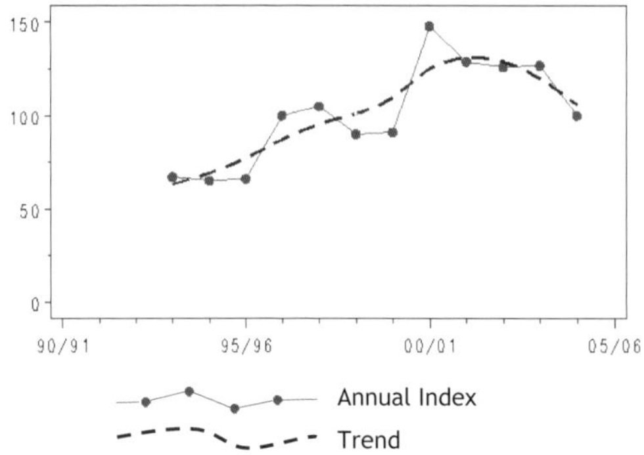

Figure 19.a, Annual indices & trend for Mandarin for GB.

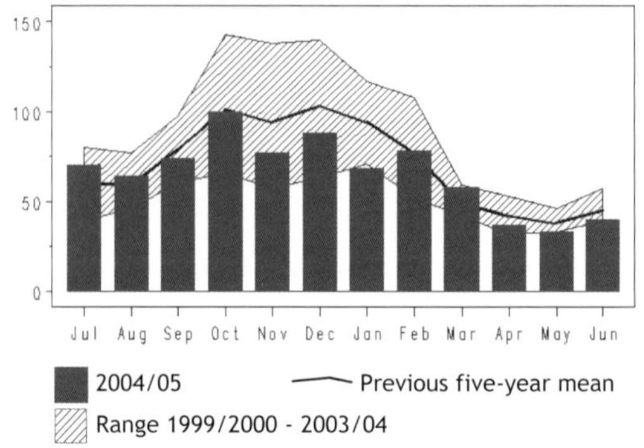

Figure 19.b, Monthly indices for Mandarin for GB.

The British annual index for Mandarin, displayed here for the first time, suggests that numbers of this species increased steadily up until 2001/02, before declining to its present level. Both the index and the counted British maximum fell substantially since 2003/04, but this is a secretive species and some of the key sites are not covered annually. In 2004/05, Mandarins were recorded from 121 British sites, just slightly down from the peak in 2003/04.

Apart from the supplementary counts provided for the combined Forest of Dean Pools, there were no three-figure counts of Mandarins in 2004/05. As usual, many sites continued to show fluctuating peaks of this species, presumably due to its unobtrusive habits and preference for heavily wooded pools. Notably higher than average counts were recorded from Longueville Marsh, Kedleston Park Lake, Stibbington Gravel Pit, Swanbourne Lake, Fonthill Lake, Osterley Park Lakes, Linacre Reservoirs and Eversely Cross & Yateley Gravel Pits. Conversely, relatively low peaks were noted from Headley Mill Pond, Passfield Pond, Darwell Reservoir, Woburn Park Lakes and the Severn Estuary, where numbers at WWT Slimbridge have fallen sharply. None were recorded in Northern Ireland; small numbers were recorded regularly from Dundrum Bay up to 2001 but none since.

	00/01	01/02	02/03	03/04	04/05	Mon	Mean
Sites with mean peak counts of 10 or more birds in Great Britain†							
Forest of Dean Ponds	171 [13]	159 [13]	120 [13]	160 [13]	221 [13]	Nov	166
Bradley Pools	43	85	55	188	65	Oct	87
Cuttmill Ponds	104	98	51	59	61	Jul	75
Wraysbury Pond	83	78	63 [12]	61	51	Dec	67
Stockgrove Country Park	80	54	70	43			62
Busbridge Lakes	57	54	47	72			58
Bough Beech Reservoir	40 [49]	77 [49]	33 [49]	56 [49]	60 [49]	Oct	53
Headley Mill Pond	16	70	76	76	23	Jan	52
Dee Flood Meadows	31	79	49	32	42	Sep	47
Passfield Pond	61	14	67	73	16	Sep	46
Connaught Water	54	26	31	44	32	Nov	37
Darwell Reservoir	46	43	25	56	13	Aug	37
Severn Estuary	72	65	28	3	4	Jan	34
Arun Valley	31	28	41	32	23	Oct	31
River Thames at Staines Bridge			31				31
Lost, Golding & Baldwins Hill Pds	45	10	78	5	12	Mar	30
Sutton Place		44	21	32	20	Aug	29
Harewood Lake	10	11	53	35	31	Oct	28
Fonthill Lake	23	17	18	20	38	Mar	23
Paultons Bird Park		21	20				21
Osterley Park Lakes	8	13	20	19	31	Jul	18
Strawberry Hill Ponds	8	7	30	23	15	Jan	17
Panshanger Estate	8	16	24	12	11	Dec	14
Woburn Park Lakes	25	12	24	8	1	Aug	14
Gatton Park	15	20	18	5	4	Oct	12
Linacre Reservoirs	3	5	14	17	23	Sep	12
Outwood Swan Sanctuary			5	19	10	Jul	11
Bramshill Park Lake	(14)	4	15 [13]	8	(9)	Oct	10
Eversley Cross / Yateley GPs	0	9	6	9	25	Jan	10
Other sites surpassing table qualifying levels in WeBS-Year 2004/2005 in Great Britain†							
Longueville Marsh	0	0	0	6	25	Jan	6
Swanbourne Lake	2	4	13	5	18	Jul	8
Kedleston Park Lake	1	0	1	0	14	Nov	3
Radnor Mere	13	0	9	12	12	Nov	9
Stibbington GP			4	5	12	Dec	7
Wraysbury Gravel Pits	6	2	(13)	12	12	Mar	9

† *as no British or All-Ireland thresholds have been set a qualifying level of 10 has been chosen to select sites for presentation in this report*

Wigeon
Anas penelope

GB max: 376,743 Dec
NI max: 9,561 Oct

International threshold: 15,000
Great Britain threshold: 4,060
All-Ireland threshold: 1,250

	S	M	L
GB change	o	o	+
NI change	o	o	-

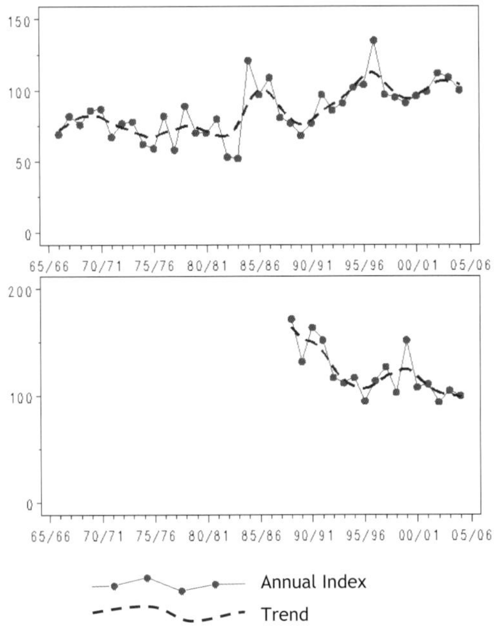

Figure 20.a, Annual indices & trend for Wigeon for GB (above) & NI (below).

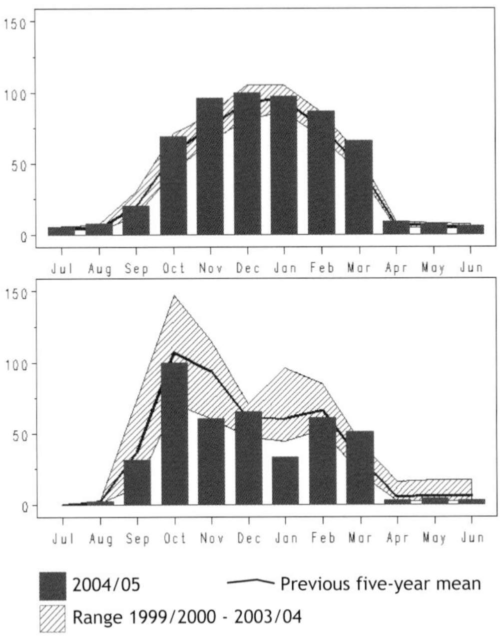

Figure 20.b, Monthly indices for Wigeon for GB (above) & NI (below).

The British annual index dropped for the second year running but overall numbers remain very healthy in Britain. This is a species that has seen major variability over time albeit with an underlying increasing trend. Although the British maximum was well below that seen in 2003/04, it was in line with the run of several winters previous to this. The monthly indices clearly show that most birds arrive in October and leave in March.

The Ribble Estuary remains by far the pre-eminent site in the UK and the peak this winter was the highest since 1998/99. As usual, the highest numbers on the Ribble were around Banks Marsh. Peak numbers were low this winter on the Somerset Levels and the Swale Estuary, but remained high at the Ouse Washes whilst Lindisfarne and the Dornoch Firth newly qualify as supporting internationally important numbers. Counts were low on the Nene Washes, in line with several other species here due to the lack of flooding. The peak on the Dee Estuary was also rather low, the lowest since 1995/96. The extraordinary build-up at Cassington/Yarnton in 2003/04 was not repeated in 2004/05, numbers dropping back to more usual levels. The peak on the Cleddau was very high for this site, most birds being found at the Pembroke River.

In Northern Ireland, there was also a small drop in the annual index but this forms a continuation of a generally declining trend over time. The majority of birds counted in Northern Ireland are found on Lough Foyle, Strangford Lough and Loughs Neagh and Beg, at all of which peaks have fluctuated in recent years.

	00/01	01/02	02/03	03/04	04/05	Mon	Mean	
Sites of international importance in the UK								
Ribble Estuary	(63,921)	68,661	75,617	(82,627)	86,157	Nov	78,266	
Somerset Levels	(28,366)	28,779	(39,546)	29,397	15,346	Jan	28,287	
Ouse Washes	14,874 [13]	26,623	26,753 [13]	33,773 [13]	30,794	Nov	26,563	
North Norfolk Coast	20,083	19,078	16,056	20,694	17,444	Jan	18,671	
Swale Estuary	17,637	15,303	22,827	20,772	13,832	Jan	18,074	
Breydon Wtr & Berney Marshes	15,700	21,700	15,999 [11]	16,811	19,019	Jan	17,846	
Lindisfarne	14,141	(12,435)	(20,016)	(12,321)	15,960	Oct	16,706	▲
Dornoch Firth	(17,445)	17,967	16,979	12,485	14,746	Oct	15,924	▲
Sites of national importance in Great Britain								
Cromarty Firth	14,027	11,987	(6,041)	12,877	13,487	Oct	13,095	
Lower Derwent Ings			11,217	13,171			12,194	
Nene Washes	10,808	5,053	11,866	8,190	4,998	Jan	8,183	
Severn Estuary	(5,789)	(5,579)	7,019	9,110	8,058	Jan	8,062	
Blackwater Estuary	6,507	(5,789)	10,976	7,057	7,385	Jan	7,981	
Alde Complex	7,145	6,647 [11]	7,387	(4,956)	7,274	Mar	7,113	
Inner Moray and Inverness Firth	7,260	7,070	7,820	(7,587)	5,595	Dec	7,066	
Morecambe Bay	7,746	5,861	5,634	7,151	8,095	Dec	6,897	
Thames Estuary	5,392	5,808	9,798	5,565	4,343	Jan	6,181	
Middle Yare Marshes	4,794	5,668	5,508	4,998	7,846	Jan	5,763	
Loch of Harray	9,476	4,255	(2,682)	4,823	4,265	Oct	5,705	
R. Avon: R'gwood-Christchurch	4,945 [13]	(1,450)	6,394	(1,783)	(999)	Feb	5,670	
Solway Estuary	(2,778)	(3,085)	(5,497)	(3,671)	(2,841)	Dec	(5,497)	
Cleddau Estuary	(3,604)	3,192	3,720	6,045	8,468	Nov	5,356	
Mersey Estuary	8,279	9,150	4,280	2,044	2,085	Oct	5,168	
Arun Valley	5,343	4,010	6,237	5,073	2,956	Nov	4,724	
Fleet and Wey	3,062	5,337	5,360	(5,105)	4,469	Nov	4,667	
Montrose Basin	3,446	4,381	4,752	5,488	(4,147)	Nov	4,517	
Dee Estuary (England & Wales)	4,681	4,941 [11]	3,979	(5,658)	2,464	Dec	4,345	
Humber Estuary	3,969	2,514	(5,513)	4,734 [11]	(3,570)	Dec	4,183	▲
Sites of all-Ireland importance in Northern Ireland								
Lough Foyle	8,051	5,696	2,609	3,978	4,589	Oct	4,985	
Strangford Lough	2,509	2,414	3,400	4,299	3,281	Oct	3,181	
Loughs Neagh and Beg	2,375	2,707	1,908	3,060	3,611	Mar	2,732	
Sites no longer meeting table qualifying levels in WeBS-Year 2004/2005								
Cassington & Yarnton GPs	(377)	(437)	(530)	4,144	576	Jan	2,360	
Other sites surpassing table qualifying levels in WeBS-Year 2004/2005 in Great Britain								
Loch of Strathbeg	611	(920)	1,094	2,058	4,376	Jan	2,035	
Exe Estuary	2,031	3,703	(2,580)	4,492	4,318	Oct	3,636	
Crouch-Roach Estuary	2,305	2,571	3,442	2,247	4,115	Jan	2,936	
Other sites surpassing table qualifying levels in WeBS-Year 2004/2005 in Northern Ireland								
Upper Lough Erne	1,858	401	797	921	1,284	Jan	1,052	

American Wigeon

Anas americana

Vagrant
Native Range: N & C America

GB max: 2 Nov
NI max: 0

Single birds were recorded at seven sites from November through to June. With the exception of the bird at the Ouse Washes during April 2005, all other records were in Scotland. Apart from the bird at Loch of Hillwell that stayed during November and December all other records were of birds in single months: Loch Bee, South Uist (November), Inner Moray and Inverness Firth (December), Lossie Estuary and Moray Coast (January) and Kinnordy Loch (June). There was no indication that any of these birds were not vagrants.

Chiloe Wigeon

Anas sibilatrix

Escape
Native Range: S America

GB max: 2 Aug
NI max: 0

Single birds were recorded from three sites between June and August. These were Harewood Lake (July and August), Blagdon Lake (June) and the Dee Estuary (August).

Gadwall

Anas strepera

International threshold: 600
Great Britain threshold: 171
All-Ireland threshold: +[†]

GB max: 15,639 Jan
NI max: 182 Dec

	S	M	L
GB change	o	(+)	++
NI change	o	(-)	o

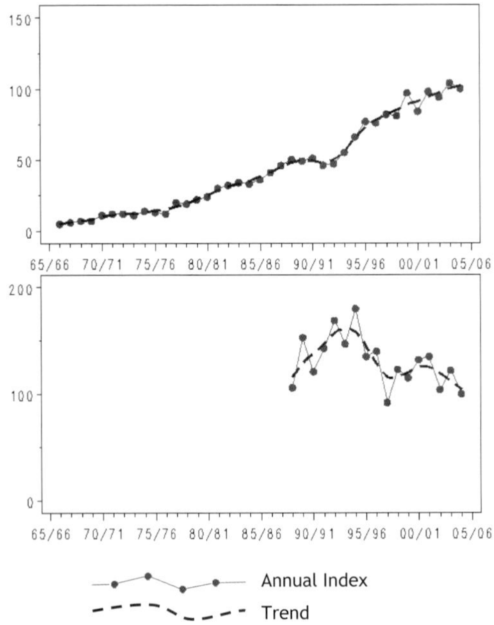

Figure 21.a, Annual indices & trend for Gadwall for GB (above) & NI (below).

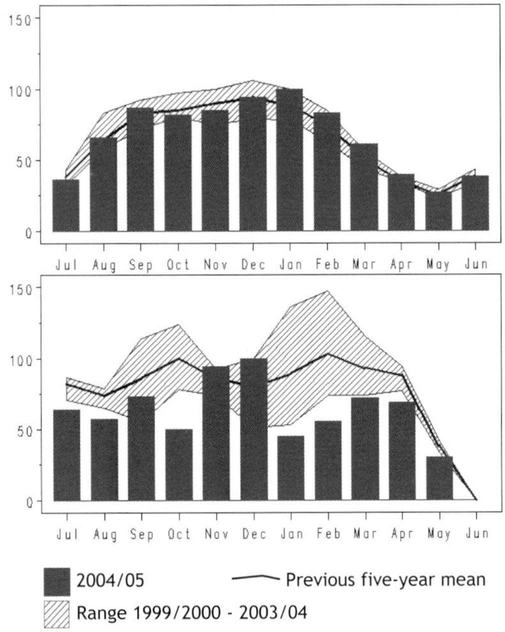

Figure 21.b, Monthly indices for Gadwall for GB (above) & NI (below).

Although the British annual index showed a slight fall in 2004/05, the long-term upwards trend appears unchanged. In Northern Ireland, the annual index has fluctuated and currently appears to be on a downwards trend, although the number of birds present in Northern Ireland is small. British monthly indices show that numbers increase rapidly to September and only gradually thereafter to mid-winter, then decline after January. However, a substantial proportion remains on WeBS sites throughout the summer.

The peak count at the Ouse Washes in March was the highest ever recorded at the site. Conversely, peak numbers at Rutland Water fell to their lowest level for 23 years. Other sites with peak counts well below expected levels were Pitsford Reservoir, Colne Valley Gravel Pits and Whitlingham Country Park, whilst five sites dropped from supporting nationally important numbers on the basis of five-year peak means. However, both North Warren & Thorpeness Mere and Staines Reservoirs newly qualify as supporting nationally important numbers, following high peaks in 2004/05. Peaks at Fen Drayton Gravel Pits, Alde Complex, Woolston Eyes, Blenheim Park Lake and Darwell Reservoir were also high in comparison to recent peaks. In Northern Ireland, Loughs Neagh and Beg remains the key site but the peak here was the lowest for the site since 1997/98.

	00/01	01/02	02/03	03/04	04/05	Mon	Mean
Sites of international importance in the UK							
Rutland Water	967	747	867	1,096	491	Jul	834
Somerset Levels	(453)	754	(1,077)	430	729	Oct	748
Ouse Washes	393	433 [13]	782	889 [13]	1,242	Mar	748
R. Avon: F'bridge-Ringwood	897 [13]	525	824	701	684	Jan	726
Lee Valley Gravel Pits	526	717	808	560	622	Dec	647

	00/01	01/02	02/03	03/04	04/05	Mon	Mean	
Wraysbury Gravel Pits	713	552	745	516	706	Sep	646	
Thames Estuary	(535)	(512)	815	(554)	471	Jan	643	
Sites of national importance in Great Britain								
Abberton Reservoir	746	173	730	519	425	Aug	519	
Loch Leven	270	320	840	635	360	Sep	485	
Pitsford Reservoir	259	581	164	898	124	Jan	405	
Hoveton Great Broad	283	310	278	667			385	
Thrapston Gravel Pits	531	(98)	(218)	207			369	
Chichester Gravel Pits	307	569	349	319	176	Feb	344	
Fen Drayton Gravel Pits	345	362	336	219	400	Dec	332	
Chew Valley Lake	310	230	360	410	315	Aug	325	
Cotswold Water Park (West)	194	267	403	375	327	Jan	313	
Minsmere	366 [13]	212	394	239	309	Oct	304	
Burghfield Gravel Pits			312	325	255	Jan	297	
Little Paxton Gravel Pits	360	287	275	339	225	Feb	297	
Orwell Estuary	150	160 [11]	465 [11]	446	234 [11]	Feb	291	
Eversley Cross & Yateley GPs	323	292	305	230	256	Jan	281	
Lower Derwent Ings			319	215			267	
Stodmarsh NNR / Collards Lgn	222	259	360	264	217	Feb	264	
Alde Complex	(106)	277	163	(244)	352	Dec	264	
Severn Estuary	298	250	253	292	194	Dec	257	
Woolston Eyes	211	124	182	297	470	Sep	257	
Middle Tame Valley Gravel Pits	(113)	(100)	(156)	(255)	(69)	Oct	(255)	
Hornsea Mere	265	240	285	219	235	Aug	249	
North Norfolk Coast	250	(221)	215	262	231	Dec	240	
Alton Water	92	268	270	360	182	Dec	234	
Sutton and Lound Gravel Pits		370	58	198	307	Feb	233	
Hickling Broad	229						229	
Colne Valley Gravel Pits	(412)	(211)	149	238	126	Dec	227	
Fairburn Ings	220	150	154	367			223	
Meadow Lane Gravel Pits	195	211	321	190	153	Jan	214	
Staines Reservoirs	(54)	101	162	126	455	Oct	211	▲
Lackford GPs		68	432	(225)	118	Sep	211	
Buckden and Stirtloe Pits	284	118	(208)				203	
Tees Estuary	83	(107)	208	231	289	Nov	203	
N. Warren & Thorpeness Mere		84 [13]	229 [13]	113	353	Feb	195	▲
Whitlingham Country Park	145	177	222	358	72	Dec	195	
Lakenheath Fen			179	263	139	Mar	194	
Blagdon Lake	178	257	17	335	148	Aug	187	
Ravensthorpe Reservoir	42	372	288	98	53	Jan	171	
Sites no longer meeting table qualifying levels in WeBS-Year 2004/2005								
Earls Barton Gravel Pits	159	140 [13]	(207)	124	(146)	Sep	155	
Dinton Pastures		291	144	97	138	Jan	168	
Brent Reservoir	306	295	109	68	69	Aug	169	
Thorpe Water Park	(157)	(55)	(74)	(74)	54	Jan	83	
Blatherwyke Lake	174	116	188		5	Mar	121	
Sites with mean peak counts of 10 or more birds in Northern Ireland†								
Loughs Neagh and Beg	155	178	149	173	130	Nov	157	
Strangford Lough	72	58	57	73	48	Sep	62	
Hillsborough Main Lake	3	27	12	41	9	Dec	18	
Other sites surpassing table qualifying levels in WeBS-Year 2004/2005 in Great Britain								
Tring Reservoirs	102	111	94	217	252	Dec	155	
Blenheim Park Lake	41	54	63	4	219	Dec	76	
Dungeness Gravel Pits	140	109	153	160	209	Sep	154	
Bainton Pits	47	76	118	168	200	Nov	122	
Alresford Pond	50	124	117	155	191	Sep	127	
Darwell Reservoir	(31)	60	24	83	191	Jan	90	
Ditchford Gravel Pits	104	118		192	180	Jan	149	
Other sites surpassing table qualifying levels in WeBS-Year 2004/2005 in Northern Ireland†								
Upper Quoile River	2	2	(0)	4	28	Dec	9	
Upper Lough Erne	0	0	4	4	10	Feb	4	

† *as no All-Ireland threshold has been set a qualifying level of 10 has been chosen to select sites for presentation in this report*

Teal

Anas crecca

International threshold: 4,000
Great Britain threshold: 1,920
All-Ireland threshold: 650

GB max: 151,875 Jan
NI max: 5,847 Jan

	S	M	L
GB change	o	o	(+)
NI change	o	o	o

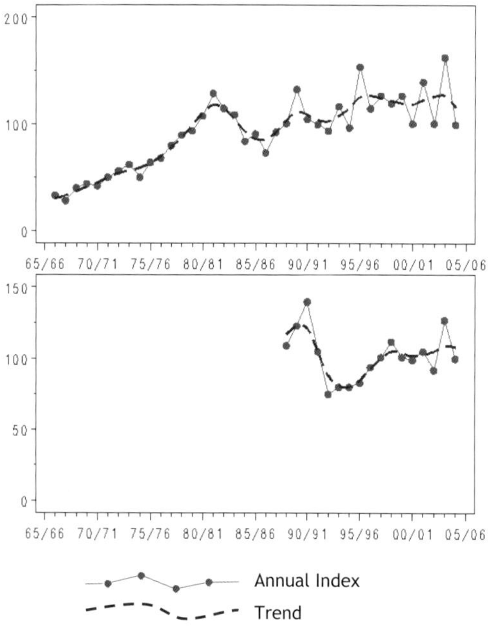

Figure 22.a, Annual indices & trend for Teal for GB (above) & NI (below).

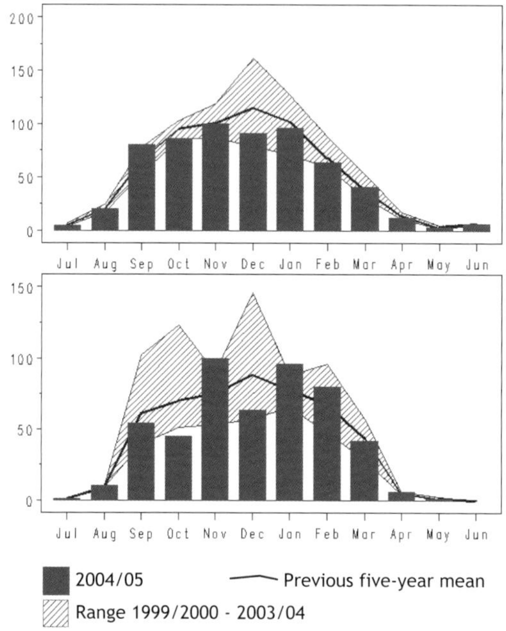

Figure 22.b, Monthly indices for Teal for GB (above) & NI (below).

In the last few years the British annual index for Teal has shown large fluctuations. In 2004/05 the index dropped by almost 40% since the previous year, but is back at the level seen in 2002/03 and the underlying trend depicts apparently healthy numbers. In 2004/05, most Teal arrived in September with relatively few more appearing later in the autumn. There were large departures between January and March. In Northern Ireland, there was also a fall in the annual index but again the underlying trend suggests numbers remain relatively high.

The peak count on the Somerset Levels was particularly low during 2004/05, due to a lack of extensive winter flooding, and some other inland wetlands dependent upon flooding also supported smaller numbers than usual, including the Arun Valley, Nene Washes and the River Avon from Christchurch to Ringwood. Conversely, numbers were high on the Ouse Washes, presumably as a result of local floodwater management. Within Liverpool Bay, numbers were low at the Mersey and Dee Estuaries but the autumn peak count on the Ribble Estuary was more than twice the previous peak count for the site, the majority of these birds favouring Banks Marsh. Nearby, the second highest site total of the winter was recorded at Martin Mere in September, conceivably involving some of the same birds. On the east coast, numbers were relatively low at almost all key coastal sites. In Northern Ireland, peak numbers were in line with their recent averages for all four key sites, although considerable fluctuation does occur between years.

	00/01	01/02	02/03	03/04	04/05	Mon	Mean
Sites of international importance in the UK							
Somerset Levels	(19,040) [13]	29,586	(33,390)	17,225	7,161	Jan	21,280
Mersey Estuary	8,777	17,660	7,855	8,364	6,023	Nov	9,736
Ribble Estuary	7,874	5,316	4,671	7,421	19,810	Oct	9,018
Thames Estuary	(4,610)	6,994	9,838	6,691	5,433	Jan	7,239
WWT Martin Mere	6,700	4,460	2,750	5,100	8,300	Sep	5,462
Dee Estuary (England & Wales)	5,622	6,887 [11]	4,361	5,459	2,752	Oct	5,016
Loch Leven	2,940	4,100	6,562	4,847	6,060	Sep	4,902
Ouse Washes	2,429	5,757	4,433 [13]	5,102	6,731	Nov	4,890
Swale Estuary	4,385	4,297	5,752	5,428	4,187	Jan	4,810
Hamford Water	2,510	9,055 [11]	3,628	6,579	2,164	Dec	4,787
North Norfolk Coast	4,186	5,718	5,281	3,436	3,730	Dec	4,470
Severn Estuary	5,151	(4,449)	3,748	(3,006)	(3,466)	Dec	4,450
Lower Derwent Ings			4,797	4,061			4,429
Breydon Wtr & Berney Marshes	4,237	6,487	3,124	1,982	4,733	Feb	4,113 ▲
Sites of national importance in Great Britain							
Blackwater Estuary	(4,867)	2,517	3,721	(2,873)	(2,064)	Dec	3,702
Otmoor	(1,313)	(856)	3,633		(2,138)	Jan	3,633
Mersehead RSPB Reserve	4,180	4,390	3,100	2,850	2,900	Jan	3,484
Alde Complex	(2,234)	3,690	2,609	(2,530)	3,028	Jan	3,109
Morecambe Bay	(2,956)	2,519	(2,261)	(2,808)	(3,699)	Nov	2,996
Inner Moray and Inverness Firth	2,794	2,289	2,948 [11]	3,439	3,397	Oct	2,973
Humber Estuary	3,370	1,284	2,681	(5,111)	2,349	Jan	2,959
Holburn Moss	3,920	1,700	2,250 [13]	3,500	3,000	Sep	2,874
Arun Valley	4,276	2,194	3,934	1,912	1,229	Nov	2,709
R. Avon: R'gwood-Christchurch	2,178 [13]	(654)	4,841	695	(309)	Dec	2,571
Dornoch Firth	2,261	2,797	2,502	2,619	2,451	Jan	2,526
Minsmere	1,700 [13]	2,227	2,189	4,381	1,984	Oct	2,496
The Wash	808	2,217	1,918	4,223	2,578	Jan	2,349
Solway Estuary	2,101	(750)	(2,813)	(1,286)	1,941	Oct	2,285
Woolston Eyes	2,100	3,675	1,320	2,072	2,170	Jan	2,267
Cleddau Estuary	2,427	(1,621)	2,095	2,129	2,269	Jan	2,230
Forth Estuary	(2,359)	(1,585)	1,984	(2,511)	1,880	Feb	2,184
Poole Harbour	2,086	1,667	2,235	(2,357)	(1,806)	Nov	2,086 ▲
Nene Washes	1,592	940	4,046	2,730	726	Nov	2,007
Sites of all-Ireland importance in Northern Ireland							
Loughs Neagh and Beg	2,007	1,633	1,887	2,732	2,019	Nov	2,056
Strangford Lough	1,189	2,121	2,177	2,232	2,015	Jan	1,947
Lough Foyle	2,888	684	2,275	582	1,038	Jan	1,493
Upper Lough Erne	308	333	1,635	407	723	Jan	681
Sites no longer meeting table qualifying levels in WeBS-Year 2004/2005							
Abberton Reservoir	488	(1,871)	736	3,863	1,224	Jan	1,636
Other sites surpassing table qualifying levels in WeBS-Year 2004/2005 in Great Britain							
Crouch-Roach Estuary	1,225	1,340	1,249	1,129	2,981 [11]	Jan	1,585
Stodmarsh NNR / Collards Lgn	1,004	1,126	1,170	1,183	2,500	Feb	1,397
Pagham Harbour	1,849	1,197	1,236	1,455	2,321	Dec	1,612

Green-winged Teal

Anas carolinensis

Vagrant
Native Range: N America

GB max:	7	Feb
NI max:	2	Nov

Green-winged Teal were recorded at 16 sites this year, a similar number to other recent years. All records involved single birds except for two at Larne Lough in November, with one also here in March. A bird at the Dee Estuary was noted monthly from November until March. Other sites at which this species was recorded include Garths Loch Scatness, Hayle Estuary, Nene Washes, Woodhorn Flashes, Langford Lowfields Gravel Pits, Montrose Basin, Steinish Canal, Bramshill Plantation Lakes, St Mary`s Island, Thanet Coast, Fleet and Wey, Poole Harbour, Blackwater Estuary and Stodmarsh NNR and Collards Lagoon.

Speckled Teal

Anas flavirostris

Escape
Native Range: S America

GB max:	5	Dec
NI max:	0	

Speckled Teal were recorded at Bramshill Park Lake between September and February, peaking at four birds in December. The only record away from this site was of three at Barcombe Mills Reservoir in December.

Mallard

Anas platyrhynchos

International threshold: 20,000**
Great Britain threshold: 3,520†
All-Ireland threshold: 500

GB max:	140,112	Dec
NI max:	6,694	Nov

	S	M	L
GB change	o	o	-
NI change	o	o	o

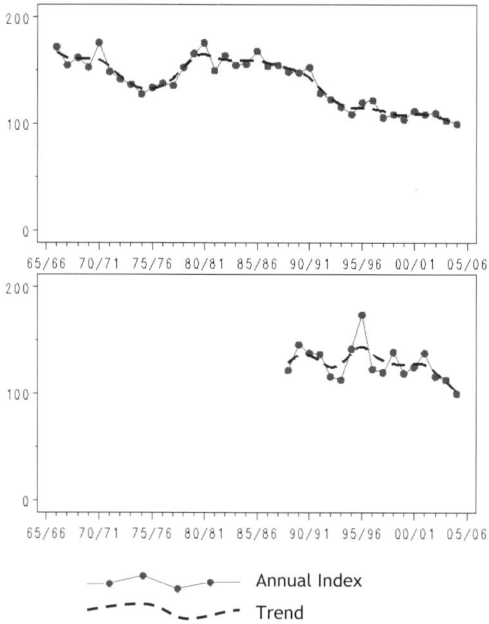

Figure 23.a, Annual indices & trend for Mallard for GB (above) & NI (below).

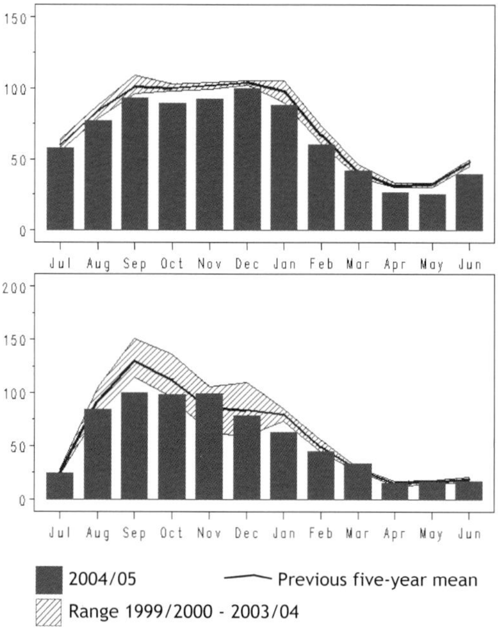

Figure 23.b, Monthly indices for Mallard for GB (above) & NI (below).

There was no sign of recovery in 2004/05 for British Mallards, the underlying trend in the annual index continuing the decline seen since the end of the 1970s, whilst the index itself reached its lowest ever value. One is left to wonder how far numbers will decline before finally stabilising. In virtually every month the monthly index values were below their recent averages. Moreover, the Northern Ireland annual index also dropped to its lowest ever value, following a pattern of relative stability until now. The counted British maximum was the second lowest since the late 1970s (when the number of sites covered was far fewer), whilst the counted maximum in Northern Ireland was the lowest since 1985/86.

The Ouse Washes remains the only site in Britain supporting nationally important numbers of Mallards, albeit with the peak here the lowest for some years. Site totals were also unimpressive in Northern Ireland, the Loughs Neagh and Beg total being the lowest since 1999/2000, although there was a record count from Lower Lough Erne. As there are so few sites supporting nationally important numbers in Great Britain, additional sites with means and peaks in excess of 2,000 Mallards are also listed. Of these other sites, many experienced small declines but there were higher than recent peaks at the Severn Estuary, Livermere, and the Swale Estuary.

However, the Mallard remains by far the most widespread and widely dispersed species of wildfowl in the UK. WeBS counts tend to reflect trends of this species more for larger wetland sites and should not be relied upon at face value to describe the numbers in the wider countryside. For example, the Breeding Bird Survey results (Raven *et al. 2005*) show a 23% increase in Mallards between 1994 and 2004, at odds with the picture from WeBS. This suggests that much of the decline is due either to a reduction in the number of autumn releases for shooting or more likely a reduction in autumn immigration.

	00/01	01/02	02/03	03/04	04/05	Mon	Mean
Sites of national importance in Great Britain							
Ouse Washes	3,657	4,457	3,580 [13]	3,988 [13]	3,505	Nov	3,837
Sites of all-Ireland importance in Northern Ireland							
Loughs Neagh and Beg	6,438	4,243	4,763	4,774	4,027	Nov	4,849
Strangford Lough	1,807	2,227	1,851	1,568	1,621	Sep	1,815
Lough Foyle	1,298	1,181	705	791	1,025	Oct	1,000
Lower Lough Erne			533	(494)	754	Dec	644
Sites with mean peak counts of 2,000 or more birds in Great Britain†							
WWT Martin Mere	2,400	3,800	3,280	3,350	2,930	Oct	3,152
Ampton Water			2,535	3,735			3,135
Severn Estuary	3,265	2,761	2,936	2,701	3,353	Sep	3,003
Humber Estuary	3,460	1,524	2,957	(2,347)	2,455	Jan	2,599
Lower Derwent Ings			2,630	2,560			2,595
The Wash	3,264	1,781	2,384	2,639	2,437	Jan	2,501
Morecambe Bay	(3,126)	(1,683)	2,455	2,208	1,891	Aug	2,420
Tring Reservoirs	(1,700)	1,834	2,800	2,000	1,557	Sep	2,048
Other sites surpassing table qualifying levels in WeBS-Year 2004/2005 in Great Britain†							
Livermere	1,010	1,146	820	440	2,517	Oct	1,187
Clifford Hill Gravel Pits Consolidated	1,620	910	1,148	2,784	2,143	Sep	1,721
Swale Estuary	1,778	1,635	1,452	1,800	2,010	Dec	1,735

† *as few sites exceed the British threshold a qualifying level of 2,000 has been chosen to select sites for presentation in this report*

Chestnut Teal

Anas castanea

Escape
Native Range: S Australia

GB max:	1	Nov
NI max:	0	

Singles were present at Salthouse, North Norfolk Coast in November and at Scotney Pit and Dungeness Gravel Pits in the following June.

Pintail

Anas acuta

International threshold:	600
Great Britain threshold:	279
All-Ireland threshold:	60

GB max:	21,745	Jan
NI max:	364	Nov

	S	M	L
GB change	o	o	o
NI change	o	(+)	(+)

Both the Great Britain annual index and counted maximum were down slightly on those of the previous year. Overall, the underlying British trend has shown a steady decline since the early 1980s but with an upturn since the low point in 2000/01. The British monthly indices show that numbers were a little higher than average in most months, particularly in October. Pintail leave rapidly in the spring, with very few remaining by April, and birds were only recorded from four sites in both July 2004 and June 2005. In Northern Ireland, the index fell a little but was still high compared to recent years, whilst the counted maximum was lower than that seen in 2003/04. The great majority of the birds recorded during the November peak were found at Strangford Lough; indeed almost all of the Strangford birds were at the far northern end of the Lough.

There were several changes in the table of important sites. The Stour Estuary fell from international importance and the Mersey fell from even supporting numbers of national importance. The latter represents a major collapse in numbers at what used to be one of the key British sites. Conversely, high counts conferred new national importance status for Wigtown Bay, Malltraeth Cob & Pools, Lindisfarne and the Crouch-Roach Estuary, whilst the Lower Derwent Valley, Mersehead RSPB and the Arun Valley are now seen as supporting internationally important numbers. At the other key resorts, numbers were on the low side at the Nene Washes, Somerset Levels, River Avon (Ringwood to Christchurch), Dee Flood Meadows, Ashleworth Ham and Coombe Hill Canal. All of these are inland sites dependent for their wildfowl interest on the degree of flooding. As also seen for Teal, however, the peak at the Ouse Washes bucked the trend of these other inland sites and the peak was the highest since 1999/2000. On the coast, whilst numbers were relatively low at the Dee Estuary they were higher than usual at the nearby Ribble Estuary, perhaps suggestive of local redistribution around Liverpool Bay. Counts were also relatively high at the Inner Moray Firth, Martin Mere, Pagham Harbour and the Duddon Estuary.

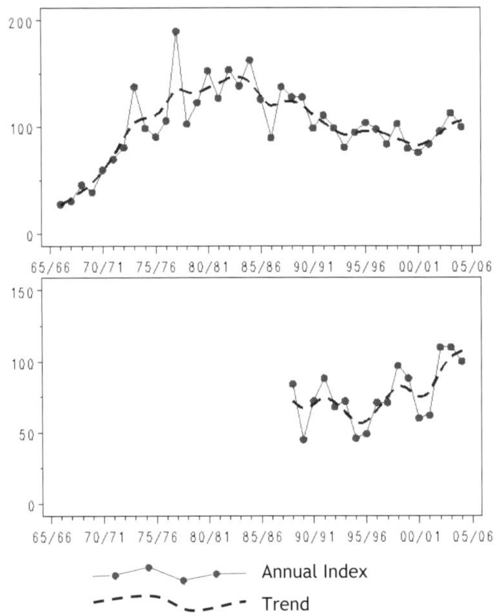

Figure 24.a, Annual indices & trend for Pintail for GB (above) & NI (below).

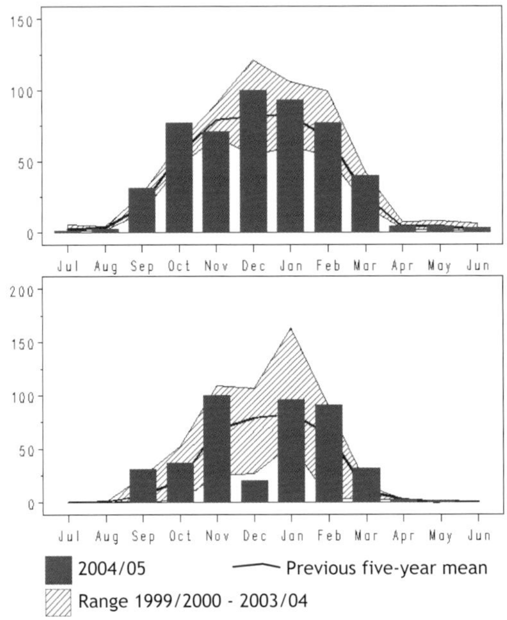

Figure 24.b, Monthly indices for Pintail for GB (above) & NI (below).

	00/01	01/02	02/03	03/04	04/05	Mon	Mean
Sites of international importance in the UK							
Dee Estuary (England & Wales)	4,216	6,023	6,000 [13]	6,317	4,312	Oct	5,374
Solway Estuary	2,818 [11]	(8,070)	(3,357)	4,183	(4,352)	Oct	4,856
Morecambe Bay	2,387	3,471	3,628	(3,942)	3,620	Oct	3,410
Burry Inlet	1,328	1,305	4,410	5,772	2,745	Dec	3,112
Ouse Washes	1,509 [13]	2,606 [13]	2,844 [13]	2,277 [13]	3,330	Jan	2,513
Severn Hams		2,000 [13]					2,000
Nene Washes	2,671	1,250	3,478	1,779	327	Jan	1,901
Ribble Estuary	819	619	1,405	(2,562)	(3,058)	Jan	1,693
Duddon Estuary	(628)	391	(415)	(1,299)	1,626 [11]	Dec	1,105
North Norfolk Coast	987	1,296	(475)	(768)	712	Oct	998
Severn Estuary	(981)	(780)	(891)	(354)	(784)	Jan	(981)
Medway Estuary	(475)	(1,118)	(333)	(95)	812 [11]	Feb	965
Somerset Levels	(1,546)	1,084	(1,315)	494	261	Jan	940
Swale Estuary	952	998	946	962	672	Jan	906
The Wash	70	516	1,253	1,086	915	Jan	768
R.Avon: R'gwood-Christchurch	1,385 [13]	280	2,013	25	46	Dec	750
Dee Flood Meadows	(990)	1,050	(628)	580	250	Jan	700
Mersehead RSPB Reserve	480	410	1,140	480	970	Dec	696 ▲
Lower Derwent Ings			660	573			617 ▲
Arun Valley	(1,171)	(413)	(775)	403	293	Jan	611 ▲

	00/01	01/02	02/03	03/04	04/05	Mon	Mean
Sites of national importance in Great Britain							
Stour Estuary	691	629	613 [11]	467	310 [11]	Dec	542 ▼
WWT Martin Mere	344	635	487	463	710 [12]	Jan	528
Pagham Harbour	340	587	304	477	834	Dec	508
Alde Complex	(506)	705	403	(330)	313	Jan	482
Blackwater Estuary	(325)	352	(498)	461	555	Jan	467
Breydon Wtr & Berney Marshes	446	329	571	271 [11]	248	Jan	373
Thames Estuary	(244)	(223)	355	(126)	(100)	Jan	355
Otmoor	396 [13]	160 [13]	481 [13]		(156)	Jan	346
Inner Moray and Inverness Firth	307	313	310	258	518	Feb	341
Wigtown Bay	115	(195)	(320)	(359)	(654)	Dec	329 ▲
Malltraeth Cob and Pools				207	421	Dec	314 ▲
Poole Harbour	296	424	191	316	338	Dec	313
Orwell Estuary	179 [11]	473 [11]	372 [11]	325 [11]	165 [11]	Feb	303
Lindisfarne	196	272 [11]	330	384	301	Nov	297 ▲
Blyth Estuary	202	368					285
Crouch-Roach Estuary	269	192	385	267	(281)	Dec	279 ▲
Sites of all-Ireland importance in Northern Ireland							
Strangford Lough	249	348	378	582	349	Nov	381
Sites no longer meeting table qualifying levels in WeBS-Year 2004/2005							
Mersey Estuary	491	134	220	152	204	Jan	240
Other sites surpassing table qualifying levels in WeBS-Year 2004/2005 in Great Britain							
Fleet and Wey	241	281	149	(281)	420	Nov	274
North West Solent	100	233	96	(391)	412	Dec	246
Chichester Harbour	43	190	69	233	297	Feb	166
Dornoch Firth	158 [11]	269	252	140	291	Dec	222

Bahama Pintail

Anas bahamensis

Escape
Native Range: S America

GB max:	14	Dec
NI max:	0	

Bahama Pintail were recorded at four sites between September and May. The single site peak of nine at Harrow Lodge Park in December was accompanied by counts of four and one at Dagenham Chase Gravel Pits and Doddington Pool respectively. Counts outside of December were all of single birds. Stanton Lake was the other site at which this species was recorded.

Garganey

Anas querquedula

International threshold:	20,000**
Great Britain threshold:	+†
All-Ireland threshold:	+†

GB max:	53	Sep
NI max:	0	

As a summer visitor, Garganey are considered here for the calendar year 2004. The first birds appeared in March at three sites, with the year's peak for Breydon Water being in this month. Birds were then more widespread in April and May, followed by a mid-summer lull, with passage birds being absent and breeding birds at their most retiring. Records increased to a September peak of 53 birds at 25 sites, a little below average for the last ten years. Late birds were still present at eight sites in October but none were seen thereafter. Overall, Garganey were recorded from 56 sites during 2004.

Garganey are never present in large concentrations in Britain (and have never been recorded at all in Northern Ireland during WeBS counts). For the third year running, the only double-figure site count was from Wraysbury Gravel Pits in September, but peaks at different sites were from a wide variety of months reflecting the variation in site usage over the season. Most sites featuring in the table below are in southeast Britain, notable exceptions being the Dee Estuary, Tees Estuary and Loch Gelly. The notable absence from Stodmarsh NNR & Collards Lagoon was due to a lack of summer counts for this site during 2004.

	2000	2001	2002	2003	2004	Mon	Mean
Sites with mean peak counts of 4 or more birds in Great Britain[†]							
Stodmarsh NNR / Collards Lgn	12	30	5	7	0		11
Thames Estuary	9	(2)	(5)	(2)	(3)	Aug	9
Wraysbury Gravel Pits	2	0	15	12	14	Sep	9
Breydon Wtr & Berney Marshes	6	4	4	4	8	Mar	5
Dungeness Gravel Pits	2	12	(0)	3	4	Aug	5
Ouse Washes	9	3	7	2	4	Sep	5
Dee Estuary (England & Wales)	(4)	(0)	(3)	(1)	(4)	Aug	(4)
North Norfolk Coast	3	(4)	(4)	(3)	4	Apr	4
Rye Harbour and Pett Level	1	7	9	2	3	Jun	4
Severn Estuary	4	(4)	(7)	2	1	Sep	4
Other sites surpassing table qualifying levels in Summer 2004 in Great Britain[†]							
Blagdon Lake	0	3	0	2	5	Sep	2
Loch Gelly	0		2	0	4	Jun	2
Staines Reservoirs	0	0	0	0	4	Aug	1
Tees Estuary	3	(1)	1	2	4	May	3

[†] *as no British or All-Ireland thresholds have been set a qualifying level of 4 has been chosen to select sites for presentation in this report*

Blue-winged Teal

Anas discors

Vagrant and escape
Native Range: Americas

GB max: 1 May
NI max: 0

One was reported from Norton Marsh in North Norfolk Coast during May.

Shoveler

Anas clypeata

International threshold: 400
Great Britain threshold: 148
All-Ireland threshold: 65

GB max: 11,141 Feb
NI max: 156 Jan

	S	M	L
GB change	o	o	o
NI change	o	o	o

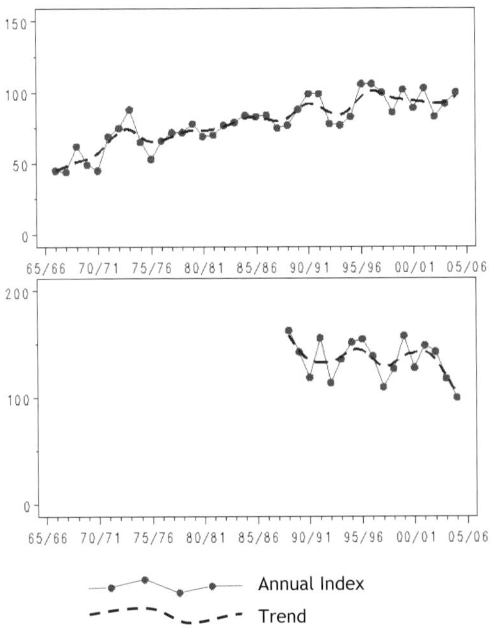

Figure 25.a, Annual indices & trend for Shoveler for GB (above) & NI (below).

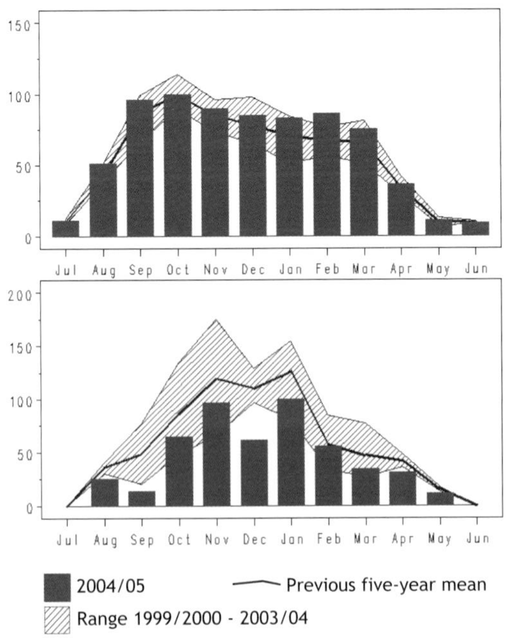

Figure 25.b, Monthly indices for Shoveler for GB (above) & NI (below).

The British annual index rose for the second year running with an underlying trend of approximate stability over the last decade following a steady increase. Additionally, the British counted maximum rose since 2003/04. The monthly indices show that British numbers have peaked already by October then decline slowly to March following which there is a larger exodus to breeding sites at home and abroad. In Northern Ireland, the annual index dropped to the lowest level so far recorded, with the counted maximum for Northern Ireland the lowest for over a decade.

At the key sites, whilst the Somerset Levels remains the top site on the basis of mean numbers, the peak at the Ouse Washes exceeded 1,000 birds for the third consecutive year. The peak autumn count at Rutland Water was the highest there since 1999. At most other sites supporting important numbers, however, peaks were lower than their recent five-year means, the main exceptions being Walthamstow Reservoirs and Colne Valley Gravel Pits. However, there were 11 sites at which counts exceeded the national 1% level in 2004/05 despite not qualifying on the basis of their five-year peak means. Many sites showed small declines, with larger declines noted from the Thames Estuary and Pitsford Reservoir, whilst four sites dropped from the list of those supporting nationally important numbers. Over 75% of the Northern Ireland January total was at Strangford Lough, where numbers were at their lowest since 1997/98.

	00/01	01/02	02/03	03/04	04/05	Mon	Mean
Sites of international importance in the UK							
Somerset Levels	(1,343)	1,170	(2,190)	784	(902)	Feb	1,372
Ouse Washes	396 [13]	968 [13]	1,125	1,104 [13]	1,173	Jan	953
Thames Estuary	564	(605)	697	415	402	Jan	537
Rutland Water	401	608	504	475	663	Oct	530
Chew Valley Lake	270	805	535	565	395	Jan	514
Breydon Wtr & Berney Marshes	620	679	415	322	468	Mar	501
Swale Estuary	511	587	440	330	292	Jan	432
Loch Leven	480	400	550	295	386	Oct	422
Abberton Reservoir	352	440	422	488	355	Oct	411
Sites of national importance in Great Britain							
Dungeness Gravel Pits	398	504	320	378	340	Oct	388
Lower Derwent Ings			442	319			381
Burry Inlet	368	215	397	327	344	Dec	330
Severn Estuary	306	366	368 [11]	325	266	Feb	326
Staines Reservoirs	(130)	356	377	261	308	Sep	326
Lee Valley Gravel Pits	374	321	(308)	(246)	275	Feb	323
Llynnau Y Fali	464	176	337	233	232	Mar	288
Alde Complex	(181)	(407)	229	(106)	175	Mar	270
Stodmarsh NNR / Collards Lgn	(409)	206	244	202	272	Feb	267
Arun Valley	392	227	259	195	175	Feb	250
Ribble Estuary	393	179	197	231	219	Dec	244
Chichester Gravel Pits	160	317	238	321	173	Jan	242
Nene Washes	190	374	262	200	177	Mar	241
North Norfolk Coast	203	289	182	212	234	Dec	224
Minsmere	241 [13]	207	233	180	227	Oct	218
Confidential SE England Site	164	520	125	120	120	Jan	210
Fairburn Ings	289	153	159	221			206
R. Avon: F'bridge-Ringwood	182	117	361	188	149	Jan	199
Morecambe Bay	57	380	(82)	184	167	Nov	197
Rye Harbour and Pett Level	160	282	167	204	162	Dec	195
Walthamstow Reservoirs	157	179	135	212	265	Sep	190
Tees Estuary	(264)	114	245	181	145	Nov	190
Blagdon Lake	145	400	75	146	160	Aug	185
Fen Drayton Gravel Pits	378	157	128	115	135	Oct	183
Wraysbury Gravel Pits	154	260 [13]	221	97	172	Mar	181
Hampton & Kempton Reservoirs	118	208	(88)	165	(134)	Dec	164
Pitsford Reservoir	114	153	91	378	70	Mar	161
Malltraeth RSPB	(157)	(145)	186	124	173	Nov	161
Dee Estuary (England & Wales)	157	(101)	(80)	(69)	(105)	Aug	157 ▲
Cotswold Water Park (West)	(188)	88	(218)	(91)	126	Mar	155 ▲
Colne Valley Gravel Pits	(62)	(120)	(38)	(96)	(154)	Mar	(154) ▲
Middle Yare Marshes	(175)	151	169	96	(111)	Dec	148 ▲

	00/01	01/02	02/03	03/04	04/05	Mon	Mean
Sites of all-Ireland importance in Northern Ireland							
Strangford Lough	159	182	199	201	119	Jan	172
Sites no longer meeting table qualifying levels in WeBS-Year 2004/2005							
Blithfield Reservoir	341	58	148	60	129	Sep	147
Brent Reservoir	183	230	125	20	129	Aug	137
Humber Estuary	146	(78)	109	(127)	99	Oct	120
Medway Estuary	(71)	(280)	(20)	(26)	19 [11]	Dec	83
Other sites surpassing table qualifying levels in WeBS-Year 2004/2005 in Great Britain							
Trinity Broads	30	(142)	55	137	304	Oct	134
Grafham Water	128	143	51	112	266	Dec	140
Chetwynd Pool	91	67	166	156	220	Oct	140
WWT Martin Mere	103	92	43	162	198	Sep	120
N. Warren / Thorpeness Mere		121	156 [13]	120 [13]	166	Feb	141
Barn Elms Reservoirs	90	154	107	132	160	Sep	129
Belvide Reservoir	81	79		50	160	Sep	93
Coombe Country Park			0	83	159	Sep	81
Woolston Eyes	210	103	71	175	157	Sep	143
Otmoor	134 [13]	(103)	130		150	Jan	138
Sutton and Lound Gravel Pits		118	10	108	150	Sep	97

Ringed Teal

Callonetta leucophrys

Escape
Native Range: S America

GB max:	1	Jun
NI max:	0	

The only record of this escape was of a single bird at Liden Lagoon in Wiltshire during June.

Red-crested Pochard

Netta rufina

International threshold:	**500**
Great Britain threshold:	?[†]
All-Ireland threshold:	?[†]

GB max:	175	Oct
NI max:	0	

Red-crested Pochards in the UK remain highly concentrated around the upper reaches of the Thames valley, most especially at the two parts of the Cotswold Water Park (where numbers have been steadily rising) but also in increasing numbers downstream at the Lower Windrush Valley Gravel Pits. A new record peak for Wraysbury Gravel Pits suggests this range may be extending even further downstream. Elsewhere, there are indications of new concentrations developing along the Trent valley. There have been no counts submitted for the Pensthorpe area of the river Wensum in Norfolk for several years now.

	00/01	01/02	02/03	03/04	04/05	Mon	Mean
Sites with mean peak counts of 10 or more birds in Great Britain[†]							
Cotswold Water Park (West)	56	58	(74)	114	81	Oct	77
Cotswold Water Park (East)	33	72	40	33	48	Oct	45
Baston and Langtoft Gravel Pits	17	16	8	(23)			16
Hanningfield Reservoir	1	6	6	(43)	2	Apr	12
Other sites surpassing table qualifying levels in WeBS-Year 2004/2005 in Great Britain[†]							
Lower Windrush Valley GPs	1	8	5	6	(19)	Feb	8
Arnot Park Lake Arnold	0	0	0	12	16	Sep	6
Sutton and Lound Gravel Pits		3	7	6	16	Nov	8
Wraysbury Gravel Pits	0	0	0	0	10	Nov	2

[†] *as no British or All-Ireland thresholds have been set a qualifying level of 10 has been chosen to select sites for presentation in this report*

Pochard

Aythya ferina

GB max:	30,553	Jan
NI max:	7,110	Feb

International threshold:	3,500
Great Britain threshold:	595
All-Ireland threshold:	400

	S	M	L
GB change	o	o	o
NI change	(--)	--	--

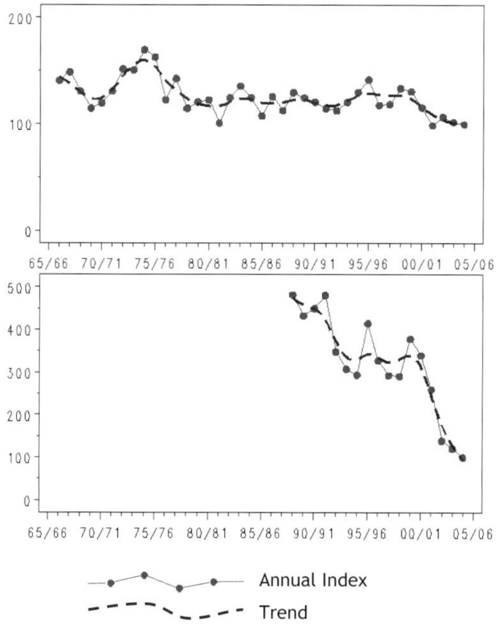

Figure 26.a, Annual indices & trend for Pochard for GB (above) & NI (below).

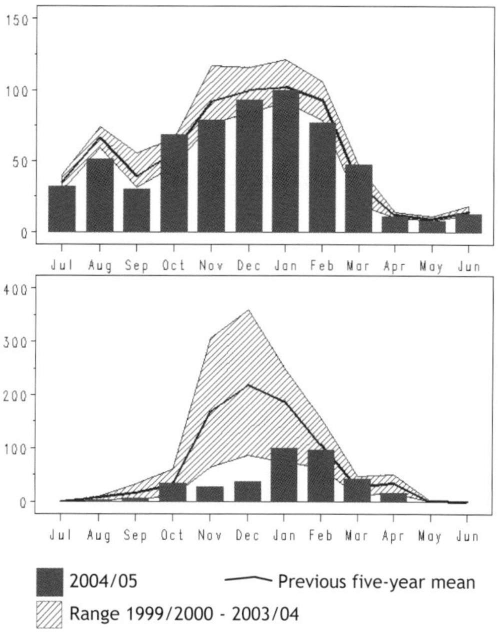

Figure 26.b, Monthly indices for Pochard for GB (above) & NI (below).

The British annual index remained similar to that of the previous year, at a historically low level, whilst the counted British max was the second lowest since 1970/71. The monthly indices in Britain showed a similar pattern to other recent years, with an August arrival, a decline in September and then a more general autumn arrival. Compared to recent years, the August arrival was lower than normal but the January peak was about average.

In Northern Ireland, the annual index slipped a further 17% to a new low, following a huge decline since 2000/01. The monthly indices show that numbers were particularly low before New Year. As usual, almost all birds in Northern Ireland were at Loughs Neagh and Beg where numbers have declined dramatically in recent years, perhaps due to increased nutrient inputs to the loughs having detrimental effects on the availability of chironomid larvae on which the species primarily feeds (Maclean *et al.* 2006). The 2004/05 peak for the site was the lowest ever.

In Britain, counts at both Abberton Reservoir and the Ouse Washes were lower than usual (the latter the lowest since 1996/97), although mean peak counts still suffice to enable these species to qualify as internationally important for the species. Peak counts were also low at most sites supporting nationally important numbers, although the very low count at the Nene Washes was to be expected given very low levels of flooding this winter at the site. Only at Chew Valley Lake were counts relatively high compared to other recent years.

	00/01	01/02	02/03	03/04	04/05	Mon	Mean
Sites of international importance in the UK							
Loughs Neagh and Beg	24,393	16,168	9,082	7,835	6,764	Jan	12,848
Abberton Reservoir	5,296	3,125	4,325	5,290	3,188	Jan	4,245
Ouse Washes	4,602 [13]	4,206	4,583	3,304 [13]	2,099	Jan	3,759

	00/01	01/02	02/03	03/04	04/05	Mon	Mean
Sites of national importance in Great Britain							
Loch Leven	1,330	4,074	2,934	2,548	2,193	Oct	2,616
Lower Derwent Ings			1,973	1,236			1,605
Middle Tame Valley Gravel Pits	1,733	1,423	(442)	(203)	(56)	Feb	1,578
Nene Washes	4,102	48	2,853	66	32	Jan	1,420
Hornsea Mere	580	1,115	1,415	1,325	1,150	Jan	1,117
Cotswold Water Park (West)	988	(512)	(377)	(499)	(571)	Jan	988
Fleet and Wey	928	1,072	926	850	746	Jan	904
Severn Estuary	1,008	1,064	772	(905)	652	Jan	880
Dungeness Gravel Pits	669	595	765	855	788	Aug	734
Loch of Boardhouse	711	822	605	705	770	Oct	723
WWT Martin Mere	861	860	750	565	472	Jan	702
Woolston Eyes	537	570	637	663	620	Feb	605
Lower Windrush Valley Gravel Pits	681	600	(384)	505	(423)	Dec	595
Sites of all-Ireland importance in Northern Ireland							
Upper Lough Erne	185	780	916	801	473	Feb	631
Sites no longer meeting table qualifying levels in WeBS-Year 2004/2005							
Cotswold Water Park (East)	723	826	371	629	410	Feb	592
Hickling Broad	390						390
Other sites surpassing table qualifying levels in WeBS-Year 2004/2005 in Great Britain							
Chew Valley Lake	290	735	475	480	635	Jan	523

Redhead

Aythya americana

Vagrant and escape
Native Range: N America

GB max:	1	Oct
NI max:	0	

A single drake was present at Kenfig Pool, Glamorgan, from October until December.

Ring-necked Duck

Aythya collaris

Vagrant
Native Range: N America

GB max:	8	Jan
NI max:	0	

Ring-necked Ducks were recorded at a total of thirteen sites this year, with a peak of eight birds in January. All records involved single birds except for two at Wimbleball Lake during December and January. Long-staying birds were at Porth Reservoir from November to March, Loch Sarclet from November to February and Lochs Davan and Kinord from November to April. Other sites hosting the species were Barrow Gurney Reservoir, Point of Ayre Gravel Pit, Attenborough Gravel Pits, Long Eaton Gravel Pits, Bredon`s Hardwick Gravel Pits, Tophill Low Reservoirs, Loch of Skene, Loch Hempriggs and Papil Water, Fetlar.

New Zealand Scaup

Aythya novaeseelandiae

Escape
Native Range: New Zealand

GB max:	1	Sep
NI max:	0	

A single New Zealand Scaup was reported from Chew Valley Lake during September and November.

Ferruginous Duck

Aythya nyroca

Vagrant and escape
Native Range: N America, Asia

GB max:	1	Aug
NI max:	0	

Ferruginous Ducks were present at only three sites compared with eight during the previous year. All records referred to single birds and included a female at Chew Valley Lake in August and September, and drakes at Welney (Ouse Washes) in November and at Abberton Reservoir in March.

Tufted Duck

Aythya fuligula

International threshold:	12,000
Great Britain threshold:	901
All-Ireland threshold:	400

GB max:	56,338	Oct
NI max:	11,536	Feb

	S	M	L
GB change	o	o	o
NI change	(-)	--	-

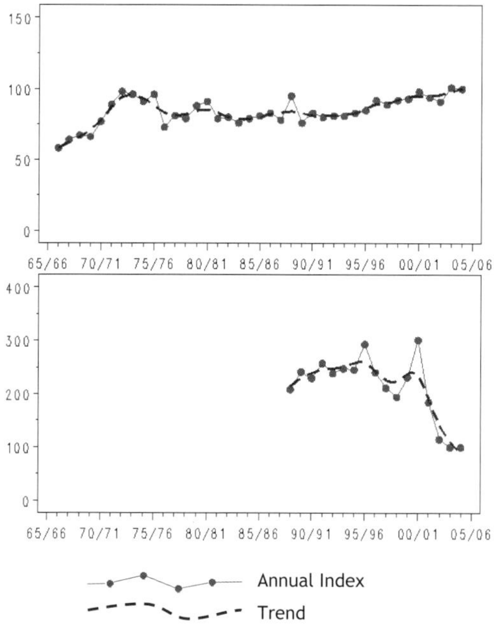

Figure 27.a, Annual indices & trend for Tufted Duck for GB (above) & NI (below).

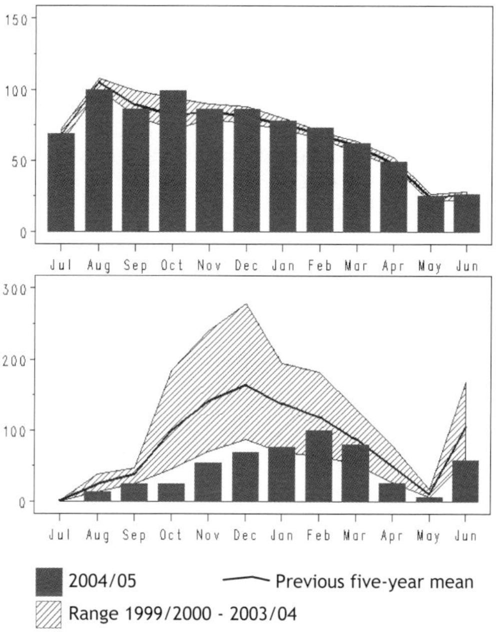

Figure 27.b, Monthly indices for Tufted Duck for GB (above) & NI (below).

The long-term upwards trend in Tufted Duck numbers wintering in Great Britain continues, although the annual index value fell just slightly. This is in line with the results of the Breeding Bird Survey which found that the species increased by 27% over the period 1994 to 2004 (Raven *et al.* 2005). As with Pochard, there is a peak in August but unlike that species there is not always a clear secondary arrival later in the autumn (although the numbers present in October 2004 were higher than usual). Conversely, in Northern Ireland, whilst the annual index did not fall any further, it remains at the very low level of the past two years. The peak on the key site, Loughs Neagh and Beg, rose slightly but still remains only a fraction of that seen as recently as 2000/01. If numbers do not increase at the site in 2005/06 they will no longer rank as internationally important. Numbers at both Upper and Lower Loughs Erne, however, remained relatively high.

At the main British sites, increasing numbers are evident at Abberton Reservoir, Pitsford Reservoir and Little Paxton Gravel Pits, whilst sustained high numbers at Cotswold Water Park (West) means that this site holds nationally important numbers of Tufted Ducks now. Particularly low counts were seen from Hanningfield Reservoir, Staines Reservoirs, Alton Water, Draycote Water and Besthorpe & Girton Gravel Pits.

	00/01	01/02	02/03	03/04	04/05	Mon	Mean
Sites of international importance in the UK							
Loughs Neagh and Beg	26,417	13,318	9,771	8,999	9,277	Feb	13,556
Sites of national importance in Great Britain							
Rutland Water	3,313	5,115	7,496	6,818	6,488	Oct	5,846
Loch Leven	3,900	3,650	4,872	3,913	3,826	Sep	4,032
Abberton Reservoir	4,414	1,418	2,487	2,067	5,112	Oct	3,100
Middle Tame Valley Gravel Pits	2,547	2,164	(915)	(325)	(129)	Sep	2,356
Pitsford Reservoir	1,202	1,263	2,441	2,226	2,506	Oct	1,928
Walthamstow Reservoirs	1,691	1,838	1,867	1,772	1,771	Aug	1,788
Hanningfield Reservoir	2,183	1,160	1,641	3,109	400	Jan	1,699
Wraysbury Gravel Pits	785	2,091	2,422	846	1,015	Feb	1,432
Ouse Washes	2,214	1,395 [13]	1,192	973 [13]	1,330	Jan	1,421
Staines Reservoirs	1,243	1,026	1,971	1,133	792	Jul	1,233
Lee Valley Gravel Pits	1,085	1,027	1,248	1,404	1,222	Nov	1,197
Chew Valley Lake	785	1,020	1,080	1,465	1,235	Oct	1,117
Alton Water	1,389	961	815	1,440	644	Feb	1,050
Cotswold Water Park (West)	(753)	736	(792)	1,199	960	Nov	965 ▲
Sites of all-Ireland importance in Northern Ireland							
Upper Lough Erne	745	998	1,065	1,236	1,295	Feb	1,068
Lower Lough Erne			635	580	674	Feb	630
Sites no longer meeting table qualifying levels in WeBS-Year 2004/2005							
Tophill Low Reservoirs	920	720	898	853	915	Aug	861
Draycote Water	744	740 [13]	1,030	1,251	122	Aug	777
Besthorpe & Girton Gravel Pits	(462)	(418)	(10)	(200)	97	Jul	294
Other sites surpassing table qualifying levels in WeBS-Year 2004/2005 in Great Britain							
Little Paxton Gravel Pits	505	531	678	653	1,137	Oct	701
Tophill Low Reservoirs	920	720	898	853	915	Aug	861

Scaup

Aythya marila

International threshold:	3,100
Great Britain threshold:	76
All-Ireland threshold:	30*

GB max: 7,599 Dec
NI max: 5,806 Feb

	S	M	L
GB change	o	o	o
NI change	o	o	+

**50 is normally used as a minimum threshold*

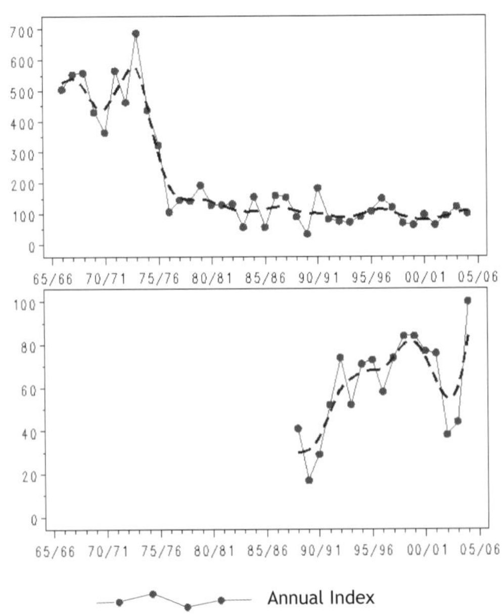

Figure 28.a, Annual indices & trend for Scaup for GB (above) & NI (below).

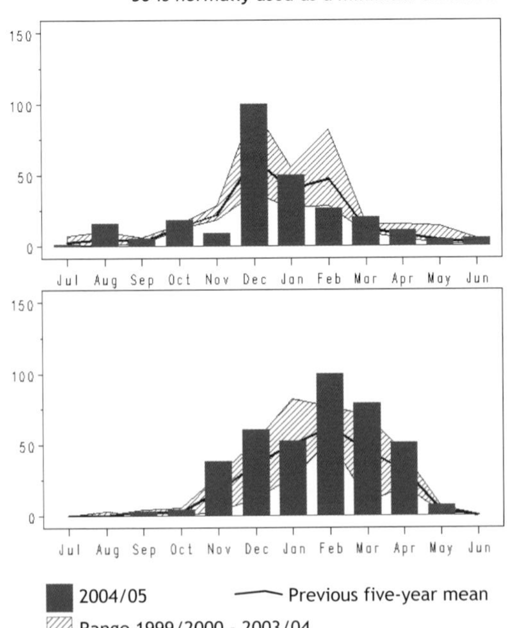

Figure 28.b, Monthly indices for Scaup for GB (above) & NI (below).

In Northern Ireland, the annual index more than doubled since 2004/05, to its highest level to date. This followed concerns about a crash in numbers at Loughs Neagh and Beg; concerns that may have proved somewhat premature (although numbers of Pochard and Tufted Duck at the lough remain low). The peak count at the site of over 5,000 Scaup was the highest single UK site count anywhere since 1993, and means that peak mean numbers at the site are once again of international importance. The monthly indices for Northern Ireland show a species that arrives later than many other wildfowl, with few before November and peaking in February. Elsewhere in the province, numbers at Carlingford Lough remained relatively low but there was a new record count in Belfast Lough at low tide. Notably, the Belfast peak was two months early than at Loughs Neagh and Beg and thus could well have involved some of the same birds.

The annual index for Great Britain saw a slight fall since 2003/04, although the counted British maximum was the highest since 1988. Moreover, this total did not include any counts for Loch Indaal, one of the key sites for the species. The Solway Estuary (not all that far from Northern Ireland) remains the key British site, and could well soon qualify as internationally important for the species with the second-highest count for the site ever. Indeed, the majority of these birds were offshore at Carse Bay where the count was considered an undercount. Elsewhere, there were high counts from the Inner Moray Firth (mostly at Milton of Culloden), Loch Ryan, Loch of Harray and Morecambe Bay, but the peak from the Firth of Forth remained low for a second year.

	00/01	01/02	02/03	03/04	04/05	Mon	Mean
Sites of international importance in the UK							
Loughs Neagh and Beg	2,633	3,389	2,565	2,674	5,144	Feb	3,281
Sites of national importance in Great Britain							
Solway Estuary	1,818 [11]	2,367 [11]	(1,077)	(1,782)	(4,610)	Dec	2,932
Loch Ryan	631	766 [13]	907 [13]	986	1,577	Dec	973
Inner Moray and Inverness Firth	313	323	923	518	2,641 [1]	Jan	944
Loch Indaal	1,200	241	755	1,003			800
Loch of Harray	311	97	(185)	420	490	Jan	330
Loch of Stenness	211	513	309	266	315	Mar	323
Cromarty Firth	424	353 [13]	160 [1]	13	47	Feb	199
Ayr to North Troon	100	200	120	(12)	(14)	Dec	140
Auchenharvie Golf Course	98			145			122
Rough Firth	204	88	0	107	204 [11]	Dec	121
Forth Estuary	240	189	130	14 [11]	22	Feb	119
Dornoch Firth	56	107	163	70	150 [13]	Feb	109
Sites of all-Ireland importance in Northern Ireland							
Belfast Lough	493	270	642	669 [11]	1,224 [11]	Dec	660
Carlingford Lough	800	618	168	(158)	233	Jan	455
Other sites surpassing table qualifying levels in WeBS-Year 2004/2005 in Great Britain							
Morocambe Bay	29	(37)	(8)	(24)	112	Jan	71

Lesser Scaup

Aythya affinis

Vagrant
Native Range: N America

GB max:	**3**	**Mar**
NI max:	**1**	**Feb**

Single birds were recorded at six sites during 2004/05. There were two long-stayers; at Abberton Reservoir in December and January and Dungeness Gravel Pits from December until March. Other records were from Upper Lough Erne in February, Beesands Ley and Vale of Coustry Complex in March and Rutland Water in May.

Eider

Somateria mollissima

GB max:	18,661	Dec
NI max:	1,712	Dec

International threshold: 15,500
Great Britain threshold: 730
All-Ireland threshold: 20*

	S	M	L
GB change	o	o	o
NI change	+	++	++

**50 is normally used as a minimum threshold*

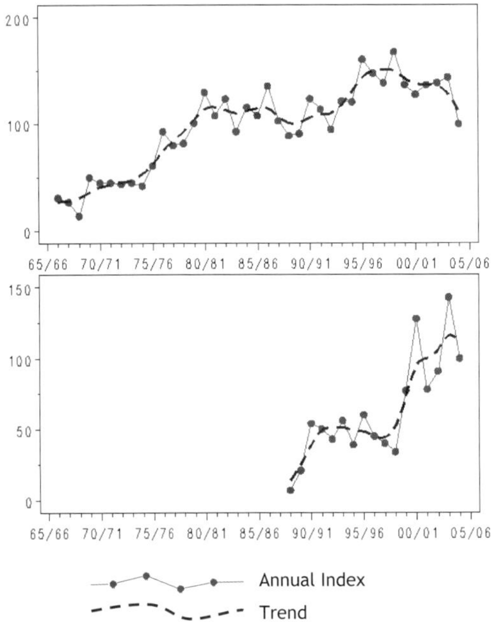

Figure 29.a, Annual indices & trend for Eider for GB (above) & NI (below).

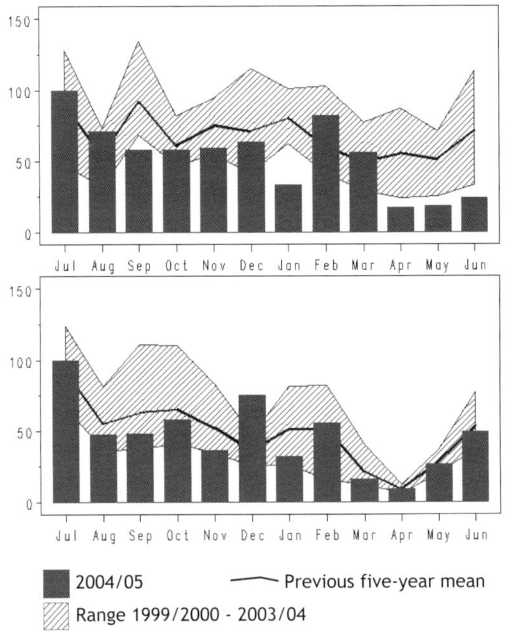

Figure 29.b, Monthly indices for Eider for GB (above) & NI (below).

The British annual index for Eider declined sharply in 2004/05 to its lowest level for more than a decade, and the counted British maximum was below 20,000 for only the second time in recent years. However, this is a species for which sizeable fluctuations in the index are frequent, and substantial parts of the range are not well covered by WeBS. Therefore, further years of data are needed to see whether this represents an actual downwards trend. In Northern Ireland, the index and counted maximum were also down on the previous year but remain relatively high in recent terms.

The Firth of Clyde remains the most important area for Eiders, although on the basis of mean numbers has fallen from supporting internationally important numbers. The excellent set of September supplementary counts for the 'greater' Clyde area was again kindly made available, with the table below showing the total for the area plus peaks for some of the individual parts (C. Waltho *pers. comm.*) The total for the Clyde was the lowest recorded of the last nine years, with particularly low counts at Gourock-Largs, Bute and Inner Loch Fyne. WeBS Core Counts for the Inner Clyde also showed numbers somewhat below average for recent years

Elsewhere, WeBS Core Counts were again deemed incomplete for the mouth of the Tay Estuary. At the Firth of Forth, in common with many other species of sea-ducks, the 2004/05 Eider peak count was low compared to recent years, in fact the lowest of the last 16 years. A supplementary count for Morecambe Bay in August 2004 was well in excess of the count in the same month during WeBS Core Counts. Recent offshore counts in Aberdeen Bay revealed the largest count of the year, at just over 6,000 birds. There is clearly likely to be overlap with the flock at the Ythan Estuary. Peak numbers at Montrose Basin were the lowest for a decade and numbers remained low also at Lindisfarne, where a low tide peak count was about twice the peak core count, which failed to surpass 600 birds. However, by far the biggest apparent decline was on the Wash where, despite recent complaints about large numbers feeding on commercial shellfish beds, the peak count of 91 was the lowest since 1987/88. However, birds can be well offshore at this site and difficult to count accurately from land. In Northern Ireland, peaks at most sites were similar to those seen in recent years, although the peak at Strangford Lough was the highest there to date.

Data are presented separately for a number of aerial surveys employing distance sampling methodology (see *Additional counts*). Whilst the derived distance estimates were not available at time of publication, it is likely that some of these (such as for the Tay) will result in estimates well in excess of the WeBS Core Counts.

Aerial surveys employing distance sampling

Area	Date	Counted	Estimate	Ref
Firth of Tay	Dec	4,378	not available	Wilson *et al.* 2006
Firth of Forth	Dec	2,033	not available	Wilson *et al.* 2006
North West Strategic Area	Oct/Nov	1,682	not available	WWT WAS Report
Outer Hebrides	Mar	621	not available	Wilson et al. 2006
Scapa Flow	Mar	599	not available	Wilson *et al.* 2006
Sound of Gigha	Mar	335	not available	Wilson *et al.* 2006
Coll & Tiree	Mar	220	not available	Wilson *et al.* 2006
Aberdeen Bay	Feb	190	not available	Wilson *et al.* 2006

	00/01	01/02	02/03	03/04	04/05	Mon	Mean
Sites of national importance in Great Britain							
FIRTH OF CLYDE	14,679 [15]	15,692 [15]	14,297 [15]	15,276 [15]	13,042 [15]	Sep	14,597 ▼
Tay Estuary	(190)	7,500 [13]	(6,000)	(4,700)	(5,636)	Dec	7,500
Forth Estuary	8,893	5,684	7,616	7,014	4,750	Mar	6,791
Inner Firth of Clyde	6,126	3,901	4,730	6,194	4,152	Aug	5,021
Morecambe Bay	5,306	3,903	(4,541)	(3,940)	5,300 [15]	Aug	4,836
Aberdeen Bay offshore				1,756 [59]	6,003 [59]	Aug	3,880
Killantringan Bay					3,600 [15]	Aug	3,600 ▲
Gare Loch	3,877 [15]	3,252 [15]	2,619 [15]	3,263 [15]	2,713 [15]	Sep	3,145
Montrose Basin	(2,500)	3,013	3,051	2,075	1,754	Jul	2,479
Farne Islands	2,434	(671)	(293)	(183)	(187)	Sep	2,434
Ythan Estuary	583	3,531	2,082	3,417	(2,029)	Jul	2,403
Lindisfarne	1,841 [11]	2,024	2,043	1,241	1,202 [11]	Dec	1,670
Moray Firth	(1,491)	(749)	(747)	1,639	1,673	Nov	1,656
Irvine Bay					1,547 [15]	Aug	1,547 ▲
Loch Ryan	2,037	(1,031)	1,188 [15]	1,803	1,150 [15]	Jul	1,545
Loch Long and Loch Goil	1,539 [15]	1,299 [15]	1,459 [15]	1,390 [15]	1,614 [15]	Sep	1,460
Gourock to Largs	908 [15]	1,097 [15]	1,773 [15]	2,220 [15]	614 [15]	Sep	1,322
Holy Loch to Toward Point	1,319 [15]	615 [15]	1,146 [15]	1,114 [15]	2,225 [15]	Aug	1,284
Girvan to Turnberry	957	(151)	1,198	(330)	1,500 [15]	Jul	1,218
Inner Loch Fyne	1,144 [15]	1,647 [15]	1,358 [15]	956 [15]	968 [15]	Sep	1,215
The Wash	1,370	1,344	2,546	703	91	Nov	1,211
Lower Loch Long					914 [15]	Aug	914 ▲
Ayr to North Troon	1,008 [15]	1,203 [15]	458 [15]	1,064	205	Sep	788
Bute	771 [15]	1,143 [15]	944 [15]	457 [15]	451 [15]	Sep	753
Sites of all-Ireland importance in Northern Ireland							
Belfast Lough	2,219 [11]	906	1,016 [11]	1,813	1,490 [11]	Dec	1,489
Lough Foyle	28	344	551	645	431	Jul	400
Outer Ards Shoreline	241		428	(256)	271	Jan	313
Strangford Lough	279	283	165	259	481	Nov	293
Larne Lough	128	107	120	55	69	Sep	96
Port Stewart - Portrush					34 [15]	Aug	34 ▲
Ballycastle - Fair Head					26 [15]	Aug	26 ▲
Other sites surpassing table qualifying levels in WeBS-Year 2004/2005 in Great Britain							
Seahouses to Budle Point	464	735	(368)	400	900	Oct	625
Sound of Barra (Barra)	242 [52]	579 [52]	398 [52]	600 [52]	886 [52]	Mar	541
Hacosay/Bluemull/Colgrave Sds	186 [10]	183 [10]	631 [10]	790 [10]	855 [10]	Mar	529

King Eider

Somateria spectabilis

Vagrant
Native Range: Arctic

GB max:	1	Dec
NI max:	0	

December saw a first-winter drake King Eider off Titchwell on the North Norfolk Coast. Elsewhere, the returning drake was present in Loch Ryan in December, January and March.

Long-tailed Duck

Clangula hyemalis

International threshold: 20,000**
Great Britain threshold: 160†
All-Ireland threshold: +†

GB max:	3,586	Dec
NI max:	15	Jan

The counted British maximum was the highest since 2000/01, and indeed the second highest on record since the 1970s. Conversely, whilst overall numbers in Northern Ireland are generally low, the peak of 15 this year was one of the lowest on record. As usual almost all records in Northern Ireland were at Belfast Lough.

The December count for the Moray Firth was the highest on record. Whether this reflects a real increase, or was due to good count conditions on the day, however, is open to question. Within the wider firth, the largest flocks were present between Nairn and Burghead. WeBS counts were low at the Forth Estuary, and the peak at Water Sound in Orkney was the lowest there in over 20 years of counts. Further south, the peak at the North Norfolk Coast was the highest since 1993/94. WeBS Core Counts were also high compared to recent averages at St Andrews Bay.

Supplementary counts supplied for the Western Isles and for Shetland continue to provide a useful picture of the status of this species away from standard WeBS count sections. In Shetland, the Hacosay/Bluemull/Colgrave area supported the highest counts since 1998/99, whilst numbers were relatively low in South Yell Sound, from Rova Head to Wadbister Ness and from Wadbister Ness to Kirkabister. In the Western Isles, there was a high count in the Sound of Harris, whilst numbers off the west coast of South Uist were lower than in recent years.

Data are presented separately for a number of aerial surveys employing distance sampling methodology (see *Additional counts*).

Aerial surveys employing distance sampling

Area	Date	Counted	Estimate	Ref
Moray Firth	Mar	787	not available	Wilson *et al.* 2006
Firth of Forth	Feb	236	not available	Wilson *et al.* 2006
Scapa Flow	Mar	146	not available	Wilson *et al.* 2006
Outer Hebrides	Mar	54	not available	Wilson *et al.* 2006
Firth of Tay	Feb	32	not available	Wilson *et al.* 2006

	00/01	01/02	02/03	03/04	04/05	Mon	Mean
Sites of national importance in Great Britain							
Moray Firth	(3,991)	(1,501)	3,585 [1]	5,446 [1]	6,402 [1]	Dec	5,144
Forth Estuary	(319)	413	435	249	(240)	Dec	366
South Uist West Coast	302 [52]		411 [52]	440 [52]	185 [52]	Mar	335
Sound of Harris	200 [52]	117 [52]		230 [52]	500 [52]	Feb	262
St Andrews Bay	7	10	97	107	232	Nov	232
Branahuie Banks (Lewis)					196	Dec	196
Hacosay/Bluemull/Colgrave Sds	169 [10]	201 [10]	59 [10]	249 [10]	303 [10]	Mar	196
Sites with mean peak counts of 50 or more birds in Great Britain†							
South Yell Sound	222 [10]	136 [10]	108 [10]	201 [10]	91 [10]	Feb	152
West Whalsay and Sounds		152 [10]					152
Broad Bay (Lewis)	35 [52]		72 [52]	300 [52]			136
Loch of Stenness	75	226	182	105 [13]	89	Jan	135
Quendale to Virkie	203 [10]	117 [10]	122 [10]	103 [10]	100 [10]	Feb	129
Sound of Barra (Barra)		150 [52]		132 [52]	80 [52]	Mar	121
Island of Papa Westray		4	182	184	102	Jan	118
Burra and Trondra		109 [10]		97 [10]	117 [10]	Feb	108
Water Sound	179	68	155	80	60	Mar	108

	00/01	01/02	02/03	03/04	04/05	Mon	Mean
West Coast (Benbecula)	145 [52]		63 [52]	92 [52]			100
Bressay Sound	31 [10]	130 [10]	176 [10]	66 [10]	90 [10]	Nov	99
Dee Mouth to Don Mouth					84	Mar	84
Sound of Taransay (Harris)			50 [52]		100 [52]	Feb	75
Allasdale Bay to Borve (Barra)		30 [52]		112 [52]	68 [52]	Mar	70
Rova Head to Wadbister Ness	84 [10]	131 [10]	63 [10]	34 [10]	21 [10]	Feb	67
Kirkabister to Wadbister Ness	74 [10]	90 [10]	21 [10]	73 [10]	(4) [10]	Feb	65
Traigh Luskentyre	49	126			21	Nov	65
Loch Indaal	231	0	6	5			61
Sites with mean peak counts of 30 or more birds in Northern Ireland†							
Lough Foyle	161	0	1	(0)	0		41
Other sites surpassing table qualifying levels in WeBS-Year 2004/2005 in Great Britain							
North Norfolk Coast	31	34	15	50	67	Mar	39

† *as few sites exceed the British threshold and no All-Ireland threshold has been set, qualifying levels of 50 and 30 have been chosen to select sites, in Great Britain and Northern Ireland respectively, for presentation in this report*

Common Scoter

Melanitta nigra

International threshold: 16,000
Great Britain threshold: 500
All-Ireland threshold: 40*

GB max: 12,197 Dec
NI max: 26 Dec

	S	M	L
GB change	++	++	+
NI change	++	++	++

**50 is normally used as a minimum threshold*

The maximum counted in Britain during WeBS Core Counts was similar to that seen in 2003/04, although clearly hugely underestimates the true total numbers present. The maximum for Northern Ireland was very low, most birds recorded in Belfast Lough. As discussed in Collier *et al.* (2005), it is known that Common Scoters, to a greater extent than almost any other species in this report, are present in substantial numbers at distances offshore which make land-based counts very partial at best. As a result, monitoring is supplemented with specialist, mostly aerial, offshore surveys. As most of these surveys use a distance-sampling methodology (see *Additional counts*), and also because the boundaries of aerial survey sites often do not closely correspond with those for ground-based surveys, the results of such aerial surveys are now presented separately.

The ground count obtained for Carmarthen Bay was substantially lower than in the past few years. However, counts were not available for the mid-winter period, and when the count was carried out on 25th February it is likely (from past experience) that many birds had already departed. Therefore, this low count does not cause undue alarm.

In the North West Strategic Area (broadly between North Wales and the Solway Firth), aerial survey resulted in an estimate of almost 50,000 Common Scoter. The highest concentrations of these were again found off Blackpool, as well as in Colwyn Bay.

On the Moray Firth, the peak count was a little lower than usual, with most birds typically found off Nairn and Culbin Bars. The total was low also for Cardigan Bay, but no dedicated aerial survey was carried out here this year. On the North Norfolk Coast, most birds were present to the west of the site, mostly off Scolt Head and Titchwell, with fewer birds in Holkham Bay than in recent years. In common with a number of other seaducks, the peak at the Forth Estuary was relatively low in 2004/05. However, counts were higher than average in St Andrews Bay, and recent shore-based offshore counts in Aberdeen Bay have confirmed the presence of substantial numbers there also.

Aerial surveys employing distance sampling

Area	Date	Counted	Estimate (confidence intervals)	Ref
North West Strategic Area	Feb/Mar	19,032	47,554 (28,592-73,122)	WWT WAS Report
Carmarthen Bay	Feb	10,104	not available	unpub. BTO data
Greater Wash / Thames Strategic Areas	Nov/Dec	3,575	4,987 (881-11,757)	WWT WAS Report
Firth of Tay	Dec	1,085	not available	Wilson *et al.* 2006
Moray Firth	Mar	693	not available	Wilson *et al.* 2006
Aberdeen Bay	Feb	255	not available	Wilson *et al.* 2006

	00/01	01/02	02/03	03/04	04/05	Mon	Mean
Sites of international importance in the UK							
Carmarthen Bay	19,506 [25]	20,078 [34]	23,288 [28]	15,446 [46]	11,100 [46]	Feb	
Sites of national importance in Great Britain							
Moray Firth	(3,848)	(3,072)	(8,351)	(7,987)	4,265 [1]	Dec	6,868
North Norfolk Coast	606	8,008	5,051	2,252	4,866	Nov	4,157
Aberdeen Bay				2,992 [59]	3,475 [59]	Aug	3,234
St Andrews Bay	2,300	1,705	(584)	1,170	2,660	Dec	1,959
Colwyn Bay	(500)	(620)	(975)	(1,737)			(1,737)
Forth Estuary	783	1,582	3,255	1,349	985	Nov	1,591
Alt Estuary	399	1,900 [13]	1,818	2,169	1,229	Nov	1,503
Cardigan Bay	(280)	(1,000)	(450)	(198)	(183)	Nov	(1,000)
The Wash	2,650	150	452	(15)	372	Jul	906
Dee Estuary (England and Wales)	24	4,000 [11]	5	26	17	Sep	814
Sites of all-Ireland importance in Northern Ireland							
Dundrum Bay	(0)	828 [24]	(0)	(0)	(0)		828
Other sites surpassing table qualifying levels in WeBS-Year 2004/2005 in Great Britain							
Lavan Sands	145	35	72	25	1,800	Nov	415

Black Scoter

Melanitta americana

Vagrant
Native Range: N America

GB max: 1 Mar
NI max: 0

A drake Black Scoter was recorded off Llanfairfechan Saltings, Lavan Sands, in March. One has wintered along this stretch of coast each winter since early 1999 and it is likely these records refer to the same individual. This is the first time that Black Scoter has featured as a full species in the Waterbird Counts following the recent taxonomic split of this species from Common Scoter (Sangster *et al.* 2005).

Surf Scoter

Melanitta perspicillata

Vagrant
Native Range: N America

GB max: 4 Jan
NI max: 0

As is often the case for this species, all records during 2004/05 came from Scottish sites. The peak count at a single site was of three at Traigh Luskentyre (Harris) in January, while two were present at Culbin Bar on the Inner Moray Firth in December and a single drake stayed throughout the winter at Shell Bay-Ruddons Point in the Forth Estuary.

Velvet Scoter

Melanitta fusca

International threshold: 10,000
Great Britain threshold: 30*
All-Ireland threshold: +

GB max: 2,014 Dec
NI max: 0

**50 is normally used as a minimum threshold*

Velvet Scoters were recorded from 26 sites in 2004/05, none of which were in Northern Ireland, and birds were recorded in every month except June. The counted British maximum of 2,014 was about average for recent years, although annual peaks vary greatly due to the concentration of the species at just a few sites, and the importance of good counting conditions offshore at these sites.

The peak count on the Moray Firth in 2004/05 was lower than in the two previous winters. The majority of birds here were off Culbin Bar, with smaller numbers off Nairn Bay and Findhorn Bay. The peak at the Forth Estuary also remained low, where the species remains widespread in the outer firth but with the highest numbers between Eastfield & Musselburgh, and at Largo Bay and Ruddons Point. There was a high count at St Andrews Bay however. Whilst the count at Lunan Bay was lower than in some recent years, it was notable that the peak occurred in August. Elsewhere, the North Norfolk Coast remains the only other site with mean numbers in excess of 30 birds.

	00/01	01/02	02/03	03/04	04/05	Mon	Mean
Sites of national importance in Great Britain							
Moray Firth	744 [1]	610	4,398	2,103 [1]	1,169 [1]	Dec	1,805
Forth Estuary	(542)	1,923	1,487	1,008	1,007	Mar	1,356
St Andrews Bay	(1,870)	800	2	90	1,050	Dec	762
Lunan Bay		400	105	(300)	125	Aug	233
North Norfolk Coast	(3)	41	55	14	45	Nov	39 ▲

Bufflehead

Bucephala albeola

Vagrant and escape
Native Range: N America

GB max: 1 Oct
NI max: 0

A single bird was present at Sea Bank (Huttoft) Clay Pits, Lincolnshire, from October to January.

Goldeneye

Bucephala clangula

International threshold: 4,000
Great Britain threshold: 249
All-Ireland threshold: 110

GB max: 12,946 Jan
NI max: 6,151 Mar

	S	M	L
GB change	o	o	o
NI change	-	-	--

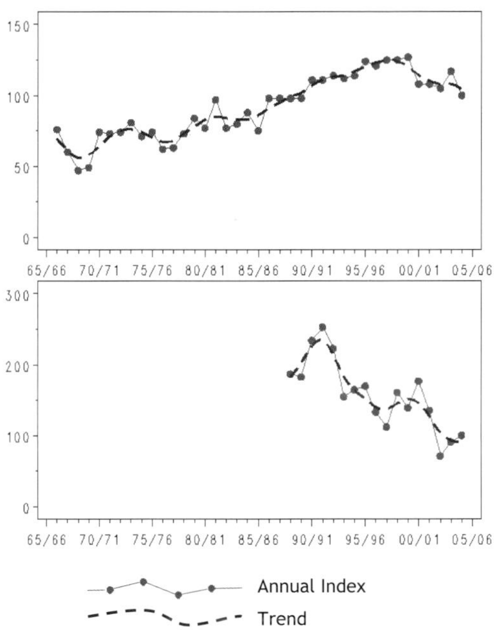

Figure 30.a, Annual indices & trend for Goldeneye for GB (above) & NI (below).

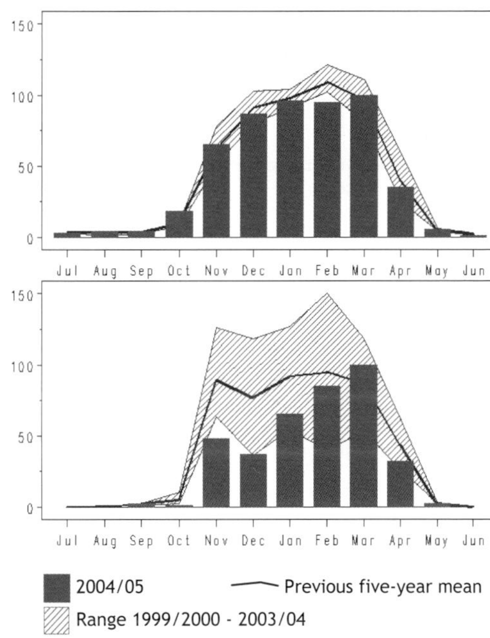

Figure 30.b, Monthly indices for Goldeneye for GB (above) & NI (below).

The British annual trend shows a decline of about 17% over the last seven years, following a long period of steady increase. However, a decline is much more evident in Northern Ireland where, although the annual index increased by 10% since 2003/04, numbers remained relatively low. The Northern Ireland picture depends greatly upon numbers at Loughs Neagh and Beg, the only site in the UK supporting internationally important numbers of Goldeneye, where something of a recovery was seen after two years of low numbers. Elsewhere in Northern Ireland, peak numbers were low at most coastal sites but higher than normal at Upper Lough Erne.

Only two sites in Great Britain regularly hold Goldeneye numbers in excess of 1,000 birds. The count at the Forth Estuary in

2004/05 was low for a second year running. Dedicated sea-duck counts in the Moray Firth found high numbers, especially concentrated around Milton of Culloden. At the other key sites, numbers were higher than normal at the Humber Estuary and Loch Leven, with lower counts at Abberton Reservoir and the Tweed between Kelso and Coldstream, the latter no longer nationally important for the species.

The monthly indices show that Goldeneye arrive in Britain much later than most of our common wintering wildfowl, with few here before November and the main departure between March and April. In early 2005 the peak occurred in March as opposed to the more usual February. A similar pattern was seen in Northern Ireland.

	00/01	01/02	02/03	03/04	04/05	Mon	Mean
Sites of international importance in the UK							
Loughs Neagh and Beg	8,583	6,454	3,661	4,497	5,787	Mar	5,796
Sites of national importance in Great Britain							
Forth Estuary	(2,414)	1,113	1,241	(753)	879	Jan	1,412
Inner Moray and Inverness Firth	1,141	993	1,352 [11]	709 [1]	1,165 [1]	Jan	1,072
Abberton Reservoir	448	(619)	469	431	394	Jan	472
Humber Estuary	498	208	618	296	595	Nov	443
Rutland Water	353	450	428	511	420	Jan	432
Inner Firth of Clyde	468	321	264	(514)	(156)	Jan	392
Stour Estuary	291	205	573	262	175 [11]	Feb	301
Hornsea Mere	85	294	(480) [13]	235	325	Mar	284
Tweed Estuary	151	312	240	390	273	Jan	273
Morecambe Bay	346	221	280	204	(297)	Feb	270
Loch of Skene	225	270	(192)	298	207	Feb	250
Sites of all-Ireland importance in Northern Ireland							
Lower Lough Erne			218	(337)	319	Dec	291
Strangford Lough	108	256	295	253	161	Mar	215
Belfast Lough	276	140 [13]	249	242 [11]	164 [11]	Feb	214
Larne Lough	136	189	130	95	73	Jan	125
Sites no longer meeting table qualifying levels in WeBS-Year 2004/2005							
River Tweed - Kelso to Coldstream	285	345	180	246	129	Jan	237
Carlingford Lough	163	(68)	103	68	102	Jan	109
Other sites surpassing table qualifying levels in WeBS-Year 2004/2005 in Great Britain							
Loch Leven	215	249	153	86	385	Dec	218
Other sites surpassing table qualifying levels in WeBS-Year 2004/2005 in Northern Ireland							
Upper Lough Erne	20	69	117	54	125	Feb	77

Smew

Mergellus albellus

International threshold:	400
Great Britain threshold:	4*
All-Ireland threshold:	+

GB max:	170	Jan
NI max:	4	Feb

**50 is normally used as a minimum threshold*

The British counted maximum for Smew was the lowest since 1994/95. Interestingly, however, the peak counted at Wraysbury Gravel Pits, the top site, was the highest ever for the site. Numbers at Wraysbury showed a strong peak in December, with birds on many of the individual pits in the complex, but declined to 25 by January and with only single figures in other winter months. Many other regular sites for wintering Smew saw low numbers in 2004/05, notably Dungeness Gravel Pits, Cotswold Water Park (West), Lee Valley Gravel Pits, Fen Drayton Gravel Pits, Rutland Water and Chew Valley Lake. Conversely, sites with peaks in excess of their five-year means included Little Paxton Gravel Pits, Abberton Reservoir, Sonning Eye/Henley Road Gravel Pits and Tophill Low. The peak total of only four in Northern Ireland, all from Loughs Neagh and Beg, was not unusually low. Overall, the vast majority of birds were recorded between October and March, but there was an unseasonal bird at Dorchester Gravel Pits in July and a late bird on the Burry Inlet in May.

	00/01	01/02	02/03	03/04	04/05	Mon	Mean	
Sites of national importance in Great Britain								
Wraysbury Gravel Pits	53	66	63	55	68	Dec	61	
Dungeness Gravel Pits	27	32	18	33	14	Jan	25	
Cotswold Water Park (West)	17	31	32	20	18	Jan	24	
Lee Valley Gravel Pits	20	23	29	23	8	Dec	21	
Rye Harbour and Pett Level	2	20	28	19	10	Jan	16	
Fen Drayton Gravel Pits	7	15	11	16	4	Dec	11	
Thorpe Water Park	6	6	11	18	10	Feb	10	
Twyford Gravel Pits		7	12				10	
Hickling Broad	10						10	
Seaton Gravel Pits and River	7	11	7 [13]	14	8	Jan	9	
Little Paxton Gravel Pits	10	4	8	4	12	Feb	8	
Rutland Water	8	12	8	8	4	Jan	8	
Middle Tame Valley Gravel Pits	8	(8)	(5)	(1)	(0)		8	
Lower Derwent Ings			7				7	
Bedfont and Ashford Gravel Pits	5	6	(6)				6	
Cassington & Yarnton GPs	(1)	(0)	(0)	10	1	Jan	6	
Marsh Lane Gravel Pits			6				6	▲
Fairburn Ings	8	6	6	4			6	
Chew Valley Lake	11	3	7	4	0		5	
Belhus Woods Country Park	0	0	7	10	6	Feb	5	▲
Abberton Reservoir	0	(9)	4	5	9	Jan	5	
Colne Valley Gravel Pits	3	0	6	(8)	7	Feb	5	
Sonning Eye & Henley Rd GPs	2	3	9	0	9	Jan	5	▲
Thrapston Gravel Pits	2	(2)	2	11			5	
Meadow Lane Gravel Pits	3	1	17	3	0		5	
Leybourne & New Hythe GPs	7	7	3	0			4	
Little Mollands Farm Pits	0	2	11	0	5	Jan	4	▲
Pitsford Reservoir	2	9	2	3	3	Feb	4	
Earls Barton Gravel Pits	(2)	0	7	7	0		4	
Eyebrook Reservoir	7	7	1	3	3	Jan	4	
Wanlip Gravel Pits				4			4	▲
Tophill Low Reservoirs	1	3	5	6	7	Feb	4	▲
Other sites surpassing table qualifying levels in WeBS-Year 2004/2005 in Great Britain								
Needingworth Quarry Lakes					10	Feb	3	
Blackwater Estuary	0	0	4 [11]	0	6	Jan	2	
Chilham & Chartham Gravel Pits	0	0	2	5	5	Feb	2	

Red-breasted Merganser

Mergus serrator

International threshold: 1,700
Great Britain threshold: 98
All-Ireland threshold: 20*

GB max: 3,232 Dec
NI max: 514 Oct

	S	M	L
GB change	o	-	o
NI change	o	o	o

**50 is normally used as a minimum threshold*

The British annual index rose by 10% since 2003/04 and the overall trend suggests winter numbers remain at a high level. The monthly British indices show that peak numbers are present at British WeBS sites between November and March, although a sizeable proportion of the peak is present year-round. Our wintering birds are known to be a mixture of local breeders and arrivals from Iceland and Greenland. The Northern Ireland index declined to its lowest level but actually shows relative stability over the last five years. In Northern Ireland there was even greater month stability than in Britain with little change between August and March.

In Britain, higher than average counts were recorded from the Fleet/Wey, Loch Ryan and the Tees Estuary. On the other hand, lower than usual counts were noted on the Forth Estuary, Poole Harbour, Inner Firth of Clyde, Tay Estuary and Exe Estuary. In Northern Ireland, counts at both Strangford and Belfast Loughs were down but peaks at Larne Lough and Carlingford Lough were both well up. The high count at Lough Foyle in 2003/04 was not sustained and a very low total at Loughs Neagh and Beg meant that the site dropped from being nationally important in an all-Ireland context for this species.

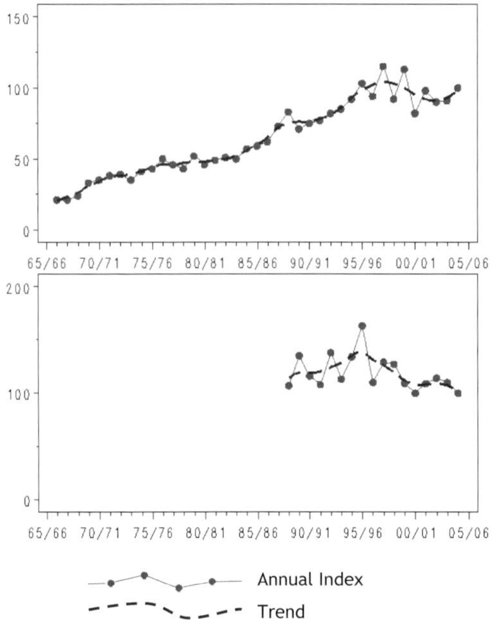

Figure 31.a, Annual indices & trend for Red-breasted Merganser for GB (above) & NI (below).

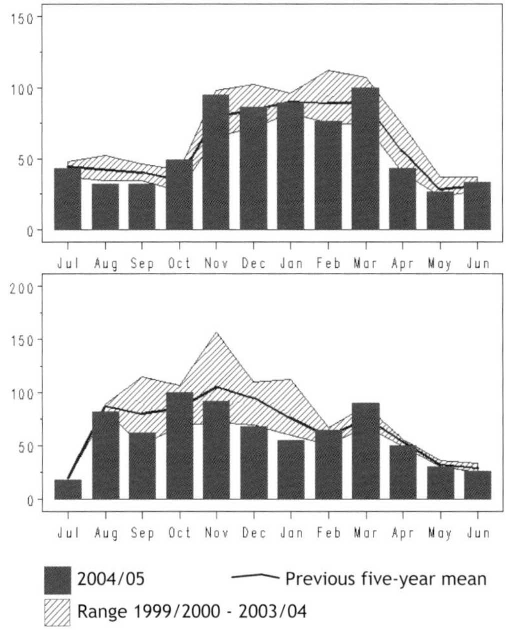

Figure 31.b, Monthly indices for Red-breasted Merganser for GB (above) & NI (below).

	00/01	01/02	02/03	03/04	04/05	Mon	Mean
Sites of national importance in Great Britain							
Forth Estuary	459	599	769	791	544	Oct	632
Poole Harbour	336	(417)	469	(392)	315	Nov	386
Fleet and Wey	283	366	358	425	413	Jan	369
Moray Firth	(295)	(234)	355	338	300	Oct	331
Morecambe Bay	338	229	(265)	(170)	(152)	Oct	284
Lavan Sands	317	164	170	264	(211)	Aug	229
Chichester Harbour	180	(159)	(184)	191	194	Mar	188
Duddon Estuary	(148)	136	220	(167)	152	Oct	169
Langstone Harbour	122	192	158	127	128	Nov	145
Inner Firth of Clyde	125	(196)	141	(164)	98	Oct	145
Loch Indaal	163	40	172	138			128
Tay Estuary	(127)	(30)	(39)	(98)	(30)	Dec	(127)
Loch Ryan	94	(113)	133 [13]	74	179	Oct	120
Cardigan Bay	(109)	118	(47)	(88)	(93)	Mar	118
Exe Estuary	139	94	112	(132)	82	Nov	112
Arran	108	(94)	(126)	103	(90)	Jul	112
St Andrews Bay	24	31	10	112	2	Oct	112
North Norfolk Coast	103	(102)	(109)	105	126	Jan	111
Solway Firth	(111)	(58)	(55)	(92)	(84)	Dec	(111)
Sites of all-Ireland importance in Northern Ireland							
Strangford Lough	148	342	187	188	189	Mar	211
Belfast Lough	169	162	228	216	75	Jan	170
Larne Lough	188	176	123	135	211	Oct	167
Carlingford Lough	44	24	106	40	154	Aug	74
Lough Foyle	15	73	37	122	52	Mar	60
Outer Ards Shoreline	(35)		62	48	54	Mar	55
Sites no longer meeting table qualifying levels in WeBS-Year 2004/2005							
Portsmouth Harbour	(63)	125	100	85	77	Nov	97
Loughs Neagh and Beg	21	27	29	14	6	Mar	19
Other sites surpassing table qualifying levels in WeBS-Year 2004/2005 in Great Britain							
Tees Estuary	34	40	36	60	113	Mar	57
Sound of Barra (Barra)	54 [52]	68 [52]	97 [52]		106 [52]	Mar	81
Goring	4	186	35	(35)	(102)	Feb	82

Goosander

Mergus merganser

International threshold: 2,700
Great Britain threshold: 161†
All-Ireland threshold: +

GB max: 2,650 Dec
NI max: 2 Dec

	S	M	L
GB change	(-)	o	+

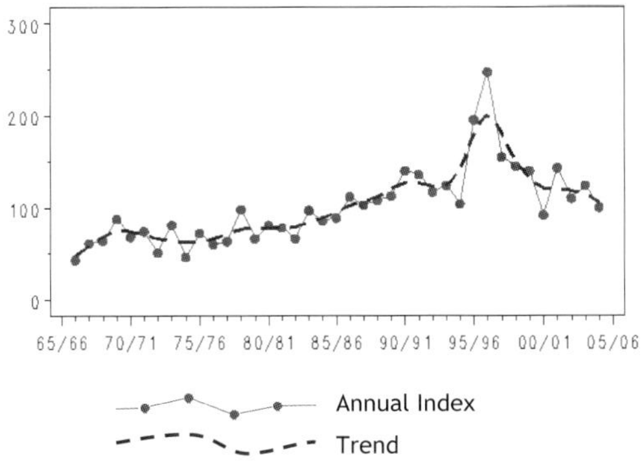

Figure 32.a, Annual indices & trend for Goosander for GB.

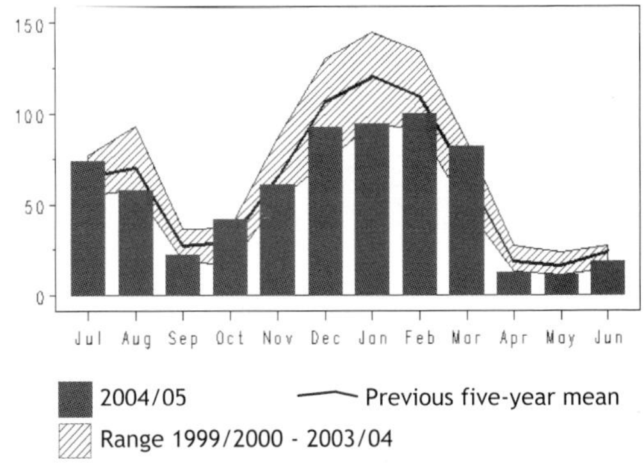

Figure 32.b, Monthly indices for Goosander for GB.

The British annual index for Goosander dropped by 19%, continuing the downwards trend seen since the peak in the mid-1990s, whilst the counted British maximum was the second-lowest seen since 1988/89. Monthly indices in 2004/05 conformed to the general pattern of occurrence seen in the past, with a mass departure in late summer to moult in Norway, then a gradual return to a mid-winter peak followed by spring dispersal. In 2004/05, numbers present in January seem to have been particularly low and there was an especially rapid drop-off between March and April.

At the key sites, the peak at the Tay Estuary was the highest since August 1999, compared to about average numbers at Tyninghame and a below-average peak on the Kelso to Coldstream stretch of the Tweed. Low numbers counted at Loch Lomond means that the site no longer supports nationally important numbers, although only a small part of this large loch is currently counted. As so few sites now support nationally important numbers, additional sites supporting a peak mean in excess of an arbitrary level of 70 birds are now also presented. All of these sites are from Yorkshire northwards. At none of these sites was the 2004/05 peak noticeably higher than average, and at Talkin Tarn and the Inner Moray Firth the peak numbers were extremely low, even during dedicated sea-duck monitoring counts at the latter site. As usual, the species was very scarce in Northern Ireland with peaks of two at Larne Lough and one at Belfast Lough.

	00/01	01/02	02/03	03/04	04/05	Mon	Mean
Sites of national importance in Great Britain							
Tay Estuary	230	245	248	192 [13]	263	Jul	236
Tyninghame Estuary	(300)	161	97	177	189	Jul	185
R. Tweed: Kelso to Coldstream	111	371	179	61	112	Dec	167
Sites with mean peak counts of 70 or more birds in Great Britain†							
Eccup Reservoir	95	103	95	137	94	Mar	105
Castle Loch (Lochmaben)	78	113	82	137	88	Oct	100
Lower Derwent Ings			154	37			96
Inner Moray and Inverness Firth	61 [13]	191 [13]	85 [13]	137 [13]	2	Oct	95
Loch Lomond	(12)	(37)	(84)	(23)	(15)	Dec	(84)
Talkin Tarn	111	141	44	(115)	5	Oct	83
Solway Estuary	61	(27)	72	(105)	84	Oct	81
Forth Estuary	86	54	(89)	53	81	Jul	73
Other sites surpassing table qualifying levels in WeBS-Year 2004/2005 in Great Britain							
Spittal to Cocklawburn		(0)	7	92	86	Jul	62
Cotswold Water Park (West)	72	(38)	(55)	50	80	Jan	67
Bala Lake		7	20	27	74	Aug	32
Blithfield Reservoir	38	74	31	51	72	Jan	53
Loch Garten				60	70	Oct	65

† *as few sites exceed the British threshold a qualifying level of 70 has been chosen to select sites for presentation in this report*

Ruddy Duck

Oxyura jamaicensis

Naturalised introduction†
Native Range: N & S America

GB max: 3,856 Jan
NI max: 33 Feb

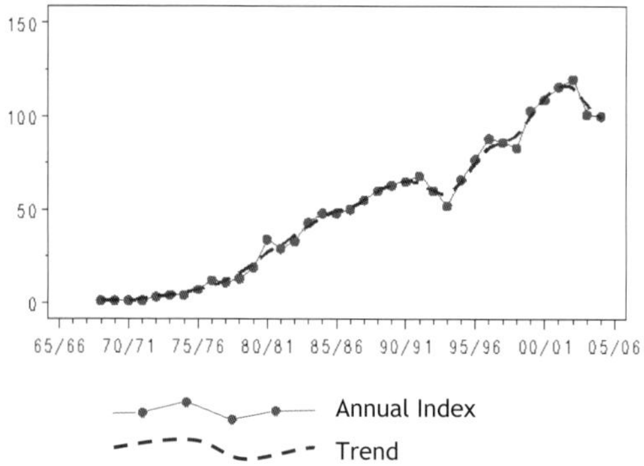

Figure 33.a, Annual indices & trend for Ruddy Duck for GB.

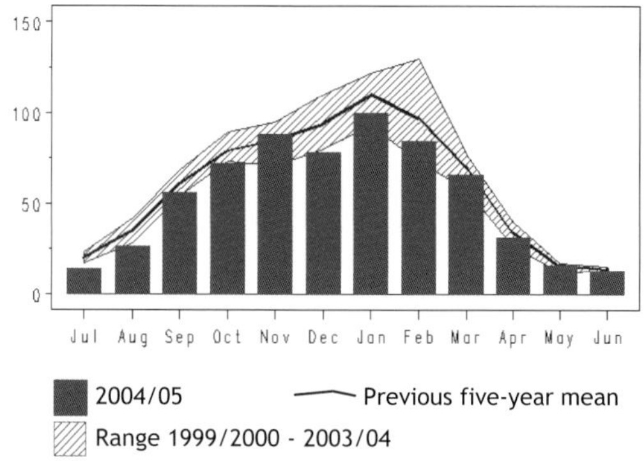

Figure 33.b, Monthly indices for Ruddy Duck for GB.

Following a major fall in the British annual index between 2002/03 and 2003/04, expectations of a further decline following the well-publicised control programme were not realised in 2004/05 and numbers appeared to remain level. The monthly indices display a pattern more expected of a winter immigrant than a resident, with increasing numbers recorded by WeBS as the autumn progresses up to a mid-winter peak. However, this presumably results from Ruddy Ducks breeding and summering on smaller waters, less well covered by WeBS than the larger sites favoured by communal winter flocks.

At the key sites, peak numbers have remained roughly stable at many. Sites at which more substantial proportional increases were seen included Dungeness Gravel Pits, Sutton & Lound Gravel Pits, Angler's Country Park, Rother Valley Country Park, Brent Reservoir and Hornsea Mere, whilst recent declines have shown signs of reversal at Rutland Water, Hanningfield Reservoir and Blithfield Reservoir. The most obvious declines in 2004/05 were noted from Chew Valley Lake, Blagdon Lake, King George V Reservoir, Eyebrook Reservoir, Clumber Park Lake, Fairburn Ings, Belvide Reservoir, Hule Moss, Loch Gelly and Kilconquhar Loch. Only three sites in Scotland supported peak numbers of Ruddy Ducks in double figures in 2004/05, whilst in Northern Ireland, the February peak was comprised entirely of birds at Loughs Neagh and Beg.

	00/01	01/02	02/03	03/04	04/05	Mon	Mean
Sites with mean peak counts of 30 or more birds in Great Britain†							
Staines Reservoirs	(244)	444	(696)	694	695 [44]	Jan	632
Rutland Water	1,187	911	482	200	251	Nov	606
Abberton Reservoir	389	456	493	678	403	Feb	484
Chew Valley Lake	647 [44]	491 [13]	427 [13]	488 [44]	220 [44]	Jan	455
Hanningfield Reservoir	287	553	(664)	285 [44]	412 [44]	Jan	440
Blithfield Reservoir	600 [44]	265	187	180 [44]	401	Nov	327
Blagdon Lake	152	463	394	249	151 [44]	Jan	282
Dungeness Gravel Pits	134	224	264	222	287	Dec	226
Pitsford Reservoir	135	293	358	103	178	Feb	213
Hilfield Park Reservoir	206	159	125	187	241 [44]	Jan	184
Stanford Reservoir	(67)	274	97	277 [44]	76	Jan	181
King George V Reservoirs	46	156	135	268	(23)	Jan	151
Holme Pierrepont Gravel Pits	106			115	189	Dec	137
Cotswold Water Park (West)	(115)	128	(60)	127	125	Nov	127
Tophill Low Reservoirs	117	173	89	110 [44]	124	Jan	123
Middle Tame Valley Gravel Pits	125	146 [44]	(120)	(96)	58 [44]	Nov	112
Eyebrook Reservoir	116 [44]	236	56	84	5	Sep	99
Sutton and Lound Gravel Pits		132	26	46	175	Feb	95
Angler's Country Park Lake	70 [13]	58	76	78	180 [44]	Dec	92
Bolton-on-Swale Gravel Pits	79	97	108	118 [44]	55 [44]	Nov	91

	00/01	01/02	02/03	03/04	04/05	Mon	Mean
Blackwater Estuary	106	152	53	69	71	Sep	90
Hollowell Reservoir	91	76	39	191	53	Mar	90
Carsington Water		141 [44]	132	0	82 [44]	Feb	89
Colwick Country Park		69 [44]		88 [44]	100 [44]	Feb	86
Walthamstow Reservoirs	(41)	40 [44]	(67)	118	90	Feb	83
Clumber Park Lake	123	122	72	76	16 [44]	Jan	82
Humber Estuary	99	45	55	116	84	Aug	80
Brent Reservoir	61	73	104	25	133	Sep	79
Fairburn Ings	100	69	94	115	5 [44]	Feb	77
Thames Estuary	63	34	106	(82)	85	Jan	74
Church Wilne Reservoir		68 [44]					68
Llyn Traffwll	61	36 [44]	80	83	78	Sep	68
Wigan Flashes	39	78	49	60	86	Oct	62
Tees Estuary	78	40	77	70	37	Sep	60
King George VI Reservoir	0	283 [13]	0	2	0		57
Great Pool Westwood Park	90	59	57	22	35	Mar	53
Llynnau Y Fali	20	32	86	57	70	Dec	53
Belvide Reservoir	170	30		6 [44]	0		52
Knight & Bessborough Res	31	122	29	23	46	Mar	50
Newsham Park				42 [44]			42
Pugneys Country Park Lakes	49	57	7	63	27	Jan	41
Barn Elms Reservoirs	24	39	36	43	59	Sep	40
Houghton Green Pool	37	36	36	40	40	Dec	38
Barrow Gurney Reservoir	16	59	38	49	19	Feb	36
Old Moor	26	40	47	28			35
Grafham Water	3	121	10	10	28	Mar	34
Loch Gelly	16 [44]	58	24 [13]	66	6	Aug	34
Swithland Reservoir	22	16	10	61	62	Feb	34
Thoresby Lake	22	30 [44]	3 [44]	69 [44]	46 [44]	Nov	34
Hule Moss	53 [13]	68 [13]	28 [13]	11 [13]	4	Sep	33
Kilconquhar Loch	80 [44]	30	42	12	5	Jun	34
Llyn Alaw	18 [44]	54 [44]	44	2	45	Jan	33
Hogganfield Loch	39	27	26	34	29	Sep	31
Woolston Eyes	43	23	29	32	27	Sep	31
Carr Vale Flash Reserve Pond	30				(4)	Mar	30
Hampton & Kempton Reservoirs	20	47	(30)	39	14	Sep	30
Sites with mean peak counts of 30 or more birds in Northern Ireland†							
Loughs Neagh and Beg	53	59	67	56	33	Feb	54
Other sites surpassing table qualifying levels in WeBS-Year 2004/2005 in Great Britain†							
Hornsea Mere	4	3	18	11	98	Feb	27
Wraysbury Gravel Pits	0	0	41	8	35	Nov	17
Colne Valley Gravel Pits	(13)	9	12	16	33	Feb	18
Swillington Ings	12	7	13	22	32	Mar	17
Rother Valley Country Park	1	14	16	14	33	Mar	16
Wintersett & Cold Hiendley Res	32	3	5	9	32	Sep	16

† *as no British or All-Ireland thresholds have been set a qualifying level of 30 has been chosen to select sites for presentation in this report*

Argentine Bluebill

Escape

Oxyura vittata

Native Range: S America

GB max: 1 Sep
NI max: 0

A single Argentine Bluebill was present on Netherfield Gravel Pits during September and October and then again in the following April and May. This is probably the same bird that has been recorded at this site for the past three years.

White-headed Duck

Escape

Oxyura leucocephala

Native Range: S Europe, W Asia

GB max: 1 Sep
NI max: 0

One bird was recorded at the Tees Estuary in September (Dormans Pool) and October (Saltholme Pools), and then one, quite possibly the same bird, was at Hilfield Park Reservoir in January and February.

Red-throated Diver

Gavia stellata

International threshold: 10,000
Great Britain threshold: 49*
All-Ireland threshold: 10*

GB max:	395	Oct
NI max:	48	Mar

**50 is normally used as a minimum threshold*

The Red-throated Diver is the most numerous of the divers around UK coasts, especially in southern and eastern Britain. The counted British maximum in October 2004 was, however, the lowest recorded by WeBS Core Counts to date. In Northern Ireland, whilst the peak total count was much lower than in 2003/04, it represented more of a return to normality after the very high numbers at Lough Foyle present that year.

The only three-figure totals obtained from WeBS Core Counts were those for the Moray Firth and the Forth Estuary; at the latter site, numbers were at their highest ever At many other sites, however, peak numbers were low, such as from Minsmere and Dengie Sands, although at these sites counts are very dependent upon sea conditions. The low count at the Inner Firth of Clyde is potentially more concerning however, as this site can be covered more consistently than many others.

In recent years, aerial surveys have taken place of many at the key shallow sea areas favoured by this species. During these aerial surveys, many divers cannot be identified to species, however, in Liverpool Bay and southeast England, the vast majority of unidentified divers are considered to be Red-throated; only small numbers of other species have been observed during flights in these areas.

The prime area for Red-throated Divers has been found to be the 'Greater Thames', the shallow offshore area off Suffolk, Essex and North Kent. In 2004/05, it was estimated that 6,437 Red-throated Divers were present at the site in January/February; this is lower than the peak estimate of 7,688 birds in February 2004 (with upper and lower confidence intervals of 5,041 and 11,725 respectively), and an estimated peak of 11,089 (8,115-15,154) in January 2003 (Webb *et al. in prep.*). In comparison, the highest ground count in 2004/05 was 423 birds, recorded at Aberdeen Bay.

The past three seasons have seen very low numbers reported from WeBS Core Counts at Cardigan Bay. However, many birds feed well offshore here and aerial survey in March 2004 gave an estimate of 1,164 Red-throated Divers in the bay (with 95% confidence intervals of 620 to 1,640 birds) (Hall *et al.* 2005).

The only three-figure totals obtained from WeBS Core Counts were those for the Moray Firth and the Firth of Forth; at the latter site, numbers were at their highest ever At many other sites, however, peak numbers were low, such as from Minsmere and Dengie Sands, although at these sites counts are very dependent upon sea conditions. The low count at the Inner Firth of Clyde is potentially more concerning however, as this site can be covered more consistently than many others.

Aerial surveys employing distance sampling

Area	Date	Counted	Estimate (ci)	Ref
Greater Thames	Jan/Feb	not available	6,437	Webb *et al. in prep.*
Greater Wash / Thames Strategic Areas	Jan/Feb	1,139	5,634 (4,485-6,595)	WWT WAS Report
North West Strategic Area	Feb/Mar	213	1,199 (879-1,571)	WWT WAS Report
Firth of Forth	Dec	57	not available	Wilson *et al.* 2006
Firth of Tay	Feb	33	not available	Wilson *et al.* 2006

	00/01	01/02	02/03	03/04	04/05	Mon	Mean
Sites of national importance in Great Britain							
Aberdeen Bay offshore				225 [31]	423 [59]	May	324
Cardigan Bay	460	732	32	22	30	Dec	255
Lade Sands	0	800	100	10	0		182
Minsmere	700 [13]	4	3	57	3	Dec	153
Moray Firth	150	74	126	166	117	Dec	127
Inner Firth of Clyde	145	112	151	126	34	Dec	114
Thames Estuary	(13)	40	(344) [32]	23	32	Jan	110
Forth Estuary	104	93	106	61	132	Oct	99
Dengie Flats	145	51	114	50	15	Mar	75
Loch Ryan	35	47	111 [13]	89	81	Oct	73

	00/01	01/02	02/03	03/04	04/05	Mon	Mean	
Pegwell Bay	4	54	215	0			68	
Lavan Sands	72 [13]	13	202 [13]	30	22	Nov	68	
Sites of all-Ireland importance in Northern Ireland								
Lough Foyle	5	7	29	147	21	Mar	42	
Strangford Lough		57 [11]	1 [11]				29	
Belfast Lough	14	41	31	13	14	Jan	23	
Outer Ards Shoreline	(18)		1	6	14	Mar	10	▲

Black-throated Diver

Gavia arctica

International threshold: 10,000
Great Britain threshold: 7*
All-Ireland threshold: 1*

GB max: 78 Mar
NI max: 1 Jan

**50 is normally used as a minimum threshold*

The British maximum count of 78 in March was the highest for four years. Conversely, whilst Black-throated Divers are scarce in Northern Ireland, the single bird at the Outer Ards in January represented a very poor year. Gerrans Bay in Cornwall remains the top site for the species, with 60% of the British total there in March. Otherwise, every other listed site for Black-throated Divers is in Scotland. Counts at both the Firth of Forth and the Moray Firth were very low in 2004/05. Counts carried out by the RAF Ornithological Society in February found the species to be widespread around the northwest coast of Scotland, with Loch Gairloch identified as holding by far the highest numbers. Many other counts in the table derive from supplementary observations submitted to help fill some of the gaps in coverage of this species. These extra counts certainly identify the Sound of Barra, Loch Slapin, Loch Roag and Broad Bay as of importance to the species.

	00/01	01/02	02/03	03/04	04/05	Mon	Mean	
Sites of national importance in Great Britain								
Gerrans Bay	35	33	53	37	47	Mar	41	
Sound of Barra (Barra)		40 [52]	37 [52]	31 [52]			36	
Loch Gairloch	23				28	Feb	26	
Loch Slapin				21 [48]	26 [48]	Apr	24	
Moray Firth	17	9	18	48	6	Dec	20	
Loch Roag		18 [52]	13 [52]				16	
Girvan to Turnberry	7	8	20	19	(9)	Mar	14	
Broad Bay (Lewis)	7 [52]	7 [52]		21 [52]			12	
Forth Estuary	(2)	24	9	5	3	Oct	10	
Little Loch Broom	17				3	Feb	10	
Arran	(0)	2	14	(1)	(0)		8	▲
Loch Ewe	(15)				0		8	
Sites of all-Ireland importance in Northern Ireland								
Outer Ards Shoreline	(3)		(0)	(0)	1	Jan	2	
Belfast Lough	(2)	2	1	4 [11]	(0)		2	
Sites no longer meeting table qualifying levels in WeBS-Year 2004/2005								
Red Point to Port Henderson	11				0		6	
Gruinard Bay	(11)				0		6	

Great Northern Diver

Gavia immer

International threshold: 50
Great Britain threshold: 30*†
All-Ireland threshold: ?†

GB max: 138 Feb
NI max: 8 Mar

**50 is normally used as a minimum threshold*

The maximum British total recorded in February was the second highest to date. However, the Northern Ireland peak of just eight was one of the lowest in recent years, due to large reductions at Lough Foyle and Carlingford Lough, following the high numbers in 2003/04. As always for this species, how representative this is of true trends is hard to judge, as many key sites are irregularly counted for WeBS, many are not covered at all and even where counts take place, they are very dependent upon sea-conditions. Moreover, the species does make substantial diurnal movements and the numbers at some sites vary depending upon the time of day that the count was made.

Given the large (and welcome) amount of supplementary data now being submitted for this species, the qualifying level for presentation within this report has been raised from a mean of five to ten birds. During WeBS Core Counts, the top site in 2004/05 was the Moray Firth, although numbers were well down on the previous year. The peak at Broadford Bay on Skye was also the lowest there of the last four years. Along the south coast, Gerrans Bay was again the top site.

The RAF Ornithological Society covered many sites in northwest Scotland, showing the continued importance of this area for the species, with double-figure counts from four sites. Other supplementary counts were kindly provided for parts of the Western Isles and Skye and include some impressive totals. However, peak counts off the west coast of South Uist appear to have declined over the last five years. The dedicated boat-based counts carried out by the Shetland Oil Terminal Environmental Advisory Group (SOTEAG) cover many important sites for this species. During 2004/05, counts were very low in Hacosay / Bluemull / Colgrave Sounds, from Kirkabister to Wadbister, and from Wadbister to Rova Head, but were closer to normal elsewhere. Whilst trends are not easy to determine, the SOTEAG team consider that numbers in Shetland have been declining in recent years (M. Heubeck *pers. comm.*).

Aerial surveys employing distance sampling

Area	Date	Counted	Estimate	Ref
Coll & Tiree	Mar	131	not available	Wilson *et al.* 2006
Sound of Gigha	Mar	104	not available	Wilson *et al.* 2006
Outer Hebrides	Mar	57	not available	Wilson *et al.* 2006
Scapa Flow	Mar	33	not available	Wilson *et al.* 2006
Mull	Mar	27	not available	Wilson *et al.* 2006

	00/01	01/02	02/03	03/04	04/05	Mon	Mean
Sites of international importance in the UK							
Deer and Shapinsay Sounds	225 [13]						225
South Uist West Coast	(173) [52]	(148) [52]	(57) [52]	(48) [52]	(63) [52]	Mar	(173)
Sound of Barra (Barra)		97 [52]	142 [52]	96 [52]	94 [52]	Mar	107
Moray Firth	38	54	60	(109)	37	Dec	60
Loch Slapin				44 [48]	59 [48]	Mar	52 ▲
Sound of Taransay (Harris)			60 [52]	70 [52]	20 [52]	Feb	50
Sites with mean peak counts of 10 or more birds in Great Britain†							
Loch Indaal	19	74	68	18			45
Kirkabister to Wadbister Ness	46 [10]	49 [10]	22 [10]	50 [10]	(2) [10]	Feb	42
Sound of Harris	60 [52]	32 [52]	35 [52]	20 [52]	42 [52]	Feb	38
East Unst		37 [10]					37
Broadford Bay		(25)	35	(43)	24	Nov	34
Whiteness to Skelda Ness	30 [10]	34 [10]	34 [10]	27 [10]	30 [10]	Jan	31
Scalloway Islands		58 [10]		19 [10]	13 [10]	Feb	30
Pontllyfni to Aberdesach				28 [13]			28
Gruinard Bay	(7)				26	Feb	26
Gualan and Balgarva				23 [52]			23
Quendale to Virkie	19 [10]	30 [10]	22 [10]	24 [10]	22 [10]	Feb	23
Rova Head to Wadbister Ness	23 [10]	38 [10]	19 [10]	30 [10]	4 [10]	Feb	23
Island of Papa Westray		6	20	22	(1)	Oct	16
Baleshare West Coast	13 [52]						13
Gerrans Bay	13	6	17	15	14	Mar	13
Loch Ewe	6				19	Feb	13
Little Loch Broom	16				8	Feb	12
Traigh Luskentyre	6	8			22	Jan	12
Island of Egilsay	28	4	3	(0)	10	Oct	11
Kirkabister to Dury Voe	11 [10]						11
Sites with mean peak counts of 10 or more birds in Northern Ireland†							
Carlingford Lough	(16)	(3)	(15)	25	2	Jan	15
Lough Foyle	0	4	26	24	5	Mar	12
Other sites surpassing table qualifying levels in WeBS-Year 2004/2005 in Great Britain†							
Red Point to Port Henderson	0				17	Feb	9
Loch Gairloch	0				11	Feb	6

† *as few sites exceed the British threshold and no All-Ireland threshold has been set a qualifying level of 10 has been chosen to select sites for presentation in this report*

Little Grebe

Tachybaptus ruficollis

International threshold: 3,400
Great Britain threshold: 78
All-Ireland threshold: ?†

GB max:	4,882	Oct
NI max:	686	Nov

	S	M	L
GB change	o	+	++
NI change	o	o	+

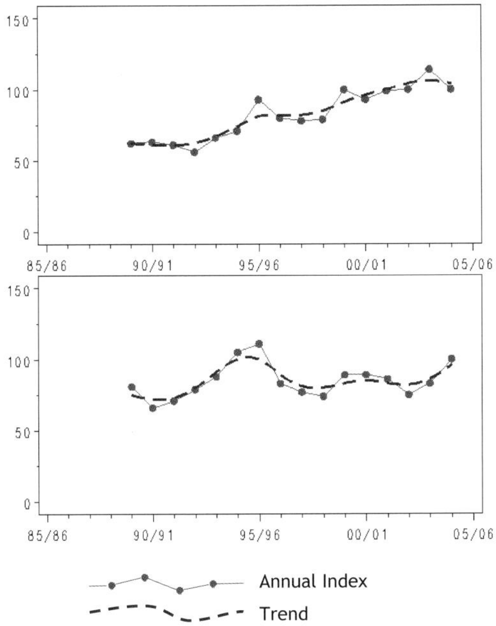

Figure 34.a, Annual indices & trend for Little Grebe for GB (above) & NI (below).

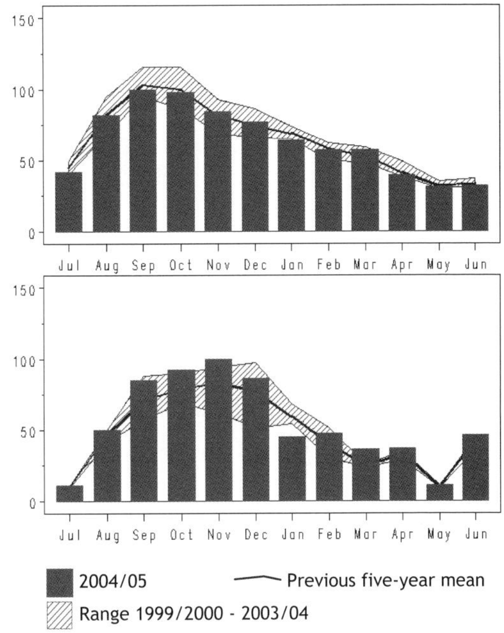

Figure 34.b, Monthly indices for Little Grebe for GB (above) & NI (below).

The British annual index for Little Grebe dropped by 13% between 2003/04 and 2004/05, but similar falls have been seen in previous years and the long-term trend is still one of steady increase. British monthly index values were similar to those of the last five years, peaking in September/October as usual, then dropping steadily through the winter, both as a result of mortality of first-year birds and then, in the spring, dispersal to breeding sites away from the larger wetlands. Whilst the counted British maximum was down a little on the previous, it was still the third highest peak to date.

Little Grebes have a very large world range, within which they tend to be dispersed in relatively small numbers and so, not surprisingly, there are no sites supporting internationally important numbers in the UK. The top site in Britain remains the Thames Estuary, where a new British site record of 444 was recorded in October. Amongst the other sites supporting nationally important numbers, there is a mixture of coastal and inland sites. The Chichester Harbour peak was the largest Core Count there to date. Conversely, the peak at the Tees Estuary was the lowest for over a decade. Numbers remain high at Chew Valley Lake but have dropped steadily at Rutland Water to leave the five-year mean just below the national qualifying level. Numbers at risen at both the Alde Complex and Blagdon Lake such that they now support nationally important numbers, whilst counts at the Lee Valley may well reach this level very soon.

Unlike a number of other waterbirds, Little Grebes seem to be doing well in Northern Ireland. The annual index for the province increased by 20% to reach its highest level since 1995/96, with a long-term picture of stability. As with many other species, numbers of Little Grebes in Northern Ireland are heavily influenced by those at Loughs Neagh and Beg, where the peak count of 466 was the highest there since autumn 1995. Elsewhere in the province there were indications of steady increase from Larne Lough and the Upper Quoile River.

	00/01	01/02	02/03	03/04	04/05	Mon	Mean	
Sites of national importance in Great Britain								
Thames Estuary	348	351	378	(198)	444	Oct	380	
Chichester Harbour	78	150 [11]	111	125	135	Dec	120	
Bewl Water	54	132	136	94	80	Sep	99	
Chew Valley Lake	95	100	70	110	110	Sep	97	
Holme Pierrepont Gravel Pits	107			55	120	Sep	94	
Swale Estuary	195	64 [11]	43	49	89	Dec	88	
Blagdon Lake	42	127	18	127	98	Aug	82	▲
Alde Complex	(58)	(76)	54	(47)	109	Dec	82	▲
Hamford Water	52	105	68	(92)	89	Jan	81	
Tees Estuary	96	(70)	104	70	54	Aug	81	
Sites no longer meeting table qualifying levels in WeBS-Year 2004/2005								
Rutland Water	87	77	58	87	70	Nov	76	
King`s Dyke Pits Whittlesey				(6)	9	Mar	9	
Sites with mean peak counts of 30 or more birds in Northern Ireland†								
Loughs Neagh and Beg	425	418	438	433	466	Nov	436	
Upper Lough Erne	94	122	75	131	104	Feb	105	
Strangford Lough	72	103	113	83	76	Oct	89	
Lower Lough Erne			(39)	(57)	(53)	Dec	(57)	
Larne Lough	36	27	32	65	77	Oct	47	
Lough Money	48	53	41	39	51	Oct	46	
Hillsborough Main Lake	40	45	37	27	28	Oct	35	
Belfast Lough	41	29	37	30	28	Sep	33	
Other sites surpassing table qualifying levels in WeBS-Year 2004/2005 in Great Britain								
Lee Valley Gravel Pits	55	59	71	83	102	Sep	74	
Other sites surpassing table qualifying levels in WeBS-Year 2004/2005 in Northern Ireland†								
Upper Quoile River	7	19	13	28	35	Jan	20	
Lough Foyle	26	12	20	31	31	Oct	24	

† *as no All-Ireland threshold has been set a qualifying level of 30 has been chosen to select sites for presentation in this report*

Great Crested Grebe

Podiceps cristatus

International threshold:	4,800	
Great Britain threshold:	159	
All-Ireland threshold:	30*	

GB max:	9,193	Oct
NI max:	2,065	Jan

	S	M	L
GB change	o	o	+
NI change	o	o	+

**50 is normally used as a minimum threshold*

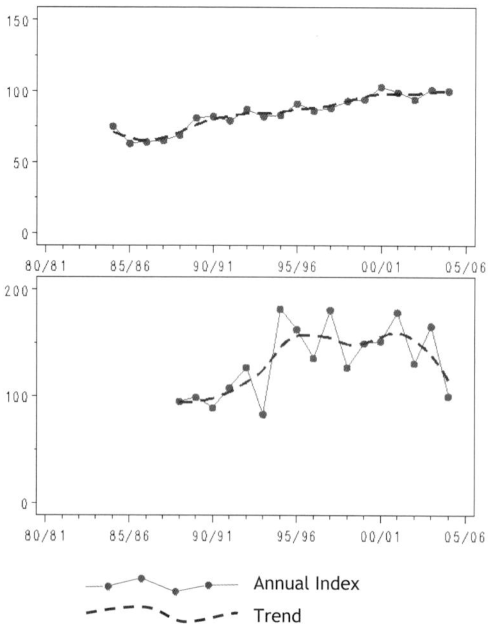

Figure 35.a, Annual indices & trend for Great Crested Grebe for GB (above) & NI (below).

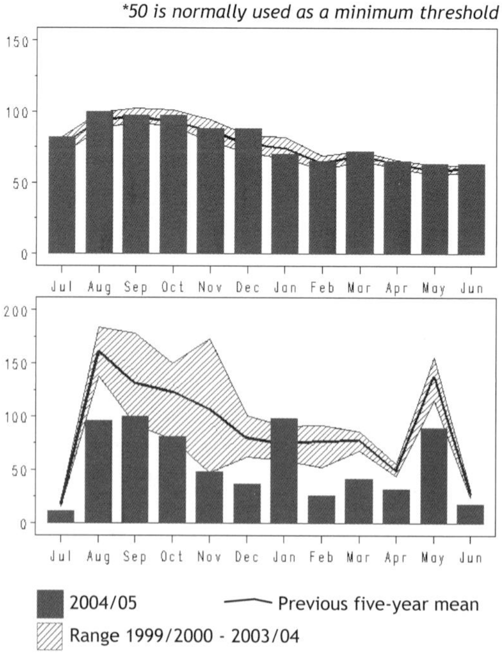

Figure 35.b, Monthly indices for Great Crested Grebe for GB (above) & NI (below).

The annual index for Great Crested Grebe in Great Britain has shown a very steady long-term rise since recording of the species began in the mid-1980s, although there was little change between 2003/04 and 2004/05. Monthly indices indicate that numbers peak in late summer. The degree of reduction in monthly indices between the peak and the following spring is much less than seen with Little Grebe, perhaps because the species remains more tied to larger waterbodies in the breeding season.

At no sites do numbers approach the level required for international importance. The nationally important sites in Great Britain are a mixture of offshore shallow sandbanks, Estuaries, inland reservoirs and gravel pit complexes. Peak numbers at Lade Sands were lower than normal, although some recent peaks have been derived from supplementary counts here. The count off Lavan Sands, however, was the highest ever recorded there by WeBS. Numbers off Minsmere were unimpressive but, as at all offshore sites, counts can very much depend upon the sea conditions on the day of the count. On the Estuaries, numbers were very low at the Swale Estuary but higher than usual at the Solway Estuary and nearby Loch Ryan; counts at these latter two peaked in different months and some of the same birds may have been involved. Inland, numbers were higher than average at Rutland Water, Grafham Water and Bewl Water and there was continued increase at Blagdon Lake, as seen also for Little Grebe. Counts were lower than the recent averages at Chew Valley Lake, Queen Mary Reservoir and Draycote Water. At Blithfield, Bough Beech and Abberton Reservoirs, peak numbers declined to such an extent that the sites no longer support nationally important numbers.

In contrast to the British index, the annual index in Northern Ireland fluctuates wildly between years; the value in 2003/04 was the lowest since 1993/94 in the province, although given this level of fluctuation it is hard to draw any firm conclusions about decline yet. Peak numbers at Belfast Lough remain high, but the peak from Loughs Neagh and Beg was well down on the previous year (although higher than in 2001/02). The massive count on Lough Foyle in 2003/04 was not repeated, the peak here crashing back down to double-figures again. At most other Northern Ireland sites, numbers show great fluctuation between years, but 2004/05 counts were disappointing from Larne Lough and the Outer Ards. Again, however, at the latter site in particular this may have been due to count conditions on the day.

	00/01	01/02	02/03	03/04	04/05	Mon	Mean
Sites of national importance in Great Britain							
Lade Sands	1,033 [47]	1,315 [47]	1,600	1,080 [13]	860	Feb	1,178
Rutland Water	997	600	607	619	815	Nov	728
Chew Valley Lake	690	480	320	330	330	Aug	430
Grafham Water	98	619	311	463	526	Dec	403
Queen Mary Reservoir	246	671	267	495	262	Dec	388
Pegwell Bay	404 [47]	354 [47]	604	20	(0)		346
Swale Estuary	528 [47]	446 [47]	(42)	316	63	Nov	338
Forth Estuary	290	224	(389)	295	(313)	Oct	302
Cotswold Water Park (West)	306	(258)	(188)	(245)	283	Oct	295
Bewl Water	261	292	356	190	330	Aug	286
Lavan Sands	388 [13]	113	308	114	446	Aug	274
Pitsford Reservoir	241	268	203	341	309	Dec	272
Loch Ryan	147	(121)	(300)	210	299	Oct	239
Morecambe Bay	245	222	187	218	(91)	Jan	218
Solway Firth	336	164	119	88	333	Dec	208
Draycote Water	219	221	255	151	98	Aug	189
Lee Valley Gravel Pits	191	181	169	204	(147)	Oct	186
Minsmere	350 [13]	5	19	463	30	Jan	173
Rye Harbour and Pett Level	68	160	48	365	186	Dec	165
Poole Harbour	151	171	127	202	(87)	Nov	163 ▲
Loch Leven	131	222	127	204	127	Oct	162
Sites of all-Ireland importance in Northern Ireland							
Belfast Lough	1,338	1,995	1,214	1,832	(1,577)	Jan	1,595
Loughs Neagh and Beg	1,547	336	930	1,695	518	Oct	1,005
Lough Foyle	38	278	782	1,030	50	Oct	436
Carlingford Lough	326	284	174	184	232	Nov	240
Upper Lough Erne	113	190	110	112	191	Feb	143

	00/01	01/02	02/03	03/04	04/05	Mon	Mean
Larne Lough	204	80	105	115	50	Sep	111
Lower Lough Erne			71	(66)	(117)	Feb	94
Outer Ards Shoreline	199		9	(7)	2	Jan	70
Strangford Lough	141	50	36	43	64	Nov	67
Sites no longer meeting table qualifying levels in WeBS-Year 2004/2005							
Blithfield Reservoir	231	110	98	202	151	Aug	158
Stour Estuary	222 [11]	139 [11]	136	139 [11]	136 [11]	Feb	154
Bough Beech Reservoir	170 [47]	146 [47]	149	196	79	Nov	148
Abberton Reservoir	77	198	144	176	52	Nov	129
Other sites surpassing table qualifying levels in WeBS-Year 2004/2005 in Great Britain							
Blagdon Lake	80	110	113	161	176	Oct	128
Attenborough Gravel Pits	90	88		107	168	Jan	113
Inner Firth of Clyde	143	121	(154)	(168)	161	Aug	149

Red-necked Grebe

Podiceps grisegena

International threshold: 1,000
Great Britain threshold: 2*
All-Ireland threshold: ?

GB max: 30 Sep
NI max: 0

**50 is normally used as a minimum threshold*

The peak British monthly total of 30 was a little higher than the previous year but still the second lowest since regular recording of this species began. The absence of the species in Northern Ireland was typical. As usual, most birds were recorded in the Firth of Forth, particularly between Port Seton and Craigielaw Point. Elsewhere, numbers were unimpressive although three at Traeth Dulas in September following two there in June 2004 were of interest. As well as the two birds at Combermere, other inland records were of single birds at Darwell Reservoir, Grafham Water, King George VI Reservoir, Queen Mother Reservoir, Stanwick Gravel Pits and Wraysbury Gravel Pits. Supplementary counts of offshore seaducks in the Moray Firth located just single Red-necked Grebes in November and December.

	00/01	01/02	02/03	03/04	04/05	Mon	Mean
Sites of national importance in Great Britain							
Forth Estuary	29	39	44	16	24	Aug	30
North Norfolk Coast	3	9	2	(2)	1	Oct	4
Lindisfarne	3 [11]	5	4 [11]	(0)	2	Dec	4
Moray Firth	1	2	(1)	1	2	Oct	2 ▲
Kirkabister to Wadbister Ness	(2) [10]						(2)
Other sites surpassing table qualifying levels in WeBS-Year 2004/2005 in Great Britain							
Traeth Dulas		0	0	2	3	Sep	1
Combermere			0	0	2	Dec	1
Hamford Water	0	1 [11]	0	0	2	Nov	1
Thames Estuary	0	0	1	1	2	Sep	1

Slavonian Grebe

Podiceps auritus

International threshold: 35
Great Britain threshold: 7*
All-Ireland threshold: ?†

GB max: 219 Oct
NI max: 10 Mar

**50 is normally used as a minimum threshold*

The peak counted total for Britain was about average for recent years and in Northern Ireland, the peak fell to more usual levels following the very high numbers at Lough Foyle the previous winter. This site still supported the highest numbers in Northern Ireland, however, with smaller counts at Larne Lough and the Bann Estuary. The Forth Estuary remains the top UK site with numbers about average for recent years, whereas numbers at the Moray Firth have shown a steady decline over the last five years. Declines were also evident at the North West Solent and the single bird on the Exe Estuary represented the lowest peak here on record. Conversely, new peak counts were recorded from Loch Ryan, Loch of Harray and Loch of Swannay. Following rises in 2003/04 at both the Blackwater Estuary and Pagham Harbour, both saw low peaks again. Low Tide Counts made at Lindisfarne revealed much higher counts than seen during WeBS Core Counts. Supplementary counts provided for Shetland recorded the highest recent counts for Whiteness to Skelda Ness, Rova Head to Wadbister Ness, and Sullom Voe. Similarly, recently provided data for the Western Isles reveals the importance of sites such as the Sound of Taransay and Sound of Harris.

	00/01	01/02	02/03	03/04	04/05	Mon	Mean	
Sites of international importance in the UK								
Forth Estuary	(44)	61	80	110	73	Feb	81	
Moray Firth	86	75	69	62	55	Dec	69	
Whiteness to Skelda Ness	43 [10]	29 [10]	55 [10]	55 [10]	59 [10]	Jan	48	
Loch Ashie	41 [13]						41	
Sound of Taransay (Harris)			50 [52]	44 [52]	15 [52]	Feb	36	
Sites of national importance in Great Britain								
Traigh Luskentyre	48	19			31	Oct	33	▼
Loch Ryan	19	31	31 [13]	32	42	Oct	31	
Loch of Harray	8	25	25	23	49	Oct	26	
Loch Indaal	23	11	31	30			24	
Inner Firth of Clyde	20	10	45	(20)	16	Oct	23	
Lindisfarne	9	14	23 [11]	(2)	30 [11]	Dec	19	
Jersey Shore	5	(31)					18	
Blackwater Estuary	9	22	9 [11]	41	11	Mar	18	
Kirkabister to Wadbister Ness	23 [10]	5 [10]	13 [10]	17 [10]	(16) [10]	Feb	15	
Gualan and Balgarva	12 [52]		12 [52]		11 [52]	Oct	12	
Sound of Harris	23 [52]		8 [52]	5 [52]	10 [52]	Feb	12	
Rova Head to Wadbister Ness	5 [10]	15 [10]	11 [10]	6 [10]	18 [10]	Feb	11	
Pagham Harbour	1	6	6	28	8	Dec	10	
Loch of Swannay	7	4	10	11	19	Mar	10	
Sound of Gigha	9 [29]						9	
Broadford Bay		8	10	6	10	Feb	9	
Vaila Sound and Gruting Voe		9 [10]					9	▲
Lavan Sands	6	(4)	15 [13]	6	5	Dec	8	
Sullom Voe	8 [10]	8 [10]	6 [10]	6 [10]	13 [10]	Feb	8	
Sites no longer meeting table qualifying levels in WeBS-Year 2004/2005								
North West Solent	8	10	(4)	5	1 [11]	Dec	6	
Exe Estuary	3	(3)	4	(9)	1	Jan	4	
Upper Loch Torridon	9				0		5	
Sites with mean peak counts of 4 or more birds in Northern Ireland†								
Lough Foyle	9	6	13	61	10	Mar	20	
Other sites surpassing table qualifying levels in WeBS-Year 2004/2005 in Great Britain								
Goring	0	0	(1)	(0)	(7)	Mar	2	

† *as no All-Ireland threshold has been set a qualifying level of 4 has been chosen to select sites for presentation in this report*

Black-necked Grebe

Podiceps nigricollis

International threshold: 2,800
Great Britain threshold: 1*†
All-Ireland threshold: ?†

GB max: 80 Dec
NI max: 1 Nov

**50 is normally used as a minimum threshold*

The British peak of 80 birds in December was the highest on record, although over 75% of these birds were at just four key wintering sites. The peak at William Girling Reservoir in 2004/05 was the highest on record whilst there was also an increase in the lower numbers on the Beaulieu Estuary. The peak for the Fleet & Wey was low however for the third year running. The species remains less than annual in Northern Ireland, the single bird at Dundrum Bay in November being the first since the 1997/98 winter.

At many of the other key sites, peak counts refer to breeding colonies and some of these have been kept confidential with the advice of the Rare Breeding Birds Panel and/or local counters. The peaks at Woolston Eyes and Kilconquhar Loch were particularly low but this was due to the lack of summer counts from both of these sites.

	00/01	01/02	02/03	03/04	04/05	Mon	Mean
Sites with mean peak counts of 5 or more birds in Great Britain†							
William Girling Reservoir	14	16	16	21	27	Feb	19
Woolston Eyes	1	41	6	23	2	Oct	15
Fal Complex	(1)	16	15	7	19	Mar	14
Langstone Harbour	15	15 [11]	15	11	16 [11]	Feb	14
Teignmouth to Berry Head (Babbacombe Bay and Tor Bay)				4	18	Dec	11
Confidential Hertfordshire Site	5	8	7	17	8	Jul	9
Confidential Northumberland Site		2	11	10	11	Jul	9
Fleet and Wey	9	13	3	1	2	Feb	6
Holme Pierrepont Gravel Pits	10			0	5	Sep	5
Kilconquhar Loch	8	8	6	2	0		5
Tamar Complex	9	3	6 [11]	2	3	Jan	5
Other sites surpassing table qualifying levels in WeBS-Year 2004/2005 in Great Britain†							
Beaulieu Estuary	3	3	4	2	6	Feb	4
Staines Reservoirs	0	3	5	1	6	Aug	3
Stour Estuary	(0)	2	6 [11]	0	5 [11]	Jan	3

† *as the British threshold is so small, and as no All-Ireland threshold has been set a qualifying level of 5 has been chosen to select sites for presentation in this report*

Cormorant

Phalacrocorax carbo

International threshold: 1,200
Great Britain threshold: 230
All-Ireland threshold: ?†

GB max: 17,535 Oct
NI max: 2,798 Oct

	S	M	L
GB change	o	o	+
NI change	o	+	+

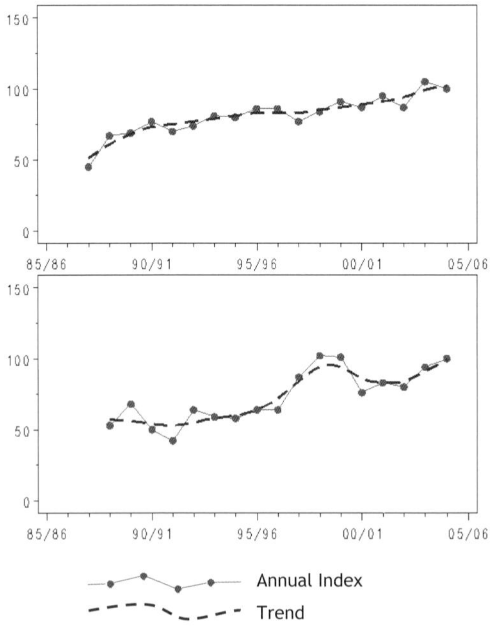

Figure 36.a, Annual indices & trend for Cormorant for GB (above) & NI (below).

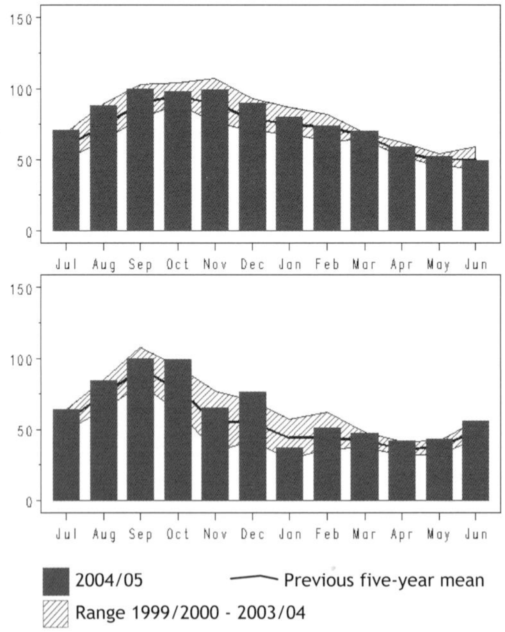

Figure 36.b, Monthly indices for Cormorant for GB (above) & NI (below).

The British annual index for Cormorant dropped slightly in 2004/05 but was still at its second highest value to date. Monthly indices show an autumn peak followed by a steady decline thereafter. At the key British sites, particularly high peaks were noted in 2004/05 at the Alt Estuary, Ribble Estuary, Dungeness Gravel Pits and Rye Harbour and Pett Level. There were also four new British sites supporting nationally important numbers, including the Alde Complex, which saw a particularly large increase; the count here was the eighth highest peak anywhere during the year. Lower peaks than expected were noted at Abberton Reservoir (where Great Crested Grebe numbers also declined sharply), Hanningfield Reservoir and the Blackwater Estuary. Most noticeable, however, was the picture in west London where peaks declined sharply at almost every site since 2003/04. The summed peaks for Queen Mary Reservoir, Wraysbury Gravel Pits, Staines Reservoirs, Queen Mother Reservoir, Wraysbury Reservoir and Queen Elizabeth II Reservoir had increased from 1,095 in 2002/03 to 4,237 in 2003/04, but in 2004/05 slumped dramatically to just 587, an 86% decline in the last year. In fact, numbers at Queen Elizabeth II remained fairly stable; without this site the decline was 93% in the last year.

In Northern Ireland, the annual index increased again, continuing a general pattern of recent increase. The peak at Loughs Neagh and Beg was the highest there since 1999/2000. There were also continued high counts at Strangford and Carlingford Loughs but lower numbers at Lough Foyle, following the short term peak there of several fish-eating species in 2003/4.

Newson *et al.* (2006) described how, whilst inland breeding used to be rare, following nesting at Abberton Reservoir in 1981 Cormorants have now nested at least once in 58 inland locations throughout England, with over 2,000 inland breeding pairs in 2005, although there has been some stabilisation in recent years. Most of the increase is due to expansion by continental *sinensis* although some *carbo* are also present in the inland colonies. This compares with a total of about 80 English coastal colonies occupied at least once between 1986 and 2005, mostly in the southwest. The coastal English population has remained stable over this period at roughly 2,000 pairs. Inland nesting in Scotland and Wales, however, is generally attributable to *carbo*, some of these colonies having been present for many years.

	00/01	01/02	02/03	03/04	04/05	Mon	Mean
Sites of international importance in the UK							
Loughs Neagh and Beg	1,416	723	1,383	1,468	1,591	Oct	1,316
Sites of national importance in Great Britain							
Morecambe Bay	1,223	398	(657)	(539)	(681)	Sep	811
Alt Estuary	574	960	569	739	984	Nov	765
Dee Estuary (England & Wales)	864	(692)	668	718	780	Oct	758
Forth Estuary	(744)	(761)	(982)	631	669	Oct	757
Thames Estuary	(740)	578	736	596	654	Sep	661
Rutland Water	425	520	529	788	697	Sep	592
Tees Estuary	649	432	438	773	471	Aug	553
Abberton Reservoir	318	780	600	480	(364)	Sep	545
Walthamstow Reservoirs	551	(531)	570	505	453	Sep	522
Solway Estuary	(678)	378	500	(594)	(454)	Sep	521
Inner Firth of Clyde	606	528	553	425	452	Oct	513
Poole Harbour	338	585	558	(412)	431	Sep	478
Dungeness Gravel Pits	294	625	235 [36]	251	870	Aug	455
North Norfolk Coast	294	(268)	581	(276)	(216)	Sep	438
The Wash	401	233	502	449	538	Nov	425
Ribble Estuary	219	358	398	(456)	543	Jan	395
Besthorpe & Girton Gravel Pits	(10)	386	415	372	336	Mar	377
Queen Mary Reservoir	112	580	342	768	44	Dec	369
Rye Harbour and Pett Level	324	218	340	382	446	Jul	342
Alde Complex	(71)	318 [11]	84	(106)	549	Dec	317 ▲
Blackwater Estuary	(209)	450	104	473	191	Nov	305
Hanningfield Reservoir	221	585	189 [36]	411	109	Mar	303
Loch Leven	488	421	68	310	222	Mar	302
Wraysbury Gravel Pits	264	306	181	607	119	Dec	295
Staines Reservoirs	499	77	41	773	21	Aug	282
Ranworth and Cockshoot Broads	298 [12]	398 [12]	270 [12]	324 [12]	115	Nov	281
Queen Mother Reservoir	360	50	91	850	25	Sep	275
Colne Estuary	103	(151)	(29)	423	297	Jan	274
Ouse Washes	197 [12]	213 [12]	347	252 [12]	294	Jan	261
Moray Firth				261 [1]	246 [1]	Dec	254
Wraysbury Reservoir	59	93	132	899	83	Mar	253

	00/01	01/02	02/03	03/04	04/05	Mon	Mean
Pagham Harbour	244	247	240	303	225	Dec	252
Herne Bay	250 [13]						250
Rostherne Mere	281	31	293	306	256	Nov	233
Lee Valley Gravel Pits	220	271	231	286	153	Nov	232
Grafham Water	71	204	349	193	344	Jan	232 ▲
Medway Estuary	220	167	(136)	305	(68)	Nov	231
Ayr to North Troon	(75)	(190)	169	(110)	292	Oct	231 ▲
Queen Elizabeth II Reservoir	90	115	308 [36]	340	295	Aug	230 ▲
Sites with mean peak counts of 130 or more birds in Northern Ireland†							
Belfast Lough	499	528	388	348	(234)	Sep	441
Outer Ards Shoreline	121		652	563	350	Oct	422
Strangford Lough	275	245	358	400	405	Sep	337
Carlingford Lough	(166)	208	206	154	221	Aug	197
Dundrum Bay	(120)	(104)	(124)	(162)	(126)	Sep	(162)
Upper Lough Erne	109	199	124	225	125	Feb	156
Lough Foyle	95	88	170	(210)	87	Jul	130
Other sites surpassing table qualifying levels in WeBS-Year 2004/2005 in Great Britain							
Middle Tame Valley Gravel Pits	116	(213)	(293)	(168)	(256)	Nov	209
Tay Estuary	(165)	197	233	236	(243)	Sep	227

† *as no All-Ireland threshold has been set a qualifying level of 130 has been chosen to select sites for presentation in this report*

Shag

Phalacrocorax aristotelis

International threshold: 2,400
Great Britain threshold: ?†
All-Ireland threshold: ?†

GB max: 2,238 Oct
NI max: 493 Oct

The Shag has only recently been added to the list of species reported on by WeBS. As yet, there is no well-defined estimate for the wintering population. Although ringing recoveries do show some movements across the North Sea, it is likely that the vast majority of Shags in winter in Britain and Ireland derive from local breeding populations. The latest estimates of breeding numbers are 27,176 pairs in Britain and 301 pairs in Northern Ireland (Mitchell *et al.* 2004). A five-year mean of peaks of 100 has been used in the table below to display key sites. In Northern Ireland, the total recorded by WeBS, peaking at 493 in October represents a relatively high proportion of the breeding population for the province. However, coverage of British Shags is proportionally much lower.

In 2004/05, Shags were recorded from 168 WeBS Core Count sites, compared to 95 in 2003/04, the increase probably more due to an continued uptake in recording Shag by counters than to any real increase. The peak British total declined since 2003/04 but that in Northern Ireland rose. This may well have been due to differences in coverage however; with further data in coming years it will be possible to generate an annual index to allow for this.

All sites with mean peaks in excess of 100 birds were in Scotland or Northern Ireland; further south, the only sites with means over 50 were the Kingsbridge Estuary and the Guernsey Coast. At the Forth Estuary, the peak count was lower than for the two previous years but coverage was incomplete Other lower than average WeBS Core Counts were seen at Anstruther Harbour, Papa Westray and Widewall Bay, whereas the count from the Moray Coast was the highest there so far. Survey work carried out by the RAF Ornithological Society on behalf of WeBS in northwest Scotland found Shags to be very widespread although generally not in very large concentrations, with the count in Loch Ewe being the highest. In Shetland, regular boat-based surveys carried out by the Shetland Oil Terminal Environmental Advisory Group found numbers to be low in South Yell Sound, Hacosay / Bluemull / Colgrave Sounds, Quendale to Virkie, and South Unst, but there was a much larger count off North Bressay than when it was last surveyed two years previously.

	00/01	01/02	02/03	03/04	04/05	Mon	Mean
Sites with mean peak counts of 100 or more birds in Great Britain†							
Forth Estuary			2,315	(1,664)	(760)	Sep	2,315
South Yell Sound	1,006 [10]	1,690 [10]	710 [10]	893 [10]	558 [10]	Feb	971
Hacosay/Bluemull/Colgrave Sds	301 [10]	1,132 [10]	423 [10]	709 [10]	232 [10]	Mar	559
Northwest Yell Sound	620 [10]	495 [10]					558
Scalloway Islands		760 [10]		424 [10]	255 [10]	Feb	480
Inner Moray and Inverness Firth			636 [1]	108 [1]	663 [1]	Dec	469
Burra and Trondra		478 [10]		476 [10]	441 [10]	Feb	465
North Bressay			53 [10]		728 [10]	Nov	391
West Whalsay and Sounds		383 [10]					383
West Yell		349 [10]					349
Kirkabister to Wadbister Ness	355 [10]	73 [10]	172 [10]	778 [10]	(97) [10]	Feb	345
Bressay Sound	282 [10]	657 [10]	114 [10]	100 [10]	272 [10]	Nov	285
Widewall Bay			68	580	140	Oct	263
Kirkabister to Dury Voe	250 [10]						250
East Unst		246 [10]					246
South Havra		428 [10]			59 [10]	Feb	244
Quendale to Virkie	98 [10]	605 [10]	123 [10]	176 [10]	97 [10]	Feb	220
South Unst		339 [10]		206 [10]	63 [10]	Mar	203
Loch Ewe					197	Feb	197
Moray Coast (Consolidated)			121	180	251	Oct	184
Inner Firth of Clyde		139	(213)	(159)	190	Oct	181
Easter Ross Coast				214 [1]	122 [1]	Dec	168
South Yell		157 [10]					157
Lunning and Lunna Holm	156 [10]						156
Broadford Bay			150	(100)	150	Dec	150
Anstruther Harbour	4	639	64	8	9	Feb	145
Rova Head to Wadbister Ness	214 [10]	83 [10]	166 [10]	132 [10]	126 [10]	Feb	144
Linga Beach		133 [13]					133
Whiteness to Skelda Ness	51 [10]	142 [10]	149 [10]	169 [10]	138 [10]	Jan	130
Ayr to North Troon		63	184	(26)	(30)	Dec	124
Southeast Yell		120 [10]					120
Arran		86	100	(151)	131	Oct	117
Girvan to Turnberry		111	80	117	115	Nov	106
Wemyss Bay to Fairlie					104	Oct	104
Island of Papa Westray		47	107	210	50	Jan	104
Sites with mean peak counts of 100 or more birds in Northern Ireland†							
Outer Ards Shoreline			227	187	280	Oct	231
Strangford Lough	17 [11]	166	193	226	218	Nov	164
Belfast Lough	39 [11]	30 [11]	215	194 [11]	90	Oct	114
Carlingford Lough		294	48	37	60	Nov	110
Other sites surpassing table qualifying levels in WeBS-Year 2004/2005 in Great Britain†							
Loch Ryan	30	(90)	(110)	79	144	Oct	91
Sullom Voe	62 [10]	50 [10]	106 [10]	104 [10]	133 [10]	Feb	91
Lindisfarne	27 [11]	118 [11]	156 [11]	48 [11]	104 [11]	Dec	91

† *as no British or All-Ireland thresholds have been set a qualifying level of 100 has been chosen to select sites for presentation in this report*

Bittern

Botaurus stellaris

International threshold:	**65**
Great Britain threshold:	**?**
All-Ireland threshold:	**?**

GB max:	**26**	**Dec**
NI max:	**0**	

It was another good year for Bittern in terms of the number of sites at which the species was recorded, with a total of 45 sites across England, Scotland and Wales. Nationally, the species was recorded in all months except June with double figures in every month from October through March. The British peak was recorded in December though was lower than in the previous few years. Peak counts of three were reported from Potteric Carr, Loe Pool and the Humber Estuary.

Squacco Heron

Ardeola ralloides

Vagrant

Native Range: Southern Europe, Middle East & Africa

GB max: 1 Oct
NI max: 0

One was present at East Chevington Pools, Northumberland, during the October count; only the fifth ever recorded during WeBS counts.

Little Egret

Egretta garzetta

International threshold: 1,300
Great Britain threshold: ?†
All-Ireland threshold: ?†

GB max: 2,780 Sep
NI max: 6 Sep

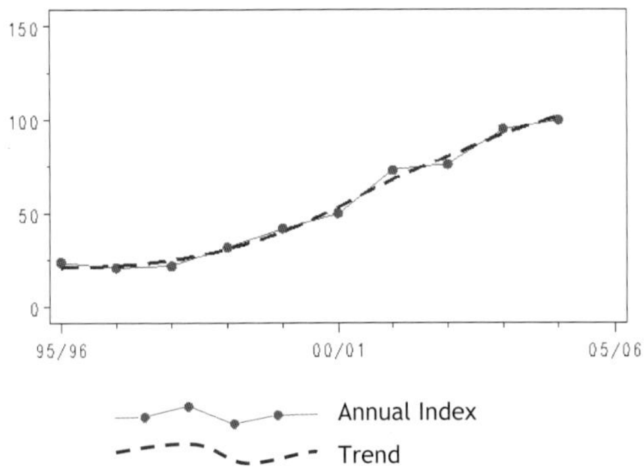

Figure 37.a, Annual indices & trend for Little Egret for GB.

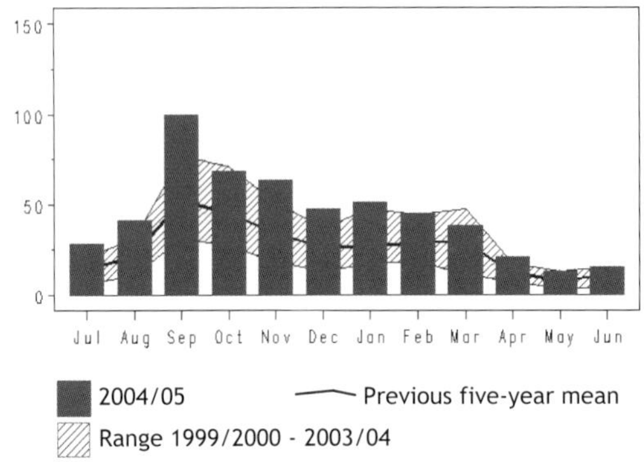

Figure 37.b, Monthly indices for Little Egret for GB.

The Little Egret continued to expand in both range and numbers in 2004/05, although not so rapidly as in some earlier years. The British annual index increased again, albeit only by 5%, whilst the British counted maximum was almost static. The monthly indices again depict a strong post-breeding peak in September and a low point from April to July, when birds probably either return to French breeding colonies or, increasingly, become more secretive around British ones. Whilst there is still no official 1% threshold value for this species, for the purposes of this report the table qualifying level has been increased from a mean of 10 to 30 birds for clarity; 68 sites supported means in excess of 10 birds in 2004/05.

Looking at counts on individual sites, it is very noticeable that almost all sites where the 2004/05 represented a new peak were towards the extremities of the British range, including most sites from the Thames to the Wash on the east coast, as well as those for Lavan Sands and the Dee Estuary. Along the south coast, peaks are more static and at some sites, including the Fowey Estuary and Portsmouth Harbour, numbers may even be declining. The northern limit to the range appears fairly static also, with peaks of just three on the Humber Estuary and four on the Ribble Estuary still surprisingly low. In Scotland, there were just scattered records of singles in Dumfries & Galloway plus one at Montrose Basin in July and August. The maximum count of six in Northern Ireland, whilst minute compared to Britain, is the highest so far for the province.

	00/01	01/02	02/03	03/04	04/05	Mon	Mean
Sites with mean peak counts of 30 or more birds in Great Britain†							
Medway Estuary	(19)	106	(126)	413 [12]	(76)	Oct	260
Chichester Harbour	220 [12]	255 [12]	218	228	129	Sep	210
Thames Estuary	83	132	201	(262)	295	Aug	195
Poole Harbour	118	197 [12]	(140)	(179)	(116)	Nov	165
Tamar Complex	121	141	129	143	120	Oct	131
Longueville Marsh	(85)	132	145	105	102	Oct	121
North Norfolk Coast	15 [12]	50 [11]	81	149 [12]	228 [12]	Nov	105
Exe Estuary	71	149	67	131	93	Aug	102
Burry Inlet	58	99	87	(141)	103	Dec	98

	00/01	01/02	02/03	03/04	04/05	Mon	Mean
Jersey Shore	64	126					95
Kingsbridge Estuary	72	100	105	(99)	86	Aug	92
Portsmouth Harbour	64	123 [12]	110 [12]	(34)	51	Sep	87
Langstone Harbour	51	99	88	90	87	Aug	83
Pagham Harbour	76	81 [12]	76	63	(60)	Sep	74
Blackwater Estuary	15	(35)	(51)	66	(159)	Aug	73
Swale Estuary	22	44	(59)	131	95	Oct	73
Fal Complex	(39)	(30)	55	(52)	89	Jan	72
Taw-Torridge Estuary	71	64	60	(74)	(56)	Aug	67
Camel Estuary	(77)	48	64	65	71	Oct	65
Colne Estuary	26	118 [12]	(2)	(35)	46	Oct	63
Severn Estuary	(13)	59	41	47	66	Sep	53
Avon Valley: Salisbury-F'bridge	(38)	49	(79)	19	57	Jan	51
The Wash	5	(6)	29	72	92	Sep	50
Cleddau Estuary	25	66	48	36	71	Sep	49
Fowey Estuary	49	79	48	35	33	Sep	49
Guernsey Shore	46	50	48	(51)	34	Aug	46
North West Solent	(30)	(44)	(25)	42	(51)	Oct	46
Stour Estuary	10	29	32	57	87	Sep	43
Lavan Sands	(6)	6	15	67	71	Sep	40
Southampton Water	25	45 [12]	(19)	(51)	(39)	Sep	40
Crouch-Roach Estuary	15	24	42	43	73 [11]	Nov	39
Hamford Water	9	31	20	53	81	Sep	39
Helford Estuary	30	33	47	35	46	Jul	38
Newtown Estuary	38	44 [12]	22	41	(21)	Aug	36
Fleet and Wey	37	(37)	38	25	32	Nov	34
Orwell Estuary	11 [12]	43 [12]	37 [12]	56 [12]	9 [11]	Nov	31
Other sites surpassing table qualifying levels in WeBS-Year 2004/2005 in Great Britain†							
Dee Estuary (England & Wales)	7	18 [11]	20	32 [12]	50 [12]	Feb	25
Pegwell Bay	17	20	23	26	(48)	Jul	27
Alde Complex	12	15	(20)	(23)	45	Nov	24
R. Avon: F'bridge- Ringwood	6	4	12	(17)	44	Jan	17
Abberton Reservoir	0	(1)	3	0	(44)	Sep	10
Breydon Wtr & Berney Marshes	2	7	19	22	42	Aug	18
Wootton Creek	13		4	10	38	Sep	16
Carmarthen Bay	(7)	(13)	(9)	23	35	Sep	29
Dart Estuary	12	9	20	44	34	Aug	24
Christchurch Harbour	(20)	24	21	38	32	Sep	29
Dengie Flats	3	18 [11]	15	(27)	31	Oct	19
Beaulieu Estuary	19	6	42	22	30	Nov	24

† *as no British or All-Ireland thresholds have been set a qualifying level of 30 has been chosen to select sites for presentation in this report*

Great White Egret

Ardea alba

Vagrant

Native Range: S Europe, Africa, Asia, N & C America

GB max: 2 Oct
NI max: 0

With the exception of September, Great White Egrets were recorded each month between August and January. The earliest record was at Fishbourne (Chichester Harbour) in August, then singles were at Machynys and Penclacwydd (Burry Inlet) and Freshwater Yantlet (Thames Estuary) in October, Pewit Island (Hamford Water) in December and Carrick Roads (Fal Complex) in January. A long staying bird remained at River Avon: Fordingbridge to Ringwood from November through January. A single spring record was at Saltholme Pools (Tees Estuary) in May.

Whilst numbers are still low (with a monthly peak of just two birds), this is a species to watch in the future. Great White Egrets have rapidly colonised the Netherlands since 1985/86. In 2003 there were an estimated 59 breeding pairs, and non-breeding waterbird counts found a peak of 445 in January 2005. This is up from the species being a rare visitor as recently as 1996/97. Colonisation of the UK therefore remains a strong possibility.

Grey Heron

Ardea cinerea

International threshold: 2,700
Great Britain threshold: ?†
All-Ireland threshold: ?†

GB max: 3,767 Oct
NI max: 405 Oct

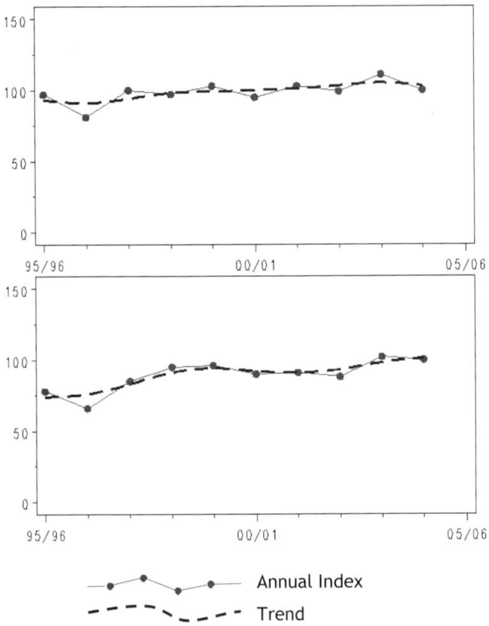

Figure 38.a, Annual indices & trend for Grey Heron for GB (above) & NI (below).

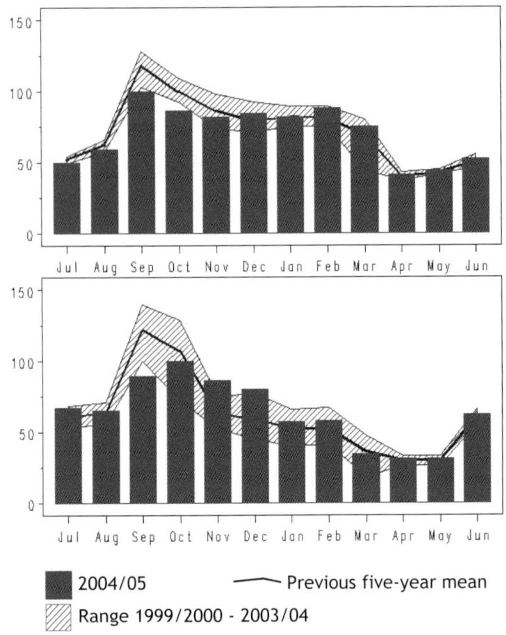

Figure 38.b, Monthly indices for Grey Heron for GB (above) & NI (below).

Both British and Northern Ireland annual indices for Grey Heron are some of the least fluctuating for any waterbird species, and although both fell slightly in 2004/05 both have shown an increase in the underlying trend over the last ten seasons, by about 11% and 36% respectively. British monthly indices show that Grey Herons peak on WeBS sites in September but remain in fairly constant numbers throughout the winter until March. As Grey Herons are typically early nesters, it initially seems strange that there is not a more rapid decline in the spring, but this is probably explained by the fact that many of the key WeBS sites contain, or are adjacent to, some of the key breeding colonies. The autumn peak was a little lower than usual with numbers later in the winter more as expected. Data from the BTO Heronries Census show that UK nest totals in 2004 were about 4% down on the 1999-2003 average, explaining at least part of this decline in autumn peak (J. Marchant *pers. comm.*).

Peak counts at a number of key sites were lower than usual, notably from the Taw-Torridge Estuary, Ouse Washes and Walthamstow Reservoirs. Higher counts than expected were noted from the Somerset Levels, Forth Estuary and Solway Estuary.

	00/01	01/02	02/03	03/04	04/05	Mon	Mean
Sites with mean peak counts of 50 or more birds in Great Britain†							
Avon Valley: Salisbury-F'bridge	326	(100)	(83)	150	(58)	Mar	238
Coombe Country Park			159	159	105	Jul	141
Somerset Levels	(148)	121	136	130	151	Jan	137
Thames Estuary	145	(129)	(124)	(94)	100	Oct	125
Walthamstow Reservoirs	117	91	133	64	60	Feb	93
Dee Estuary (England & Wales)	124	63 [11]	(111)	87	67	Aug	90
Inner Firth of Clyde	81	90	87	81	90	Aug	86
Ouse Washes	70 [13]	100	104 [13]	78	66	Oct	84
Severn Estuary	(67)	69	104 [11]	73	69	Oct	79
Morecambe Bay	69	51	101	91	68	Sep	76
Forth Estuary	59	47	62	(78) [11]	104	Nov	70
Tees Estuary	85	58	66	64	56	Aug	66

	00/01	01/02	02/03	03/04	04/05	Mon	Mean
Inner Moray and Inverness Firth	44	56	91	67	55	Dec	63
Solway Estuary	(48)	28 [11]	(69)	(70)	72	Oct	60
Taw-Torridge Estuary	69	(20)	(41)	77	30	Aug	59
The Wash	56	49	54	76	50	Sep	57
Colne Valley Gravel Pits	(21)	(36)	(33)	68	44	Mar	56
Montrose Basin	87	55	24	54	(40)	Aug	55
Cromarty Firth	41	48	44	73	47	Oct	51
Cotswold Water Park (West)	(33)	(34)	(12)	(37)	(50)	Feb	(50)
Sites with mean peak counts of 50 or more birds in Northern Ireland†							
Loughs Neagh and Beg	267	87	226	208	172	Oct	192
Strangford Lough	92	113	103	102	90	Nov	100
Other sites surpassing table qualifying levels in WeBS-Year 2004/2005 in Great Britain†							
Wraysbury Gravel Pits	27	32	34	22	58	Feb	35
R. Avon: F'bridge-Ringwood	22	19	46	28	56	Jan	34
Swale Estuary	36	43	35	26	51	Oct	38
Kingsbridge Estuary	39	26	22	25	50	Sep	32
Rutland Water	42	29	52	49	50	Aug	44
Other sites surpassing table qualifying levels in WeBS-Year 2004/2005 in Northern Ireland†							
Lough Foyle	20	45	31	45	54	Jul	39
Outer Ards Shoreline	10		18	22	52	Oct	26

† *as no British or All-Ireland thresholds have been set a qualifying level of 50 has been chosen to select sites for presentation in this report*

White Stork

Ciconia ciconia

Vagrant and escape
Native Range: Europe, Africa & Asia

GB max: 4 Aug
NI max: 0

The usual free-flying White Storks were present at Harewood Lake throughout the year, peaking at four in August. These birds, originating from the captive collection here but sometimes wandering widely, have been recorded during WeBS counts since 1999.

Sacred Ibis

Threskiornis aethiopicus

Escape
Native Range: Africa & Middle East

GB max: 2 Apr
NI max: 0

The long-staying bird at Outwood Swan Sanctuary was recorded in six months between July and April. One was also recorded at nearby Gatton Park in April, probably the same individual.

Spoonbill

Platalea leucorodia

International threshold: 100

GB max: 12 Nov
NI max: 0

Spoonbills were recorded at 17 sites across Britain, about average in recent years. None have ever been recorded during WeBS counts in Northern Ireland. Birds were reported in every month and double figures were reached during November and April with 12 and ten birds respectively. The top site was the now regular wintering site of the Taw-Torridge Estuary, which held up to six birds between November and March.

African Spoonbill

Platalea alba

Escape
Native Range: Africa

GB max: 1 Oct
NI max: 0

A lone African Spoonbill was reported at Cardiff Bay on the Severn Estuary in October. This individual then reappeared during the November count on the Burry Inlet, where it was recorded monthly along the Machynys to Penclacwydd stretch through until May. This bird was initially recorded in Norfolk and Shropshire before appearing in Wales.

Water Rail

Rallus aquaticus

International threshold: ?
Great Britain threshold: ?†
All-Ireland threshold: ?†

GB max: 509 Nov
NI max: 7 Dec

Due to their liking for dense waterside vegetation, Water Rails recorded during WeBS counts are assumed to represent only a small proportion of the true number actually present in the country. The British maximum of 509 was one of the higher peaks recorded by WeBS, and the seven in Northern Ireland were the largest number recorded in the province. However, it is unclear the extent to which this says anything about trends in numbers in the UK. Nonetheless, the sites from which the highest numbers are recorded tend to be the same each year. The count of 63 on the Somerset Levels was the second highest site total ever recorded by WeBS, with over half the total being present on Westhay Moor. Particularly low counts were recorded from North Warren, North Norfolk Coast, Rye Harbour / Pett Level and Kilconquhar Loch, although these may have been more connected to local weather conditions or count methods than relating to the actual numbers of birds present.

	00/01	01/02	02/03	03/04	04/05	Mon	Mean
Sites with mean peak counts of 10 or more birds in Great Britain†							
Somerset Levels	43	45	(45)	45	63	Dec	49
Grouville Marsh	(30)	30	25	20	20	Oct	25
Kenfig Pool	7	30	39	27	17	Mar	24
Stodmarsh NNR & Collards Lgn	25	27	28	20	15	Mar	23
Middle Yare Marshes	(4)	23	17	18	(4)	Nov	19
Poole Harbour	15	15	24	(10)	(12)	Nov	18
Lee Valley Gravel Pits	18	18	18	12	(7)	Jan	17
Longueville Marsh	(10)	15	15	15	20	Nov	16
Southampton Water	13	20	18	(7)	11	Feb	16
Rye Harbour and Pett Level	31	13	8	16	6	Dec	15
Thames Estuary	(6)	12	21	(8)	11	Jan	15
Burry Inlet	9	16	10	18	16	Feb	14
Dee Estuary (England & Wales)	16	8 [11]	13	(5)	16	Feb	13
Fleet Pond	25		6	10	12	Dec	13
North Norfolk Coast	13	22	(10)	10	7	Oct	13
N. Warren & Thorpeness Mere		4	2 [13]	44	1	Jan	13
Chichester Harbour	12	14	16	6	13	Feb	12
Kenfig NNR	12 [21]						12
Ingrebourne Valley	14	10	14	12	7	Feb	11
Doxey Marshes SSSI	8	8	(14)	7	12	Nov	10
Kilconquhar Loch	15	8	14	11	0		10
Loe Pool	4	1	19	(16)	10	Dec	10
River Cam - Kingfishers Bridge	6	6	22 [13]	7	8	Jan	10
Other sites surpassing table qualifying levels in WeBS-Year 2004/2005 in Great Britain†							
Fleet and Wey	8	(4)	5	4	13	Nov	8
Llynnau Y Fali	8	5	7	4	12	Dec	7
Cotswold Water Park (West)	(10)	(4)	(1)	1	10	Mar	7

† *as no British or All-Ireland thresholds have been set a qualifying level of 10 has been chosen to select sites for presentation in this report*

Spotted Crake

Porzana porzana

Scarce

GB max: 2 Aug
NI max: 0

Two Spotted Crakes were recorded at Tophill Low Reservoirs during August, while single birds were at the same location as well as at Exton on the Exe Estuary during September.

Moorhen

Gallinula chloropus

International threshold: 20,000**
Great Britain threshold: 7500†
All-Ireland threshold: ?†

GB max: 13,464 Jan
NI max: 239 Jan

Whilst Moorhens are one of the most widely recorded species during WeBS counts, the numbers recorded are assumed to represent only a tiny proportion of the true country total. The peak counted by WeBS in Britain was roughly in line with recent years, although the peak in Northern Ireland was the lowest total for over a decade, largely as a result of low counts at Loughs Neagh and Beg. The Severn Estuary remains top site based on the five-year mean counts. A particularly high peak was recorded at the Grand Western Canal: Greenway Bridge to North Devon Link Road, while relatively low counts were recorded at the North Norfolk Coast, Pitsford Reservoir, Bewl Water and the Dee Estuary. There were no sites with counts that were substantially higher than average. Breeding Bird Survey data has shown an increase of around 20% in the past 10 years; however, there have been small declines over the past two years.

	00/01	01/02	02/03	03/04	04/05	Mon	Mean
Sites with mean peak counts of 100 or more birds in Great Britain†							
Severn Estuary	735	557	476	443	409	Oct	524
WWT Martin Mere	510	485	490	440	420	Nov	469
Lower Derwent Ings			463	444			454
Thames Estuary	(268)	345	472	324	371	Jan	378
Lee Valley Gravel Pits	315	357	312	340	301	Mar	325
Somerset Levels	310	308	325	276	327	Feb	309
North Norfolk Coast	309	334	243	280	186	Mar	270
Pitsford Reservoir	175	267	209	326	133	Aug	222
Bewl Water	230	200	254	215	165	Aug	213
Durham Coast	240	225	160	(0)	(158)	Nov	208
Rutland Water	252	211	189	191	192	Sep	207
Burry Inlet	196	209	175	169	202	Dec	190
Chew Valley Lake	185	165	105	245	125	Sep	165
Arun Valley	202	148	172	163	128	Dec	163
Humber Estuary	87	101	215	224	(170)	Jan	159
Chichester Gravel Pits	132	157	149	161	167	Feb	153
Avon Valley: Salisbury- F'bridge	(69)	(79)	(56)	143	(112)	Sep	143
Colne Valley Gravel Pits	(51)	(149)	(110)	(58)	120	Mar	135
Dee Estuary (England & Wales)	134	199	(116)	121	86	Feb	135
Barn Elms Reservoirs	95	170	131	137	135	Nov	134
Orwell Estuary	160 [11]	117 [11]	100 [11]	164 [11]	109 [11]	Feb	130
Tring Reservoirs	149	110	106	115	135	Jan	123
Sutton and Lound Gravel Pits		(160)	118	112	94	Feb	121
Thanet Coast	87	95	123	169	133	Oct	121
Hamford Water	68	(72)	134	(156)	90	Jan	112
River Wye - Bakewell to Haddon	101	89	131	126	109	Jan	111
Fairburn Ings	(115)	86	73	154			107
Old Moor	59	122	(131)	116			107
Ouse Washes	206 [13]	61	70	95	92	Mar	105
Southampton Water	(102)	(81)	(81)	125	83	Jan	104
Ingrebourne Valley	76	122	116	121	77	Dec	102
Rye Harbour and Pett Level	99	107	71	116	117	Jan	102
Sites with mean peak counts of 30 or more birds in Northern Ireland†							
Loughs Neagh and Beg	244	380	211	177	124	Mar	227
Upper Lough Erne	59	60	(46)	(32)	46	Jan	55
Belfast Lough	44	47	62	27	51	Jan	46
Other sites surpassing table qualifying levels in WeBS-Year 2004/2005 in Great Britain†							
Grnd W. Cnl: G'wy Br-N. Dvn Rd	62	73	80	103	132	Sep	90
R. Cam: Owlstone Rd-Baits B Lk	80	99	93	76	117	Jan	93
Tees Estuary	77	73	115	110	103	Dec	96
Other sites surpassing table qualifying levels in WeBS-Year 2004/2005 in Northern Ireland†							
Lower Lough Erne			26	20	(30)	Dec	25

† *as few sites exceed the British threshold and no All-Ireland thresholds have been set qualifying levels of 100 and 30 have been chosen to select sites, in Great Britain and Northern Ireland respectively, for presentation in this report*

Coot

Fulica atra

International threshold: 17,500
Great Britain threshold: 1,730
All-Ireland threshold: 250

GB max: 109,032 Nov
NI max: 3,758 Dec

	S	M	L
GB change	o	o	o
NI change	(-)	-	-

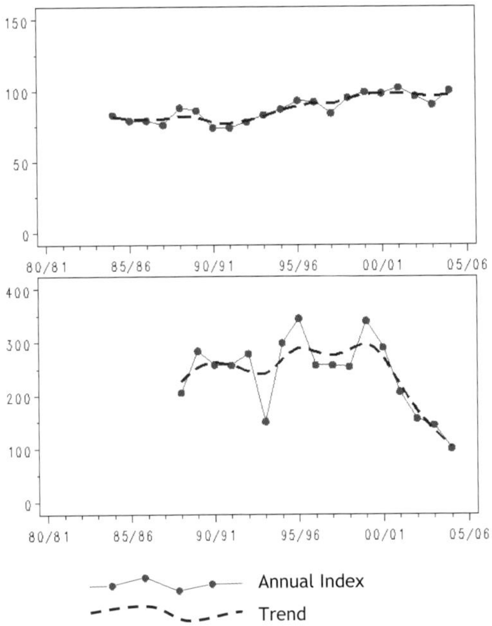

Figure 39.a, Annual indices & trend for Coot for GB (above) & NI (below).

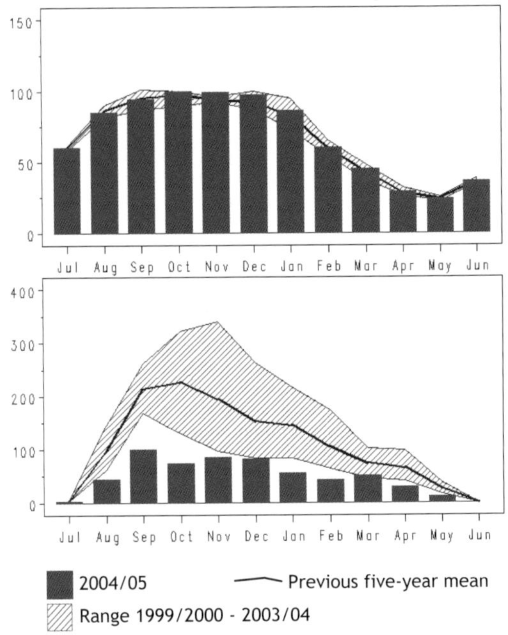

Figure 39.b, Monthly indices for Coot for GB (above) & NI (below).

Coot numbers in Britain are currently relatively stable, following a steady increase during the 1990s. In Northern Ireland, on the other hand, a major decline has taken place since the 2000/01 winter; the peak counted total for the province was the lowest to date. As with other species, including Mute Swan, Pochard and Tufted Duck, this decline is due almost entirely to a reduction in the numbers counted at Loughs Neagh and Beg. The Northern Ireland monthly indices show an all-year reduction in numbers, suggesting a local reason for the decline rather than one of reduced immigration. At the UK scale, the Breeding Bird Survey found a 77% increase in Coot over the period 1994-2004 (Raven *et al.* 2005).

Following a run of declining numbers, the peak at Abberton Reservoir increased again and this site remains by far the most important in the UK. Other proportionally high peaks were noted at Rutland Water, Ouse Washes, Carsington Water and Blagdon Lake. Conversely, peaks at Hanningfield Reservoir and Alton Water were much smaller than usual; a large decline in Tufted Duck numbers was also seen at Hanningfield Reservoir during 2004/05.

	00/01	01/02	02/03	03/04	04/05	Mon	Mean
Sites of national importance in Great Britain							
Abberton Reservoir	11,645	7,610	6,885	6,166	9,697	Dec	8,401
Cotswold Water Park (West)	(3,806)	4,161	(2,528)	4,042	4,077	Oct	4,093
Rutland Water	3,375	3,283	3,969	4,021	4,733	Oct	3,876
Cheddar Reservoir	3,410	2,950	2,975	3,100	3,873	Nov	3,262
Chew Valley Lake	3,500	2,360	3,715	3,285	3,335	Sep	3,239
Lee Valley Gravel Pits	2,751	3,245	3,250	3,213	3,435	Dec	3,179
Fleet and Wey	2,346	3,418	2,353	(2,923)	3,275	Oct	2,863
Hanningfield Reservoir	4,282	1,369	3,426	3,791	463	Jun	2,666
Loch Leven	2,100	1,818	3,205	2,650 [13]	2,375	Oct	2,430
Ouse Washes	2,062 [13]	2,488 [13]	1,349	2,039	(3,877)	Feb	2,363
Cotswold Water Park (East)	2,227	2,634	2,365	2,296	1,850	Nov	2,274

	00/01	01/02	02/03	03/04	04/05	Mon	Mean	
Pitsford Reservoir	2,331	2,746	1,949	1,823	2,354	Nov	2,241	
Lower Windrush Valley GPs	2,274	1,720	2,016	2,341	(2,075)	Nov	2,088	
Middle Tame Valley Gravel Pits	1,674	(2,106)	(1,284)	(559)	(393)	Feb	1,890	
Little Paxton Gravel Pits	3,014	1,679	1,831	1,334	1,422	Dec	1,856	
Sites of all-Ireland importance in Northern Ireland								
Loughs Neagh and Beg	6,645	2,555	4,344	4,124	1,890	Nov	3,912	
Upper Lough Erne	899	1,660	1,447	2,062	1,462	Feb	1,506	
Strangford Lough	400	581	420	230	223	Jan	371	
Lower Lough Erne			272	197	308	Feb	259	▲
Sites no longer meeting table qualifying levels in WeBS-Year 2004/2005								
Dungeness Gravel Pits	1,564	1,573	1,528	1,943	1,651	Nov	1,652	
Alton Water	655	2,536	2,491	649	142	Dec	1,295	
Hickling Broad	1,021						1,021	
Other sites surpassing table qualifying levels in WeBS-Year 2004/2005 in Great Britain								
Blagdon Lake	1,012	2,846	628	1,993	2,080	Aug	1,712	
River Avon: F'bridge-Ringwood	1,439	1,628	1,069	1,494	1,765	Oct	1,479	
Carsington Water		1,690	1,332	526	1,731	Jan	1,320	

Coot (Richard Vaughan)

American Coot

Fulica americana

Vagrant
Native Range: N America

GB max:	1	Dec
NI max:	0	

The single bird recorded at Benston Loch during December probably refers to the same individual recorded on nearby Loch of Clickimin during the previous winter.

Crane

Grus grus

Scarce

GB max:	2	Sep
NI max:	0	

Four single birds were recorded in late summer and autumn, all in northern Britain. One was at Loch of Strathbeg in August, then singles at Carsebreck & Rhynd Lochs and on the Solway Estuary at Calvo Marsh during September, with the last on the Humber Estuary in November.

Oystercatcher

Haematopus ostralegus

GB max:	234,868	Dec
NI max:	20,366	Oct

International threshold:	10,200
Great Britain threshold:	3,200
All-Ireland threshold:	500

	S	M	L
GB change	o	o	o
NI change	o	o	o

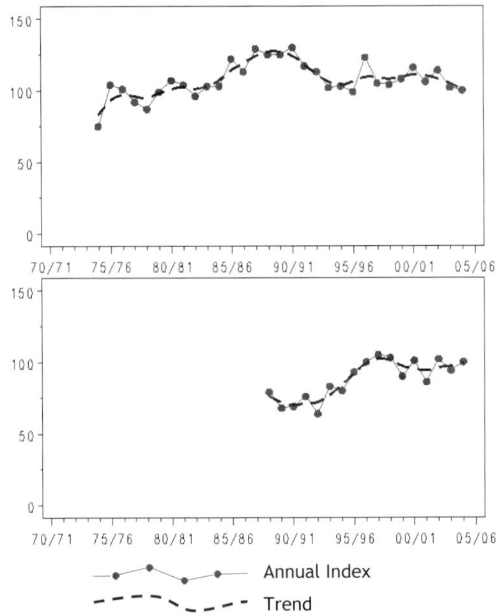

Figure 40.a, Annual indices & trend for Oystercatcher for GB (above) & NI (below).

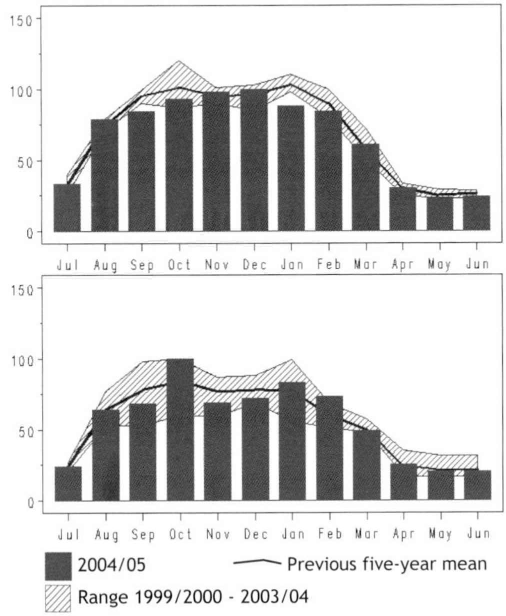

Figure 40.b, Monthly indices for Oystercatcher for GB (above) & NI (below).

The Oystercatcher remains one of the most numerous and widespread wintering wader species in the UK. Although the British annual index dropped slightly, the underlying trend seems fairly stable at the moment. Similarly, numbers wintering in Northern Ireland also seem stable.

There was no change in the sites qualifying as nationally or internationally important for Oystercatcher. At the top of the list remains Morecambe Bay, the 55,000 counted in December being the highest winter count here since 1997 and representing 23% of the British total in that month. Within the site, Oystercatchers are widespread but with a peak sectional count of nearly 20,000 birds between Morecambe and Arnside in December. Recent aerial surveys have looked into the low tide distribution of waterbirds within Morecambe Bay and confirmed the species to be widely dispersed at low tide too, except for a gap around mouth of the Kent Estuary. At the two other major west coast sites, numbers were lower than usual on the Solway but a little higher than normal on the Dee. Amongst the other listed sites, proportionally high peaks were recorded from Swansea Bay and Lough Foyle but peaks at the Ribble Estuary, Duddon Estuary, Dengie Flats and Belfast Lough were on the low side.

During the autumn, most key sites supported similar peak numbers to those seen in winter, although substantially higher numbers were noted from Strangford and Belfast Loughs, and the Thames Estuary.

	00/01	01/02	02/03	03/04	04/05	Mon	Mean
Sites of international importance in the UK							
Morecambe Bay	50,831	47,286	48,752	(48,600)	55,072	Dec	50,485
Solway Estuary	(34,196)	(35,035)	(47,415)	(34,099)	(30,397)	Nov	(47,415)
Dee Estuary (England & Wales)	21,326	31,851 [11]	20,373	23,906	25,956	Dec	24,682
Thames Estuary	(14,938)	18,814	26,803	23,858	(14,907)	Dec	23,158
Ribble Estuary	(17,784)	23,072	(12,395)	19,915	12,953	Mar	18,647
The Wash	13,457	13,371	16,760	14,684	16,395	Feb	14,933
Burry Inlet	13,347	15,253	14,570	13,831 [11]	16,219	Nov	14,644

	00/01	01/02	02/03	03/04	04/05	Mon	Mean
Sites of national importance in Great Britain							
Forth Estuary	6,814	(6,631)	9,279	7,834	6,569	Dec	7,624
Lavan Sands	6,897	7,831	7,612	6,796	5,718	Nov	6,971
Duddon Estuary	(4,867)	6,907	(6,476)	8,683	5,272	Nov	6,954
Carmarthen Bay	(4,154)	(5,575)	(4,530)	(4,597)	6,736	Feb	6,736
Inner Moray and Inverness Firth	6,049	5,153	6,087	7,624	4,681	Dec	5,919
Swale Estuary	5,427	6,270	5,058	5,858	5,225	Jan	5,568
Inner Firth of Clyde	5,060	5,488	5,386	4,627	4,737	Dec	5,060
Humber Estuary	3,834	(3,318)	(2,963)	3,305 [11]	(4,582)	Jan	3,907
North Norfolk Coast	3,755	3,990	3,011	3,858	3,778	Feb	3,678
Swansea Bay	3,500	3,563	3,797	2,857 [11]	4,605	Jan	3,664
Dengie Flats	2,033	7,061 [11]	3,034	(1,450)	1,865	Jan	3,498
Sites of all-Ireland importance in Northern Ireland							
Strangford Lough	7,149	8,298	8,625	7,412	6,454	Feb	7,588
Belfast Lough	5,647	4,276 [13]	5,542 [11]	4,248 [11]	3,909 [11]	Nov	4,724
Lough Foyle	2,730	2,294	2,326	2,231	3,095	Jan	2,535
Outer Ards Shoreline	1,621		1,968	1,812	1,740	Jan	1,785
Dundrum Bay	1,707	(1,428)	(1,250)	(1,425)	(1,252)	Dec	1,707
Carlingford Lough	(1,184)	(986)	1,289	(1,414)	1,410	Dec	1,371

Sites surpassing international passage threshold in the UK in 2004/2005					
Morecambe Bay	40,978	Sep	Dee Estuary (England and Wales)	19,423	Oct
Solway Estuary	30,961	Oct	The Wash	15,699	Aug
Thames Estuary	20,393	Oct	Ribble Estuary	14,095	Oct
Sites surpassing national passage threshold in Great Britain in 2004/2005					
Forth Estuary	8,213	Sep	Inner Moray and Inverness Firth	5,376	Oct
Duddon Estuary	6,241	Oct	Inner Firth of Clyde	4,759	Oct
Burry Inlet	6,148	Aug	Swale Estuary	4,700	Aug
Carmarthen Bay	6,009	Sep	Lavan Sands	4,232	Aug
Sites surpassing national passage threshold in Northern Ireland in 2004/2005					
Strangford Lough	9,018	Oct	Outer Ards Shoreline	1,666	Oct
Belfast Lough	5,299	Oct	Carlingford Lough	1,419	Sep
Lough Foyle	1,930	Oct			

Black-winged Stilt

Himantopus himantopus

Vagrant
Native Range: Worldwide

GB max: 1 Jul
NI max: 0

The long-staying bird at Titchwell on the North Norfolk Coast was recorded in each month from July to December, and then again in April. This bird, which first appeared here in 1993, finally disappeared for good on 21st May 2005 and is presumed dead.

Avocet

Recurvirostra avosetta

International threshold: 730
Great Britain threshold: 35*

GB max: 5,760 Jan
NI max: 0

	S	M	L
GB change	+	++	++

**50 is normally used as a minimum threshold*

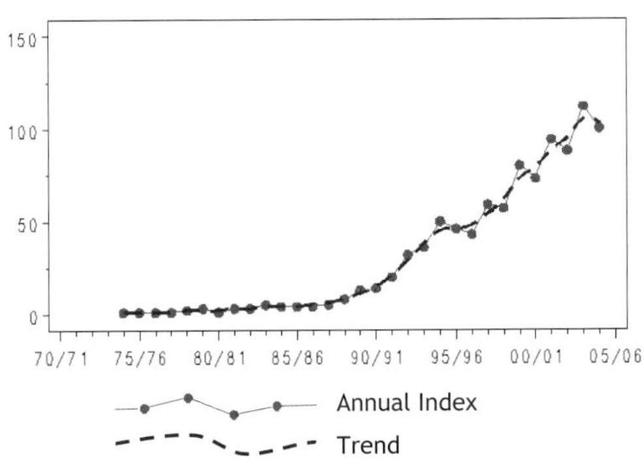

Figure 41.a, Annual indices & trend for Avocet for GB.

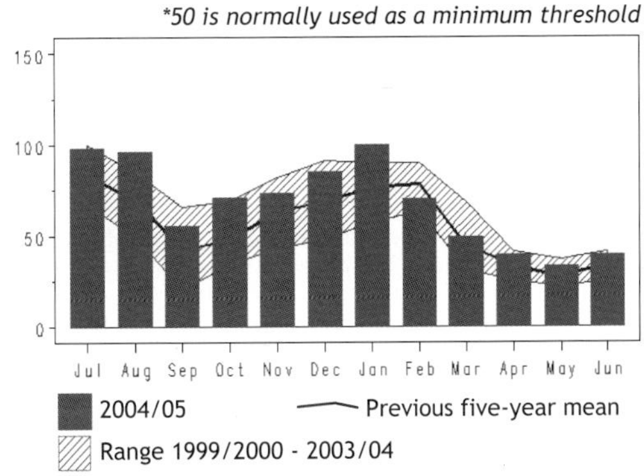

Figure 41.b, Monthly indices for Avocet for GB.

The British maximum in 2004/05 was slightly lower than during the previous year yet still represents the third highest ever total for this species. The national index witnessed a slight fall during 2004/05 but a pattern of 'stepped' increase has been evident during the past ten years and there is no reason yet to consider that the species will not continue to increase in the coming years. Monthly indices were above average throughout the year and reached their highest for over five years in August and January.

Internationally important numbers of Avocets were recorded from the same three sites as last year. Peak numbers were similar to recent levels at Poole Harbour and the Alde Complex and exceeded 1,000 for the second time ever at the Thames Estuary. However, the second highest site total of the year was made at the nearby Swale Estuary, with the highest ever count for this site. Elsewhere, higher than usual peaks were recorded at Hamford Water, Deben Estuary, Blackwater Estuary and Stour Estuary, whilst Low Tide Counts at the Crouch-Roach and Stour Estuaries were way in excess of anything recorded there during Core Counts previously. The Crouch-Roach Estuary does, however, have coincident boundaries with the Foulness area of the Thames Estuary and movement presumably occurs between the two sites. Conversely, low peak counts were noted from the Exe Estuary and the Wash.

As seen during the last few years, the importance of Breydon Water for Avocets in late summer was highlighted, the August count being about four times the winter peak. Other sites substantially more important outside the winter period were the North Norfolk Coast, the Wash and the Humber Estuary, representing the northern edge of the British range and largely reflecting breeding populations. The count of over 400 on the Humber is particularly remarkable at a site that only eight years ago held the species as only an occasional visitor. However, it is notable that the listed sites are still confined to the south and southeast, between the Tamar and Humber. Away from here, there were double-figure peaks of 26 on the Severn Estuary (mostly from Bridgwater Bay and Goldcliff Saline Lagoons), 24 on the Ribble Estuary and 12 at WWT Martin Mere, numbers at the Lancashire sites peaking in the summer and representing breeding colonies.

	00/01	01/02	02/03	03/04	04/05	Mon	Mean
Sites of international importance in the UK							
Poole Harbour	1,491	1,893	1,007	(1,493)	1,480	Jan	1,473
Alde Complex	1,007	1,174	1,089	1,073	1,058	Mar	1,080
Thames Estuary	563	1,447	839	658	1,153	Jan	932
Sites of national importance in Great Britain							
Swale Estuary	(145)	532	318	451	1,290	Mar	648
Medway Estuary	301	860	(650)	(615)	309 [11]	Feb	547
Blyth Estuary	524	463					494
Hamford Water	242	485	(406)	433	520	Nov	420
Exe Estuary	366	528	436	353	297	Feb	396
Tamar Complex	452	277	317 [11]	394	438	Feb	376
Colne Estuary	351	465	(383)	205 [13]	(90)	Jan	351
North Norfolk Coast	(72)	228	334	508	283	Mar	338
Deben Estuary	165	193	170	353	323	Jan	241
Breydon Wtr & Berney Marshes	272	172	224	268	232	Mar	234
Blackwater Estuary	167	125	151 [11]	295	428	Mar	233
Humber Estuary	126	121	281	(271)	215	Mar	203
Crouch-Roach Estuary	0	(43)	(9)	(17)	288 [11]	Nov	144 ▲
The Wash	6	347	130	180	37	Mar	140
Minsmere	120	10	1	107	86	Mar	65

Sites surpassing international passage threshold in the UK in 2004/2005					
Breydon Wtr & Berney Marshes	1,012	Aug	Poole Harbour	864	Oct
Alde Complex	1,004	Oct			
Sites surpassing national passage threshold in Great Britain in 2004/2005					
North Norfolk Coast	712	Apr	Swale Estuary	363	Oct
Hamford Water	663	Oct	Tamar Complex	216	Oct
Thames Estuary	571	Oct	Deben Estuary	183	Oct
The Wash	532	Aug	Exe Estuary	83	Oct
Medway Estuary	490	Oct	Ouse Washes	71	Apr
Humber Estuary	425	Apr	Orwell Estuary	68	Oct
Blackwater Estuary	373	Oct	Colne Estuary	64	Oct

Stone-curlew

Scarce

Burhinus oedicnemus

GB max: 2 May
NI max: 0

One was seen on the North Norfolk Coast in August, whilst two birds were recorded close to a known breeding location in eastern England during May.

Little Ringed Plover

Charadrius dubius

International threshold: 2,400
Great Britain threshold: ?†
All-Ireland threshold: ?†

GB max: 225 Apr
NI max: 0

The 2004 British peak of just over 200 Little Ringed Plovers was about average for the past ten years. As usual, there were none recorded in Northern Ireland. Note that as a summering species, this account refers to the calendar year 2004 rather than the 2004/05 WeBS year. During 2004 the birds were recorded from 149 sites. Most records were of singles or groups of up to three, although ten or more were recorded at ten sites during single months. The peak spring at the Ouse Washes represents the highest ever count for the site, as does the peak for Nosterfield Gravel Pits, which is three times that of the previous year's peak.

Sites with ten or more birds in Great Britain in 2004†

Ouse Washes	18	Apr
Nosterfield Gravel Pits	18	Jun
Barn Elms Reservoirs	15	Jun
Belvide Reservoir	14	Jul
Thames Estuary	12	Jul
Upton Warren Local Nature Reserve	12	Jun
Barton Pits	10	Jun
Sutton and Lound Gravel Pits	10	Jun
Sandbach Flashes	10	Jul
Rutland Water	10	Jul

† *as no British or All-Ireland thresholds have been set a qualifying level of 10 has been chosen to select sites for presentation in this report*

Ringed Plover

Charadrius hiaticula

International threshold: 730
Great Britain winter threshold: 330
Great Britain passage threshold: 300
All-Ireland threshold: 125

GB max: 18,317 Aug
NI max: 747 Oct

	S	M	L
GB change	o	o	o
NI change	o	o	-

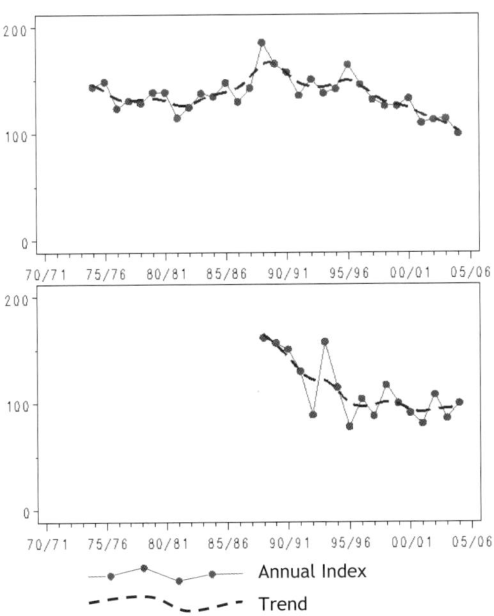

Figure 42.a, Annual indices & trend for Ringed Plover for GB (above) & NI (below).

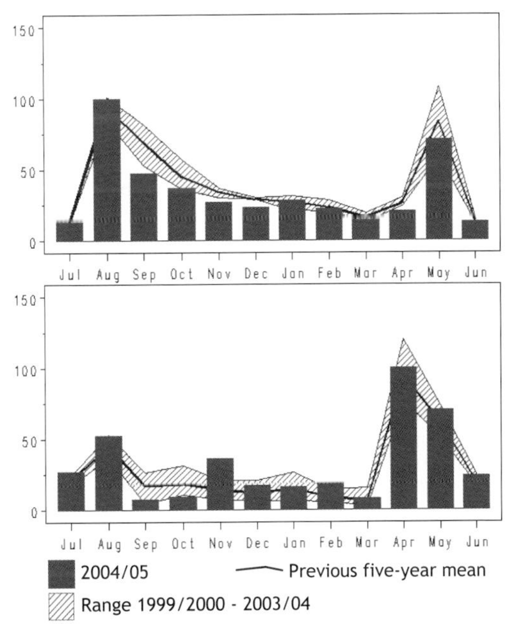

Figure 42.b, Monthly indices for Ringed Plover for GB (above) & NI (below).

Continuing the decline seen in the British Ringed Plover population since the late 1980s, the annual index for Great Britain dropped to a new all-time low. However, the counted British maximum was average for the past 15 years. The monthly indices suggest that August passage numbers were a little higher than usual but numbers in most months then below average. May passage was about as expected. In Northern Ireland, the index remained more stable, although overall numbers involved are very much lower. An interesting feature of the Northern Ireland monthly indices is that spring passage seems to peak in April rather than May, perhaps as this involves more birds breeding in Iceland and Greenland than British wintering birds that also includes Fennoscandian breeders.

The Thames Estuary remains the top site, both in winter and during passage, and is joined in the list of sites supporting internationally important wintering numbers by Hamford Water, although on the basis of mean numbers rather than by a particular increase in 2004/05. Elsewhere, higher than usual peaks were recorded at South Ford, Swansea Bay, Colne Estuary and the Duddon Estuary, the latter as a result of Low Tide Counts at the site. Particularly low site peaks were noted from the Thanet Coast, Outer Ards, Belfast Lough and the Forth Estuary.

	00/01	01/02	02/03	03/04	04/05	Mon	Mean	
Sites of international importance in the UK								
Thames Estuary	954	765	794	(654)	872	Nov	846	
Hamford Water	(678)	1,302 [11]	201	(576)	(333)	Nov	752	▲
Sites of national importance in Great Britain								
Solway Estuary	(330)	(289)	(599)	(286)	(305)	Nov	(599)	
North Norfolk Coast	369	(471)	262	464	411	Feb	395	
Thanet Coast	528	407	412	389	123	Nov	372	
Humber Estuary	409	350	225	418 [11]	(194)	Dec	351	
South Ford	341		373	250	400	Feb	341	
Swansea Bay	214	436	269	330	431	Jan	336	▲
Morecambe Bay	(473)	298	246	303	357	Jan	335	
Medway Estuary	(126)	(89)	(249)	(136)	332 [11]	Dec	332	
Sites of all-Ireland importance in Northern Ireland								
Strangford Lough	494 [11]	618 [11]	236 [11]	277 [11]	342	Nov	393	
Outer Ards Shoreline	313		315	(198)	142	Jan	257	
Carlingford Lough	(116)	(203)	(240)	161	223	Dec	207	
Belfast Lough	142	188	189	234 [11]	109 [11]	Nov	172	
Sites no longer meeting table qualifying levels in Winter 2004/2005								
Langstone Harbour	413	268	394	201 [11]	254	Nov	306	
Forth Estuary	(356)	266	343	303 [11]	172	Jan	288	
Other sites surpassing table qualifying levels in Winter 2004/2005 in Great Britain								
Colne Estuary	224	244	(15)	212	466	Dec	287	
Duddon Estuary	232	232	(227)	222	350 [11]	Feb	259	

Sites surpassing international passage threshold in the UK in 2004/2005					
Thames Estuary	1,998	Aug	Severn Estuary	1,111	Aug
Ribble Estuary	1,459	Aug	Solway Estuary	970	May
North Norfolk Coast	1,301	Sep	Morecambe Bay	944	Aug
Humber Estuary	1,277	Aug	Dee Estuary (England and Wales)	756	Aug
The Wash	1,223	May			
Sites surpassing national passage threshold in Great Britain in 2004/2005					
Pegwell Bay	640	Aug	Ardivachar Point (South Uist)	360	Aug
Dyfi Estuary	600	Aug	Hayle Estuary	350	Aug
Tay Estuary	568	Oct	Swale Estuary	343	Aug
Duddon Estuary	567	Aug	Hamford Water	327	Sep
Stour Estuary	496	Aug	Forth Estuary	324	Oct
South Ford	466	Aug	Burry Inlet	321	Aug
Tyninghame Estuary	373	May			
Sites surpassing national passage threshold in Northern Ireland in 2004/2005					
Outer Ards Shoreline	302	Oct	Carlingford Lough	251	Oct
			Belfast Lough	206	Sep

Killdeer

Charadrius vociferus

Vagrant
Native Range: N America

GB max: 0
NI max: 1 Feb

A single bird was present in February at Upper Lough Erne (east of Durnish Island). This is the first ever example of this transatlantic vagrant recorded by WeBS.

Kentish Plover

Charadrius alexandrinus

Scarce

GB max: 2 Apr
NI max: 0

Two records of single birds were both in April, with one at Breydon Water and the other on the Exe Estuary at Dawlish Warren.

Dotterel

Charadrius morinellus

Scarce

GB max: 2 May
NI max: 0

There were two records of Dotterel and as expected both referred to passage birds. Autumn passage was marked by a single bird in the Thames Estuary during October, and spring passage by two at Traeth Dulas during May.

Golden Plover

Pluvialis apricaria

International threshold: 9,300
Great Britain threshold: 2,500
All-Ireland threshold: 2,000

GB max: 208,610 Jan
NI max: 16,252 Feb

	S	M	L
GB change	o	+	++
NI change	o	o	o

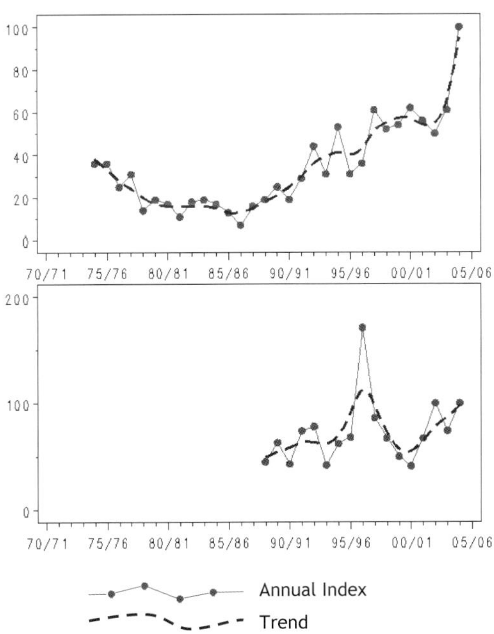

Figure 43.a, Annual indices & trend for Golden Plover for GB (above) & NI (below).

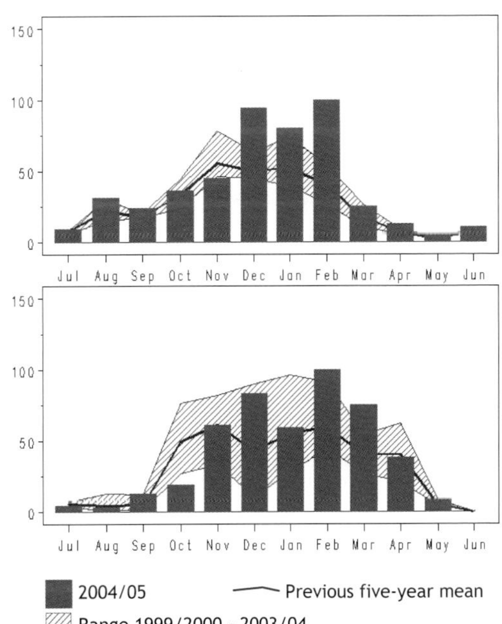

Figure 43.b, Monthly indices for Golden Plover for GB (above) & NI (below).

The number of Golden Plover recorded at WeBS Core Count sites in Britain has shown a dramatic increase over the past couple of years. The British maximum surpassed 200,000 birds for the first time. Monthly indices were the highest in the past five years between December and February, yet were around average for the rest of the year. Numbers in the Netherlands fell during December and January (van Roomen 2005), coinciding with the rise in British indices and suggesting a response to cold weather. This corresponds with high counts at several east coast sites such as the Humber Estuary, The Wash, Breydon Water and Berney Marshes and the Blackwater Estuary.

Numbers remained high through to February, and into March in Northern Ireland; this matched the timing of wintry weather and might highlight local movements away from inland areas to wetland areas counted for WeBS. The forthcoming Winter Plover Survey will aim to assess the size of the population of this species on a range of habitats, including those not covered by WeBS.

The peak count from the Nene Washes was the highest ever recorded at the site, being twice that of the previous highest count. Other record counts were recorded at Breydon Water and Berney Marshes, North Norfolk Coast, Chichester Harbour, Middle Yare Valley, Cleddau Estuary, Dengie Flats and the Alde Complex. Peak counts at Hamford Water and Lough Foyle were the second highest for the sites, as was the Taw-Torridge Estuary count, which was the highest for over 30 years. Lower than average counts were received from Strangford Lough, Ribble Estuary and Clifford Hill Gravel Pits.

In general, the wintering populations surpass passage numbers; however, numbers at Loch of Strathbeg peaked during September with wintering numbers at the site averaging around 12% of the autumn peak. Both October and September counts at Breydon Water and Berney Marshes were the highest passage counts at the site to date, as was the Carmarthen Bay count. Passage numbers at Loughs Neagh & Beg were slightly above the average for the last ten years.

	00/01	01/02	02/03	03/04	04/05	Mon	Mean
Sites of international importance in the UK							
Humber Estuary	25,133	29,607	40,585	(50,662)	43,217	Jan	37,841
The Wash	13,740	14,109	19,089	25,817	34,900	Dec	21,531
Breydon Wtr & Berney Marshes	13,280	10,200	8,900	10,464	30,940	Feb	14,757
Blackwater Estuary	18,826	(8,082)	12,455 [11]	6,986	12,747	Jan	12,754
Carmarthen Bay	(5,001)	(800)	(500)	(9,832)	(7,661)	Jan	(9,832)
Strangford Lough	6,948 [11]	11,726 [11]	8,766	15,988 [11]	4,578	Feb	9,601
Swale Estuary	(6,217)	13,898	3,282	10,935	(6,560)	Jan	9,372 ▲
Sites of national importance in Great Britain							
Morecambe Bay	(4,121)	(5,649)	(3,349)	(7,304)	(4,431)	Jan	(7,304)
Blyth Estuary	10,000	3,510					6,755
Pegwell Bay	4,000	7,000	7,229 [11]	8,000			6,557
Old Moor	4,700	5,500	(7,700)	7,000			6,225
Solway Estuary	8,065 [11]	(3,333)	(3,708)	4,459	6,145 [11]	Nov	6,223
Thames Estuary	7,911	3,538	(3,268)	(1,823)	6,440	Jan	5,963
Somerset Levels	(5,077)	5,169	1,260	8,609	8,136	Jan	5,794
Lynemouth Ash Lagoons				5,700			5,700
Stour Estuary	6,620 [11]	8,531 [11]	2,567 [11]	7,083 [11]	2,012 [11]	Nov	5,363
Taw-Torridge Estuary	(1,900)	(4,500)	(2,612)	3,300	(6,000)	Jan	4,600
Nene Washes	500	4,440	4,320	650	13,000	Feb	4,582 ▲
North Norfolk Coast	3,386	4,917	1,919	5,039	5,975	Dec	4,247
Hamford Water	4,164	2,464	2,384	3,204	5,606	Feb	3,564
Lindisfarne	(3,598)	2,881	(3,383)	3,822 [11]	3,920	Nov	3,555
Forth Estuary	1,027	2,419	(4,632)	6,940 [11]	2,658	Nov	3,535
St Mary`s Island	(1,000)	(2,000)	3,000	3,200	3,000	Nov	3,067
Ribble Estuary	(4,341)	3,075	(2,671)	(3,300)	1,705	Jan	3,018
Crouch-Roach Estuary	3,889	2,602	2,165	1,354	4,771 [11]	Nov	2,956 ▲
Colne Estuary	4,045	1,820	(82)	(1,480)	(1,450)	Mar	2,933
Clifford Hill Gravel Pits	4,500	3,560	2,500	2,740	1,000	Jan	2,860
Chichester Harbour	2,941	(2,436)	2,237	(2,822)	3,048	Feb	2,762 ▲
Ouse Washes	216	4,035	2,828 [13]	2,844	3,456	Dec	2,676

	00/01	01/02	02/03	03/04	04/05	Mon	Mean
Sites of all-Ireland importance in Northern Ireland							
Loughs Neagh and Beg	(7,621)	(2,817)	4,631	7,091	(3,447)	Nov	6,448
Lough Foyle	2,590	4,100	3,320	5,719	7,372	Feb	4,620
Sites no longer meeting table qualifying levels in Winter 2004/2005							
Stanwick Gravel Pits Consolidated		4,504		880	2,001	Dec	2,462
Mersey Estuary	(2,227)	(2,000)	(600)	(4,200)	1,000	Jan	2,357
Outer Ards Shoreline	1,411		3,164	1,369	206	Mar	1,538
Medway Estuary	(30)	(14)	(75)	(15)	(22)	Dec	(75)
Other sites surpassing table qualifying levels in Winter 2004/2005 in Great Britain							
Confidential SE England Site	1,600	2,600	500	600	6,500	Jan	2,360
Camel Estuary	1,500	800	727	515	4,750 [11]	Jan	1,658
Middle Yare Marshes	(95)	1,945	85	(96)	4,400	Mar	2,143
Cleddau Estuary	860	2,240	1,060	(2,664)	(4,273)	Dec	2,219
Dengie Flats	1,090	1,900	3,288	2,275	3,660	Mar	2,443
Alde Complex	648	793	1,444	(696)	3,346	Jan	1,558
Severn Estuary	(1,754)	806 [13]	1,215 [11]	2,060	3,100	Jan	1,795
Wigtown Bay	1,134	2,000	(602)	(3,604)	2,500	Dec	2,310
Other sites surpassing table qualifying levels in Winter 2004/2005 in Northern Ireland							
Bann Estuary	807	1,660 [13]	1,400	2,265	2,100	Mar	1,646

Sites surpassing international passage threshold in the UK in 2004/2005					
Humber Estuary	24,571	Oct	The Wash	12,945	Aug
Sites surpassing national passage threshold in Great Britain in 2004/2005					
Breydon Wtr & Berney Marshes	6,100	Oct	Solway Estuary	3,593	Oct
Blackwater Estuary	4,298	Oct	Carmarthen Bay	2,900	Oct
Loch of Strathbeg	3,623	Sep	Forth Estuary	2,624	Oct
Sites surpassing national passage threshold in Northern Ireland in 2004/2005					
Loughs Neagh and Beg	2,832	Oct			

Grey Plover

Pluvialis squatarola

International threshold: 2,500
Great Britain threshold: 530
All-Ireland threshold: 40*

GB max: 42,690 Oct
NI max: 977 Nov

	S	M	L
GB change	o	o	++
NI change	-	o	+

**50 is normally used as a minimum threshold*

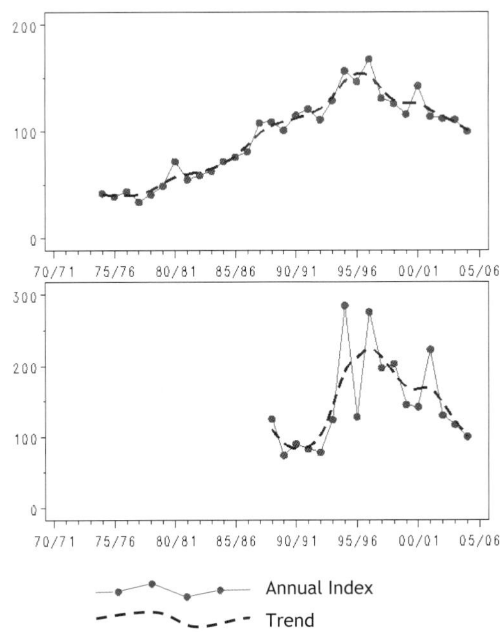

Figure 44.a, Annual indices & trend for Grey Plover for GB (above) & NI (below).

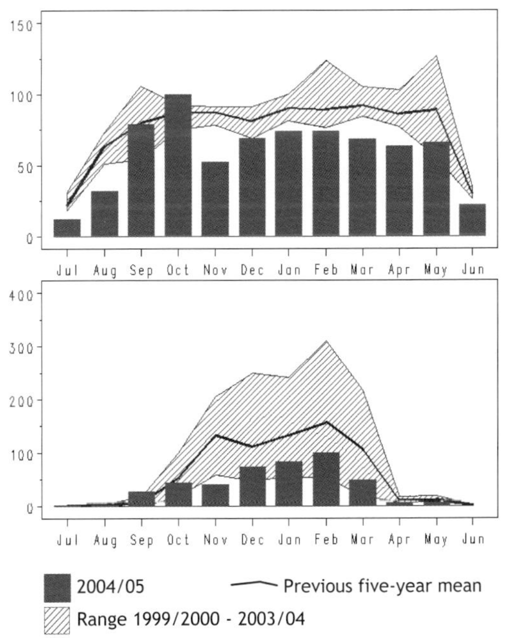

Figure 44.b, Monthly indices for Grey Plover for GB (above) & NI (below).

Grey Plover have shown further decline on top of those of the past few years. The British index has shown signs that the species has been declining for the last eight years; current levels being similar to 18 years ago. A very similar pattern has been evident in Northern Ireland. British monthly indices were the lowest for the past 5 years in all except three months, and in only October was the value higher than average. Monthly indices in Northern Ireland were again below average throughout most of the year.

Much of these declines can be attributed to the eastwards shift in the wintering population as a result of milder winters, as has been highlighted by Austin & Rehfisch (2005). In agreement with this opinion numbers of birds wintering in the Netherlands have shown a steady increase over recent years (van Roomen *et al.* 2005).

A number of key sites saw their lowest winter peaks of at least the last five years. On the Wash, the winter peak was the second-lowest for 20 years, despite both autumn and spring passage numbers here being about average for recent years. Numbers at the Orwell Estuary have fallen below the national qualifying threshold whereas an exceptionally high count at the Deben Estuary has contributed to the site's inclusion in the table.

	00/01	01/02	02/03	03/04	04/05	Mon	Mean
Sites of international importance in the UK							
The Wash	7,495	8,395	7,778	10,447	6,605	Mar	8,144
Thames Estuary	6,923	(5,160)	3,195	(3,812)	(2,681)	Dec	5,093
Ribble Estuary	5,139	6,285	1,658	5,568	3,529	Jan	4,436
Dengie Flats	7,826	3,640	3,768	2,943	2,466	Feb	4,129
Hamford Water	(2,803)	3,267	2,984	(1,746)	(2,915)	Dec	3,126
Blackwater Estuary	2,920	2,228	3,230	2,011	4,043	Nov	2,886
Stour Estuary	(3,130)	3,084	3,013	2,975	2,128	Mar	2,866
Sites of national importance in Great Britain							
Swale Estuary	(2,992)	1,745	2,181	1,892	1,389	Nov	2,040
Chichester Harbour	2,180	(3,180)	1,700	1,515	1,420	Mar	1,999
Alt Estuary	1,538	2,500	1,099	3,098	1,501	Nov	1,947
Humber Estuary	(1,320)	1,567	(1,300)	2,285 [11]	(964)	Jan	1,926
Medway Estuary	3,221	1,616	938	1,544	733	Nov	1,610
Dee Estuary (England & Wales)	(823)	2,201 [11]	966	(1,851)	758	Dec	1,444
North Norfolk Coast	1,382	1,720	1,374	1,316	1,386	Feb	1,436
Colne Estuary	1,331	1,357	(141)	(705)	(623)	Feb	1,344
Lindisfarne	(1,230)	1,016	(635)	(656)	775	Jan	1,007
Morecambe Bay	1,288	1,043	657	778	1,001	Mar	953
Langstone Harbour	1,405	504	982	1,119	701	Feb	942
Pagham Harbour	979	713	704	1,348	873	Dec	923
Eden Estuary	646	812	690	371	450	Dec	594
Solway Estuary	520 [11]	(482)	(466)	509	602 [11]	Nov	544
Deben Estuary	308	340	344	656	1,037	Jan	537 ▲
Sites of all-Ireland importance in Northern Ireland							
Strangford Lough	268	273	398 [11]	137	114	Feb	238
Carlingford Lough	(17)	45	52	(57)	33	Dec	47
Dundrum Bay	28	(19)	(72)	(27)	(32)	Dec	44 ▲
Sites no longer meeting table qualifying levels in Winter 2004/2005							
Orwell Estuary	484	323 [11]	413	710	350 [11]	Dec	456
Other sites surpassing table qualifying levels in Winter 2004/2005 in Great Britain							
Beaulieu Estuary	600	708	188	46	560	Feb	420

Sites surpassing international passage threshold in the UK in 2004/2005					
The Wash	13,480	Oct	Ribble Estuary	2,895	Apr
Thames Estuary	3,975	Oct	Stour Estuary	2,507	Oct
Dengie Flats	2,912	May			
Sites surpassing national passage threshold in Great Britain in 2004/2005					
Alt Estuary	2,234	Sep	Dee Estuary (England and Wales)	1,214	Oct
Chichester Harbour	2,140	Oct	Lindisfarne	962	Aug
Hamford Water	1,940	Oct	Morecambe Bay	904	Oct
Humber Estuary	1,901	Oct	Langstone Harbour	782	Oct
Swale Estuary	1,451	Sep	Medway Estuary	762	Oct
Blackwater Estuary	1,441	Aug	Pagham Harbour	587	Oct
North Norfolk Coast	1,359	Sep			

Lapwing

Vanellus vanellus

International threshold: 20,000**
Great Britain threshold: 20,000**†
All-Ireland threshold: 2,500

GB max:	410,308	Jan
NI max:	21,055	Dec

	S	M	L
GB change	o	o	++
NI change	-	o	-

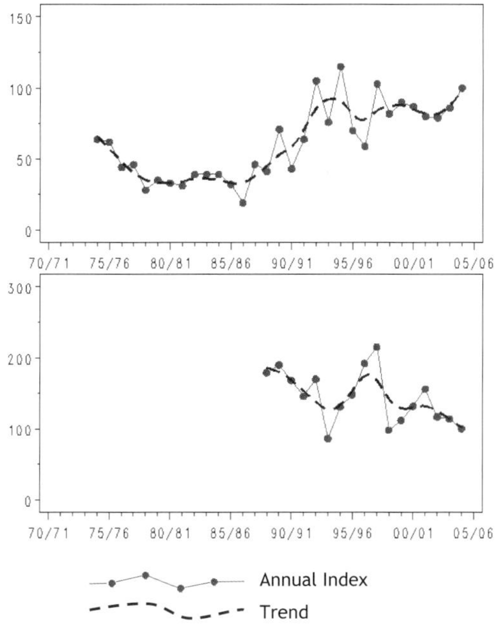

Figure 45.a, Annual indices & trend for Lapwing for GB (above) & NI (below).

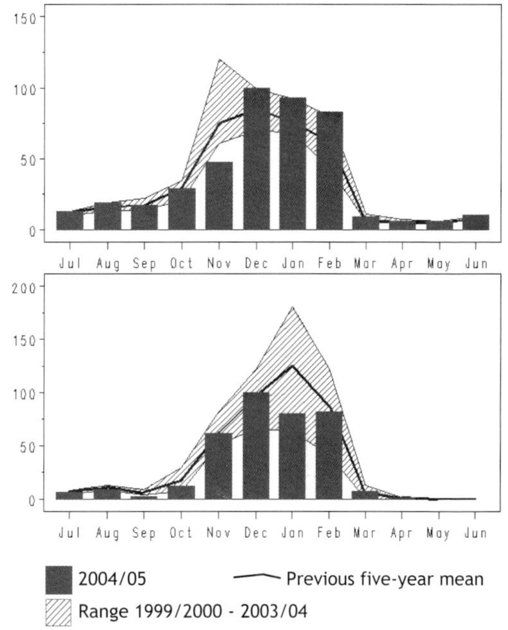

Figure 45.b, Monthly indices for Lapwing for GB (above) & NI (below).

The British index underwent a definite increase during 2004/05 with the underlying trend reaching its highest ever level. However, it is important to remember that this represents numbers of birds on wetland sites and does not take into account birds using agricultural or other types of habitats. In line with index figures the British maximum was the highest for five years and, typically for this species, was recorded during the core winter period. As the British index rose a very different picture was witnessed in Northern Ireland with the trend reaching its lowest ever level. This could perhaps indicate a shift eastwards in the wintering populations as suggested for other species (Austin & Rehfisch 2005) or perhaps instead a change in habitat use.

The forthcoming Winter Plover Survey will aim to assess the size of the population of this species on a range of habitats, including those not covered by WeBS.

In accordance with national figures counts at individual sites were on the whole higher than average. The peak at the Somerset Levels was the highest single site count since November 1999 and sees the Somerset Levels overtake The Wash at the top of the list of key sites. Numbers at the Ribble Estuary reached their second highest ever level, leading to this site surpassing the international qualifying level.

	00/01	01/02	02/03	03/04	04/05	Mon	Mean	
Sites of international importance in the UK								
Somerset Levels	(50,328)	41,675	16,053	23,641	60,834	Jan	38,506	
The Wash	31,165	43,558	43,672	29,350	43,822	Dec	38,313	
Ribble Estuary	(12,405)	(9,579)	(14,500)	(15,374)	25,991	Jan	25,991	▲
Humber Estuary	16,870	10,719	(36,309)	(39,865)	(16,856)	Dec	24,124	
Sites of all-Ireland importance in Northern Ireland								
Strangford Lough	6,214	10,527	6,977	8,884 [11]	5,792	Dec	7,679	
Loughs Neagh and Beg	(6,281)	(4,264)	3,090	6,282	7,584	Dec	5,809	
Lough Foyle	(2,277)	(3,320)	2,629	4,240	3,606	Nov	3,492	

	00/01	01/02	02/03	03/04	04/05	Mon	Mean
Sites with mean peak counts of 5,000 or more birds in Great Britain†							
Breydon Wtr & Berney Marshes	18,300	19,380	15,230	15,890	29,136	Jan	19,587
Morecambe Bay	16,213	13,504	(13,714)	(20,750)	16,701	Dec	16,792
Swale Estuary	(13,585)	14,804	14,974	16,523	(13,270)	Jan	15,434
Thames Estuary	(19,073)	(10,282)	16,036	10,229	14,657	Jan	14,999
Mersey Estuary	(1,930)	(5,284)	(5,675)	(12,150)	(9,370)	Jan	(12,150)
Severn Estuary	(9,817)	(7,439)	12,129 [11]	(6,889)	11,312	Jan	11,721
Blackwater Estuary	20,309	(9,005)	11,053 [11]	7,472	6,785	Jan	11,405
Ouse Washes	1,289	19,219 [13]	8,125 [13]	13,577	12,240	Dec	10,890
Nene Washes	7,100	4,230	21,016	3,870	7,050	Feb	8,653
Solway Estuary	8,596 [11]	(5,211)	(7,340)	8,218	(5,989)	Dec	8,407
Pegwell Bay	5,900	6,000	10,282 [11]	10,000			8,046
Dee Estuary (England & Wales)	6,270	9,206	6,470	7,853	7,512	Jan	7,462
North Norfolk Coast	5,799	7,830	5,124	7,358	7,833	Feb	6,789
Colne Estuary	6,430	(2,182)	(765)	(1,950)	(2,402)	Dec	6,430
Crouch-Roach Estuary	6,537	3,697	4,939	5,386	11,288 [11]	Nov	6,369
Confidential SE England Site	5,000	11,000	1,800	1,700	10,000	Jan	5,900
Tees Estuary	5,597	(3,196)	6,017	6,623	4,571	Dec	5,702
Other sites surpassing table qualifying levels in Winter 2004/2005 in Great Britain†							
Fiddlers Ferry Power Station Lgn	4,000	4,000	3,000	3,500	6,000	Jan	4,100
Camel Estuary	(1,405)	3,000	689	1,851	5,511 [11]	Jan	2,763
Alde Complex	4,026	2,241 [11]	4,358	(3,841)	5,472	Dec	4,024
Stodmarsh NNR and Collards Lagoon	3,375	4,000	1,200	3,400	5,000	Feb	3,395

† *as few sites exceed the British threshold a qualifying levels of 5,000 has been chosen to select sites for presentation in this report*

Knot

Calidris canutus

International threshold: 4,500
Great Britain threshold: 2,800
All-Ireland threshold: 375

GB max: 265,938 Nov
NI max: 5,919 Feb

	S	M	L
GB change	o	o	o
NI change	(--)	--	+

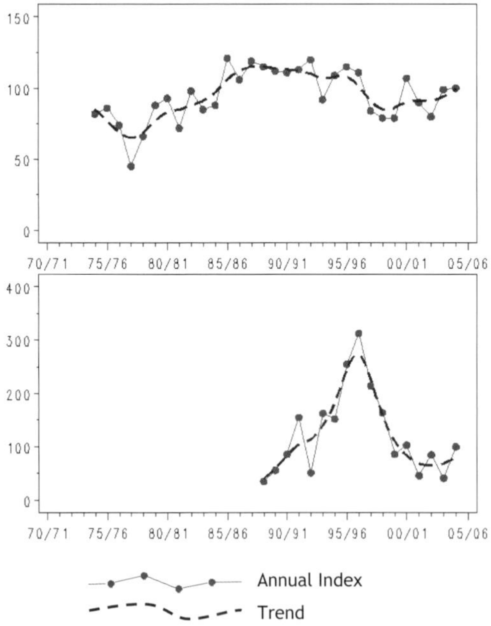

Figure 46.a, Annual indices & trend for Knot for GB (above) & NI (below).

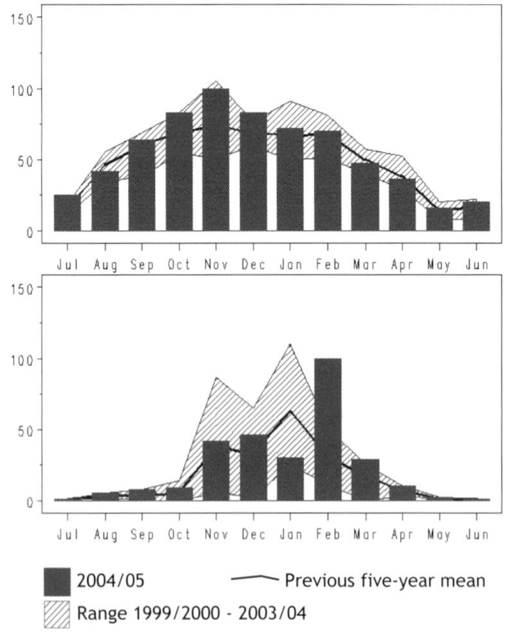

Figure 46.b, Monthly indices for Knot for GB (above) & NI (below).

Following a moderate decline in 1997/98 the underlying trend in the British index has increased for the sixth year running and is showing signs of recovery towards levels more akin to those of ten years ago. Over the past 15 years the British trend has shown a reverse pattern to that of the Dutch wintering population, providing evidence for interchange of birds between these areas. Furthermore, over the past two years monthly indices have been above average between October and December, a time when numbers in the Netherlands were lower than normal (van Roomen 2005).

The Northern Irish index showed a slight rise, largely due to higher than average numbers during February. Peak numbers at Strangford Lough, the most important site for Knot in Northern Ireland by far, were similar to the five-year mean of the site; however, this was the highest Core Count since 2000/01.

Peak numbers at the Wash were the highest recorded since December 1992. Numbers at Dengie Flats were the highest ever recorded at the site while the Solway peak was the highest for ten years. Conversely, numbers at eight of the 11 remaining sites holding internationally important numbers have declined, some such at the Dee Estuary by almost 75%. The current peak was the highest recorded at the Blackwater Estuary, which surpassed the international threshold for this species for the first time.

Only at North Norfolk Coast and Ribble Estuary did passage numbers exceed the wintering populations; both being slightly above average compared to the past five years.

	00/01	01/02	02/03	03/04	04/05	Mon	Mean
Sites of international importance in the UK							
The Wash	72,939	80,452	51,642	48,372	105,912	Nov	71,863
Morecambe Bay	72,908	66,031	(61,968)	67,959	29,596	Dec	59,692
Ribble Estuary	(20,331)	36,202	(23,691)	44,947	(21,540)	Mar	40,575
Humber Estuary	34,888	49,991	18,936	50,557 [11]	(37,015)	Feb	38,593
Thames Estuary	(38,357)	27,425	30,060	43,873	33,024	Nov	34,548
Alt Estuary	31,219	44,012	25,045	30,000	19,006	Nov	29,856
Dee Estuary (England & Wales)	5,672	52,792	26,769	38,070	10,243	Jan	26,709
Dengie Flats	19,400	13,600	10,550	8,000	22,700	Nov	14,850
North Norfolk Coast	29,636	16,214	9,224	7,523	6,735	Dec	13,866
Solway Estuary	9,159 [11]	(3,784)	(9,620)	8,725	13,142	Nov	10,342
Stour Estuary	8,036 [11]	6,998 [11]	8,648 [11]	6,564 [11]	8,454	Nov	7,740
Forth Estuary	5,807	7,232	8,936	6,907 [11]	5,077	Dec	6,792
Strangford Lough	5,863	4,000 [11]	10,340 [11]	4,058	5,730	Feb	5,998
Blackwater Estuary	(4,470)	(2,495)	1,700 [11]	(5,982)	6,273	Nov	4,606 ▲
Sites of national importance in Great Britain							
Burry Inlet	4,800	2,000	3,800	3,500	8,259	Feb	4,472
Lindisfarne	3,130 [11]	2,858	(4,512)	(6,751)	4,197	Jan	4,290
Cromarty Firth	5,050	2,621	3,132	4,932	5,000	Feb	4,147
Montrose Basin	2,800	5,000	5,800	(2,562)	1,990	Feb	3,898
Hamford Water	5,431	1,957	2,935	4,160	(2,481)	Jan	3,621
Swale Estuary	(4,200)	2,900	1,500	4,050	2,538	Mar	3,038
Inner Moray and Inverness Firth	3,373	1,980	1,873	3,663	3,446	Dec	2,867
Sites of all-Ireland importance in Northern Ireland							
Dundrum Bay	(981)	(555)	(603)	320	(475)	Dec	587
Lough Foyle	490	20	345	942	470	Mar	453
Sites no longer meeting table qualifying levels in Winter 2004/2005							
Medway Estuary	1,370	1,950	4,085	1,817	3,024 [11]	Jan	2,449
Tees Estuary	2,356	4,416	2,604	3,012	1,221	Nov	2,722
Other sites surpassing table qualifying levels in Winter 2004/2005 in Great Britain							
Carmarthen Bay	15	80	267	135	5,475	Jan	1,194
Medway Estuary	1,370	1,950	4,085	1,817	3,024 [11]	Jan	2,449

Sites surpassing international passage threshold in the UK in 2004/2005					
The Wash	92,938	Oct	Morecambe Bay	11,348	Oct
North Norfolk Coast	38,714	Oct	Thames Estuary	7,320	Oct
Ribble Estuary	36,200	May	Dee Estuary (England and Wales)	6,269	Oct
Humber Estuary	16,078	Aug	Solway Estuary	5,249	Oct
Alt Estuary	12,661	Sep			
Sites surpassing national passage threshold in Great Britain in 2004/2005					
Stour Estuary	3,410	Oct	Dengie Flats	3,200	May
Sites surpassing national passage threshold in Northern Ireland in 2004/2005					
Strangford Lough	484	Oct			

Sanderling

Calidris alba

GB max:	11,836	Aug
NI max:	282	Apr

International threshold:	1,200
Great Britain winter threshold:	210
Great Britain passage threshold:	300
All-Ireland threshold:	35*

	S	M	L
GB change	o	o	o

**50 is normally used as a minimum threshold*

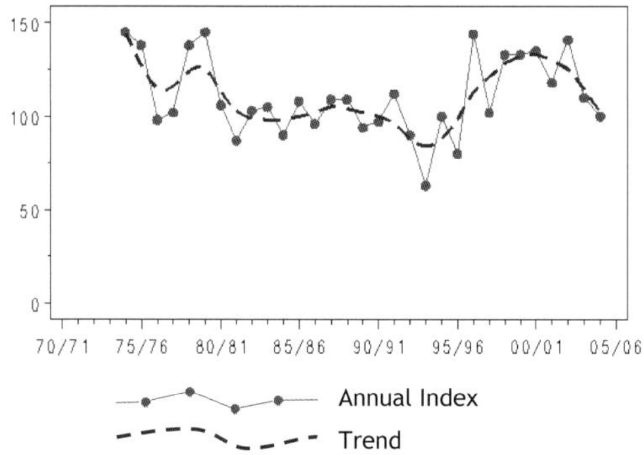

Figure 47.a, Annual indices & trend for Sanderling for GB.

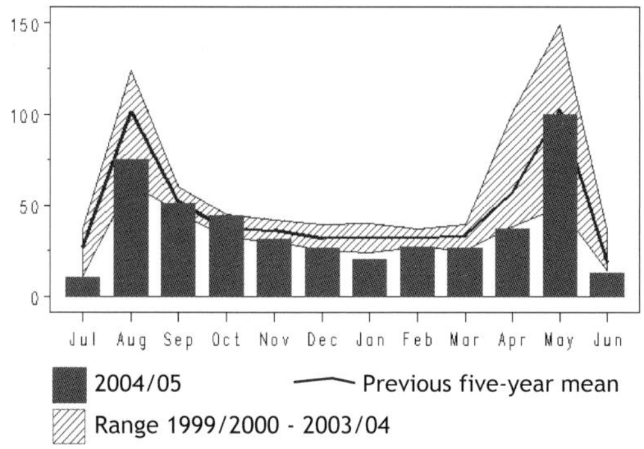

Figure 47.b, Monthly indices for Sanderling for GB.

Sanderling numbers in Britain continued to decline, the counted maximum being the lowest for four years. Nationally, numbers were below average throughout most of the year, only just surpassing the average in October. In contrast, numbers in the Netherlands have previously shown a strong increase and counts were particularly high from August to the following March (van Roomen 2005). The Northern Ireland maximum was the highest ever recorded and as usual this was during spring passage. Wintering numbers in Northern Ireland peaked at 56 in December, most of which were at Dundrum Bay and at the Bann Estuary.

Numbers at key sites were unexceptional; peak counts at the top four key sites being well below average. Additionally, the winter peak for the Alt Estuary was the lowest for nearly ten years with the five-year mean falling below the international threshold. To compensate, numbers at Lindisfarne and North Bay (South Uist) have now surpassed the national threshold. Peaks recorded by Core Counts at these two sites are typically variable; numbers at Lindisfarne have been supplemented by recent Low Tide Counts.

Passage numbers were again slightly down on the previous year; as usual peak numbers were witnessed during August and then in the following May. Numbers at The Wash were the lowest since 2000/01 and were well below average, whereas numbers at Thames Estuary and Carmarthen Bay were above average. Passage numbers at Dee Estuary (England & Wales) were the highest for over 20 years.

	00/01	01/02	02/03	03/04	04/05	Mon	Mean	
Sites of international importance in the UK								
Ribble Estuary	1,290	3,004	2,680	2,400	(1,453)	Nov	2,344	
Sites of national importance in Great Britain								
Carmarthen Bay	730	1,600	1,770	833	769	Mar	1,140	
Alt Estuary	967	1,556	1,431	913	815	Mar	1,136	▼
North Norfolk Coast	1,179	1,319	1,150	601	889	Nov	1,028	
Thames Estuary	334	552	875	385	562	Jan	542	
Duddon Estuary	606	486	287	585	361	Mar	465	
Thanet Coast	677	434	444	342	418	Nov	463	
Humber Estuary	546	358	440	370 [11]	(96)	Dec	429	
The Wash	317	504	496	317	395	Mar	406	
Solway Estuary	(117)	(218)	(266)	(370)	(302)	Mar	(370)	
Jersey Shore	253	391					322	
Dee Estuary (England & Wales)	100	550	286	(379)	274	Nov	318	
Ardivachar Point (South Uist)	0		398	460	400	Nov	315	
Howmore Estuary SSSI Coast				312 [55]			312	
Pegwell Bay	375	123	373 [11]	(115)			290	
Morecambe Bay	391	275	240	306	225	Dec	287	

	00/01	01/02	02/03	03/04	04/05	Mon	Mean	
Swansea Bay	234	356	410	200	234	Jan	287	
Forth Estuary	(262)	274	389	269 [11]	181	Dec	278	
Tees Estuary	373	259	280	240	199	Feb	270	
South Ford	228		120	250	430	Feb	257	
Lindisfarne	58	321 [11]	283 [11]	221	388 [11]	Dec	254	▲
North Bay (South Uist)	302 [13]		67	235	340	Nov	236	▲
Lade Sands	320 [13]	236	140	118	350	Mar	233	
Sites of all-Ireland importance in Northern Ireland								
Dundrum Bay	(132)	(0)	(30)	(0)	(48)	Dec	(132)	
Killough Harbour		76 [11]					76	
Sites no longer meeting table qualifying levels in Winter 2004/2005								
Durham Coast	(0)	(0)		(0)	(54)	Feb	(54)	

Sites surpassing international passage threshold in the UK in 2004/2005					
Ribble Estuary	4,830	Apr	Alt Estuary	2,278	May
The Wash	2,386	Aug	Thames Estuary	1,269	Sep
Sites surpassing national passage threshold in Great Britain in 2004/2005					
Dee Estuary (England and Wales)	1,026	Oct	Carmarthen Bay	499	Sep
North Norfolk Coast	1,021	Apr	Solway Estuary	416	Apr
Morecambe Bay	765	Sep	Thanet Coast	404	Oct
Humber Estuary	589	Aug	Tees Estuary	307	May
Ardivachar Point (South Uist)	550	Sep			
Sites surpassing national passage threshold in Northern Ireland in 2004/2005					
Bann Estuary	282	Apr			

Little Stint
Calidris minuta

International threshold: 2,000
Great Britain threshold: ?[†]
All-Ireland threshold: ?[†]

GB max: 66 Sep
NI max: 0

The national peak for Britain was slightly below that of the previous year and was the second lowest value since 1997/98. In line with this fall just three sites supported counts in excess of ten Little Stints compared to seven in 2003/04. For the second year running no Little Stints were recorded by WeBS in Northern Ireland, which might be due to short-staying passage birds being absent on WeBS count days.

Little Stint were reported from 49 sites, 16 more than during the previous year. As usual peak numbers were during autumn passage and winter numbers peaked at 12 in January; the spring peak exceeded this by one, with 13 in April.

Sites with more than 10 birds during passage periods in Great Britain in 2004/2005[†]		
Ribble Estuary	12	Apr
Severn Estuary	11	Oct
Thames Estuary	11	Sep

[†] *as no British or All-Ireland thresholds have been set a qualifying level of 10 has been chosen to select sites for presentation in this report*

Temminck's Stint
Calidris temminckii

Scarce

GB max: 2 May
NI max: 0

No Temminck's Stints were recorded during WeBS counts in autumn 2004, but there were records from three sites in spring 2005, with singles at the Camel Estuary and Dungeness Gravel Pits in May, and at Lindisfarne in June.

White-rumped Sandpiper
Calidris fuscicollis

Vagrant
Native Range: America

GB max: 1 Sep
NI max: 0

This species was recorded twice, with singles at Cley Marshes on the North Norfolk Coast in September and at Minsmere in October. All individuals of this species recorded by WeBS have fallen in the period July to December.

Baird's Sandpiper

Calidris bairdii

Vagrant
Native Range: America

GB max:	1	Oct
NI max:	0	

A single bird was recorded at Loch Paible, North Uist during October.

Pectoral Sandpiper

Calidris melanotos

Vagrant
Native Range: America, N Siberia, Australia

GB max:	4	Jul
NI max:	0	

Pectoral Sandpipers were present at both ends of the country during July, with two at the Forth Estuary, one at Loch of Strathbeg and one at Dungeness. Four singles in September were at Blagdon Lake, Mersey Estuary, Siblyback Reservoir and Thames Estuary. The latest record for 2004/05 was at Blackwater Estuary in October.

Curlew Sandpiper

Calidris ferruginea

International threshold:	7,400
Great Britain threshold:	?†
All-Ireland threshold:	?†

GB max:	316	Sep
NI max:	1	Sep

As usual peak numbers across Britain and Northern Ireland were recorded during September. The British peak was twice that of the previous year. However, as with all species that are routinely recorded during passage, numbers are subject to great variation due to daily turnover and a single monthly WeBS count cannot be considered a complete assessment of the total number of individuals at a site during the month.

Site peaks of 10 or more birds were recorded at 11 sites, all of which, except Seven Estuary and Morecambe Bay, were in on the east coast. The latest records of the autumn were of single birds at Burry Inlet, Tamar Complex and Swale Estuary in November. None were seen then throughout the winter, until the spring when five were present in March.

Sites with more than 10 birds during passage periods in Great Britain in 2004/2005†

Breydon Wtr & Berney Marshes	39	Sep
Severn Estuary	37	Sep
Swale Estuary	29	Sep
Thames Estuary	28	Sep
Forth Estuary	26	Sep
North Norfolk Coast	20	Sep
Blackwater Estuary	15	Sep
Humber Estuary	13	Sep
Morecambe Bay	12	Sep
Tees Estuary	10	Sep
The Wash	10	Aug

† *as no British or All-Ireland thresholds have been set a qualifying level of 10 has been chosen to select sites for presentation in this report*

Purple Sandpiper

Calidris maritima

International threshold:	750
Great Britain threshold:	180†
All-Ireland threshold:	10*

GB max:	1,298	Jan
NI max:	86	Mar

	S	M	L
GB change	o	-	--

**50 is normally used as a minimum threshold*

With a drop of 8% from the previous winter, the British annual index for Purple Sandpiper continued the downward trend that has been apparent since the late 1980s. Examination of monthly indices suggests that numbers were especially low early in the winter, but picked up after New Year. Perhaps birds remaining further north later into the autumn were forced south as the winter progressed. The maximum number of counted birds in Britain was down on the same figure for the previous year, whilst the maximum count in Northern Ireland dropped for the second year running.

Only two sites now support numbers in excess of the threshold for national importance in Britain. The peak count for Papa Westray was the highest yet recorded there. No birds were picked up at Balranald during the winter. In Northern Ireland, whilst counts held up on the Outer Ards, being average compared to the past 20 years, the peak count at Belfast Lough fell to an all-time low.

As so few sites qualify for national importance, additional sites supporting a mean peak of over 100 birds are also listed. 2004/05 peak counts were low at Egilsay, Forth Estuary and between Seahouses and Budle Point, but were higher than average at the Dee Estuary (Aberdeen) and Ardivachar Point on South Uist, as well as at Thurso Bay. Supplementary counts carried out for specific monitoring projects are also included in the table, drawing attention to other noteworthy concentrations. However, as always this is a species that is very under-represented by WeBS and it will be extremely interesting to see what the next Non-estuarine Coastal Waterbird Survey (NEWS) in January 2007 reveals about its current status.

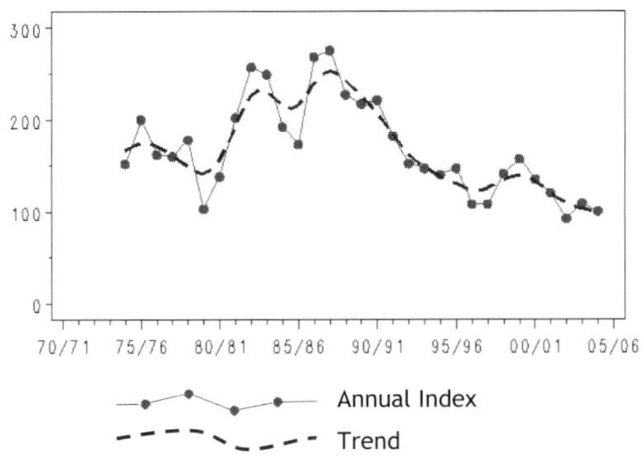

Figure 48.a, Annual indices & trend for Purple Sandpiper for GB.

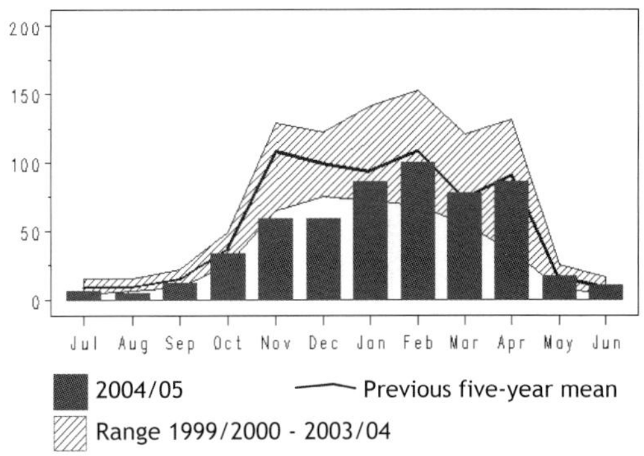

Figure 48.b, Monthly indices for Purple Sandpiper for GB.

	00/01	01/02	02/03	03/04	04/05	Mon	Mean
Sites of national importance in Great Britain							
Island of Papa Westray		330	120	216	385	Jan	263
Farne Islands	207	194	(185)				201
Sites of all-Ireland importance in Northern Ireland							
Outer Ards Shoreline	82		122	83	84	Mar	93
Belfast Lough	(13)	16	15	17	6	Jan	14
Sites with mean peak counts of 100 or more birds in Great Britain†							
Island of Egilsay	334	4	141	195	81	Jan	151
Forth Estuary	(159)	172	248	72	93	Jan	149
Moray Coast	(158)	144	89	127	134	Dec	130
Balranald RSPB Reserve			190	180	0		123 ▼
Howmore Estuary SSSI Coast				120 [55]			120
Ardivachar Point (South Uist)	100		120	110	144	Dec	119
Bornish & Ormiclate Machairs				112 [55]			112
East Unst		110 [10]					110
Dee Estuary (Scotland)	84	71	92	81	185	Feb	103
Seahouses to Budle Point	205	151	25	64	63	Nov	102
Other sites surpassing table qualifying levels in Winter 2004/2005 in Great Britain†							
Thurso Bay			37	97	120	Nov	85

Sites surpassing national passage threshold in Great Britain in 2004/2005		
Farne Islands	375	May
Sites surpassing national passage threshold in Northern Ireland in 2004/2005		
Outer Ards Shoreline	15	Oct

† *as few sites exceed the British threshold a qualifying level of 100 has been chosen to select sites for presentation in this report*

Dunlin

Calidris alpina

GB max:	334,093	Jan
NI max:	9,185	Feb

International threshold:	13,300		
Great Britain winter threshold:	5,600		
Great Britain passage threshold:	2,000		
All-Ireland threshold:	1,250		
	S	M	L
GB change	o	o	o
NI change	(-)	-	o

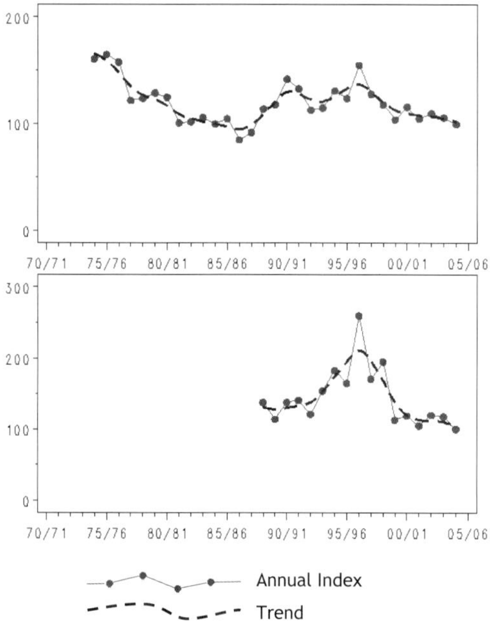

Figure 49.a, Annual indices & trend for Dunlin for GB (above) & NI (below).

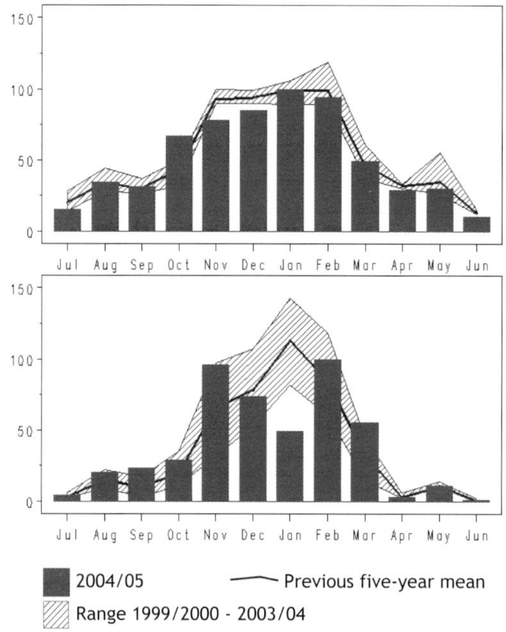

Figure 49.b, Monthly indices for Dunlin for GB (above) & NI (below).

The recent steady decline in numbers of British wintering Dunlin continued in 2004/05, the annual index reaching its lowest point for 17 years, albeit only declining 6% since 2003/04. The same was seen in Northern Ireland, the index here reaching its lowest ever value although numbers in the province have been roughly stable since the turn of the century. However, at only one site, the Stour Estuary, have mean peak numbers fallen below the international qualifying threshold. Amongst these key sites, the majority saw peak 2004/05 numbers lower than recent averages, with particularly low counts from the Dee Estuary, Thames Estuary, Lindisfarne, Dengie Flats (for the second year running) and the Colne Estuary. The site with the highest peak compared to normal was Langstone Harbour, where numbers were the highest since 1992/93. Numbers in Langstone Harbour have increased over recent years, whereas at nearby Portsmouth and Chichester Harbours numbers have shown some declines; however, these numbers refer to roosting birds, which regularly move between the harbours during the tidal cycle (Bill & Hollins, 1989 in Clark & Eyre, 1993).

In Northern Ireland, the Core Count at Strangford Lough was the highest since February 1999 but the peak at Lough Foyle was very low.

Most key sites supported wintering populations well in excess of their peak passage numbers, whereas passage numbers at the Mersey, Solway and Thames Estuaries were much higher than normal.

	00/01	01/02	02/03	03/04	04/05	Mon	Mean
Sites of international importance in the UK							
Mersey Estuary	60,330	45,756	58,463	40,170	43,020	Dec	49,548
Thames Estuary	44,907	48,104	54,205	(27,318)	40,838	Dec	47,014
The Wash	35,080	31,069	42,794	31,624	39,041	Feb	35,922
Dee Estuary (England & Wales)	41,656	34,448 [11]	21,266	41,679	16,878	Feb	31,185
Severn Estuary	(17,417)	20,401	25,734	23,801	(16,069)	Dec	23,312

	00/01	01/02	02/03	03/04	04/05	Mon	Mean
Langstone Harbour	23,790	17,500	17,320	24,286	28,239	Feb	22,227
Humber Estuary	18,502	24,378	24,168	19,182 [11]	(14,733)	Mar	21,558
Ribble Estuary	(36,473)	11,141	11,423	24,445	24,024	Jan	21,501
Morecambe Bay	27,645	18,947	18,214	18,847	(17,848)	Feb	20,913
Blackwater Estuary	(37,550)	15,004	18,806	13,958	(16,007)	Dec	20,265
Chichester Harbour	16,773	17,947 [11]	15,661	12,552	12,651	Dec	15,117
Solway Estuary	15,093	12,861	12,850	17,576	(14,628)	Nov	14,602
Sites of national importance in Great Britain							
Stour Estuary	15,822	16,469 [11]	12,863 [11]	9,268	11,974 [11]	Jan	13,279 ▼
Forth Estuary	11,900	13,296	12,143	7,840 [11]	9,132	Dec	10,862
Swale Estuary	(7,795)	11,280	14,761	5,034	9,181	Feb	10,064
Alt Estuary	5,026	8,438	6,885	12,743	8,540	Nov	8,326
Lindisfarne	5,777	9,085	(9,991)	(9,503)	5,885	Jan	8,048
Dengie Flats	9,700	15,720	7,710	2,700	3,040	Nov	7,774
Medway Estuary	(5,118)	5,872	6,901	8,086	9,373 [11]	Dec	7,558
Burry Inlet	5,401	6,654	4,955	10,150	6,318	Dec	6,696
Poole Harbour	4,852	(6,929)	(6,323)	(5,463)	(7,026)	Jan	6,119
Duddon Estuary	(4,258)	5,415	3,942	7,680 [11]	6,970 [11]	Nov	6,002
Colne Estuary	9,100	6,823	(350)	4,411	3,359	Jan	5,923
Hamford Water	5,625	10,686 [11]	3,064	(3,476)	4,290	Dec	5,916
Sites of all-Ireland importance in Northern Ireland							
Strangford Lough	2,733 [11]	3,352	4,408 [11]	4,967 [11]	4,934	Feb	4,079
Lough Foyle	5,800	2,804	4,209	4,212	1,688	Nov	3,743
Carlingford Lough	(1,390)	(2,090)	(2,872)	(2,339)	2,238	Jan	2,483
Belfast Lough	1,366 [11]	1,278	1,193	1,461 [11]	1,136 [11]	Dec	1,287

Sites surpassing international passage threshold in the UK in 2004/2005					
The Wash	42,361	Sep	Thames Estuary	27,003	Oct
Mersey Estuary	29,600	Oct	Solway Estuary	13,324	Oct
Ribble Estuary	27,847	May			
Sites surpassing national passage threshold in Great Britain in 2004/2005					
Morecambe Bay	12,094	Oct	Forth Estuary	5,255	Oct
Humber Estuary	10,814	Jul	Chichester Harbour	4,690	Oct
Severn Estuary	10,812	Oct	Lindisfarne	4,047	Oct
Dee Estuary (England and Wales)	10,407	Oct	Solway Estuary	4,005	May
Langstone Harbour	9,580	Oct	Duddon Estuary	3,152	Oct
Dengie Flats	8,254	May	Swale Estuary	2,921	Oct
Blackwater Estuary	7,968	Oct	North Norfolk Coast	2,613	Aug
Alt Estuary	7,588	May	Breydon Wtr & Berney Marshes	2,230	May
Stour Estuary	6,621	Oct	Medway Estuary	2,142	Oct

Ruff

Philomachus pugnax

International threshold: ?
Great Britain threshold: 7*
All-Ireland threshold: +†

GB max: 884 Jan
NI max: 24 Oct

**50 is normally used as a minimum threshold*

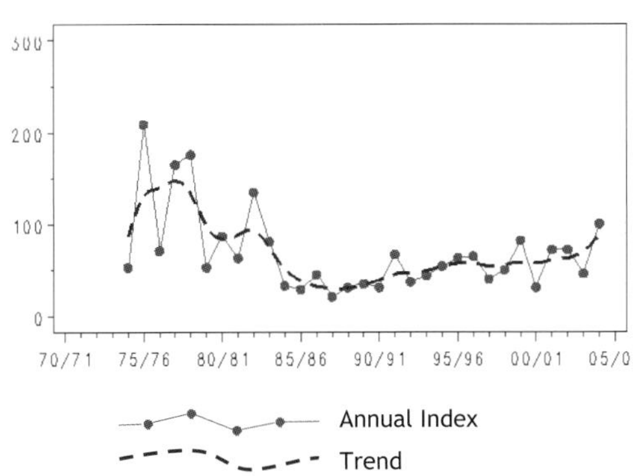

Figure 50.a, Annual indices & trend for Ruff for GB.

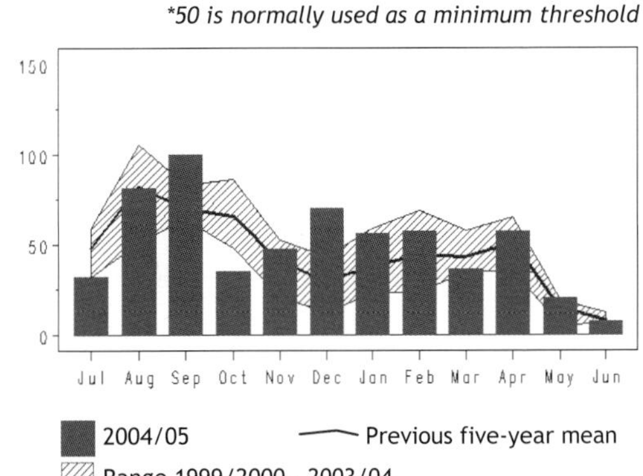

Figure 50.b, Monthly indices for Ruff for GB.

The British annual index for wintering Ruff rose during 2004/05, largely due to high numbers on the Ouse Washes during January (416) and February (431). Unusually, the

British maximum was recorded in January, the summed total being two-thirds higher than during the autumn peak. Low autumn numbers were mostly due to site coverage, as the monthly indices show a higher than usual occupancy in September, although falling rapidly in October. Winter indices were also above average between November and February.

Typically peak numbers are recorded during autumn passage, which can start as early as the end of June. Ruff colour-marked in the Netherlands during spring have been sighted in the Britain during autumn passage, mostly on the east coast although one bird turned up in northwest England as early as late June, and one was in Kent during the following winter.

The highest site counts were at the Ouse Washes, where counts exceeded 100 birds in every month from October though until March. Other three-figure counts were made at the North Norfolk Coast and the Swale Estuary. High numbers recorded during Low Tide Counts at the Crouch-Roach Estuary helped the site attain national importance status for this species. The highest count in Northern Ireland was of 16 on Lough Foyle in October.

	00/01	01/02	02/03	03/04	04/05	Mon	Mean
Sites of national importance in Great Britain							
Ouse Washes	189	334	359	232 [13]	(431)	Feb	309
Lower Derwent Ings			179	99			139
WWT Martin Mere	116	190	151	86	83	Mar	125
North Norfolk Coast	103	66	105	155	140	Nov	114
Nene Washes	38	30	275	128	16	Dec	97
Breydon Wtr & Berney Marshes	52 [13]	155	55	43	86	Dec	78
Needingworth Quarry Lakes			(106)		2	Jan	69
RSPB Hanson Wetland Creation	126			42			69
Swale Estuary	29	46	95	41	128	Mar	68
Middle Yare Marshes	(33)	37	82	(17)	53	Nov	57
Blackwater Estuary	10	49	82	19	24	Mar	37
Ribble Estuary	63	5	76	21	5	Feb	34
Arun Valley	28	29	22	52	24	Jan	31
Thames Estuary	7	34	35	43	28	Dec	29
Somerset Levels	(15)	(15)	29	33	10	Jan	24
Dungeness Gravel Pits	55	0	42	7	10	Jan	23
Crouch-Roach Estuary	2	(0)	(4)	(0)	42 [11]	Jan	22 ▲
Holland Marshes	23	41	6	12	14	Mar	19
Hamford Water	12	6 [11]	26	20	17	Dec	16
Sandbach Flashes	8		26	11	13	Dec	15
Humber Estuary	4	20 [13]	25	5 [11]	(7)	Mar	14
Confidential SE England Site	9	32	0	7	16	Mar	13
Dee Estuary (England & Wales)	8	(12)	5	10	29	Jan	13
Stodmarsh NNR & Collards Lgn	25	15	11	7	0		12
Cresswell Pond	15	(32)	1	12	0		12
Abberton Reservoir	0	0	2	51	2	Nov	11
Fen Drayton Gravel Pits	46	0	0	7	0		11
Severn Estuary	1 [13]	3	21	(0)	13	Jan	10
Hardley Flood	0		33	7	0		10
Hagnaby Lock Fen	0	6	19	9	11	Feb	9
Fairfield SSSI	12	9	12	0	0		7
Minsmere	8	8	5	6	9	Mar	7
Rutland Water	9	8	4	10	2	Nov	7
Tees Estuary	10	6	8	9	4	Jan	7
East Chevington Pools	1	14	(0)	7	4	Feb	7
Sites no longer meeting table qualifying levels in Winter 2004/2005†							
Druridge Pool	0	21	0	8	0		6
Eyebrook Reservoir	25	2	0	0	0		5
Other sites surpassing table qualifying levels in Winter 2004/2005 in Great Britain							
North West Solent	0	4	(3)	3	9	Mar	4
Rye Harbour and Pett Level	0	0	12	5	9	Jan	5

Sites with more than 50 birds during passage periods in Great Britain in 2004

North Norfolk Coast	Sep	270	Breydon Wtr & Berney Marshes	Jul	69
Ouse Washes	Oct	181			

† *as no All-Ireland threshold has been set a qualifying level of seven has been chosen to select sites for presentation in this report*

Jack Snipe

Lymnocryptes minimus

International threshold: ?
Great Britain threshold: ?†
All-Ireland threshold: 250†

GB max: 183 Dec
NI max: 1 Sep

The number of Jack Snipe recorded during WeBS Core Counts is widely variable as the species is difficult to detect and often overlooked. Nethertheless, national maxima were not too disimilar to previous years. Chat Moss remained top of the table, largely due to supplementary counts of this species supplied for this site, as well as for Bickershaw Colliery Area. In total 21 sites held mean peaks of five or more, one less than in 2003/04. Numbers at other sites were unexceptional, although 2004/05 peaks of five or more were recorded at four additional sites.

	00/01	01/02	02/03	03/04	04/05	Mon	Mean
Sites with mean peak counts of 5 or more birds in Great Britain†							
Chat Moss		68 [27]	46 [27]	28 [27]	34 [27]	Dec	44
Doxey Marshes SSSI	6	64	(30)	16	61	Dec	37
Dornoch Firth		25 [11]					25
Chichester Harbour	31 [13]	16	39	7	6	Feb	20
Lower Derwent Ings			11	22			17
Bickershaw Colliery Area		10 [27]	17 [27]	11 [27]	14 [27]	Nov	13
Dee Estuary (England & Wales)	9	22 [11]	13	(9)	3	Dec	12
Inner Moray and Inverness Firth	(3)	19	13	8	5	Feb	11
Waulkmill Glen & Littleton Res	15	6	10	12	10	Dec	11
Humber Estuary	(4)	5	13	(5) [11]	(2)	Dec	9
Kemerton Lake				9			9
Severn Estuary	12	9	7 [13]	5	10	Mar	9
Fiddlers Ferry Power Station Lgn	4	32	6	0	0		8
Ardrossan-West Kilbride	(2)	6	8	2	6	Nov	6
Boat of Garten Pools					6	Nov	6
Morecambe Bay	(2)	10	5	3	6	Jan	6
Stour Estuary	6	12	2	6 [11]	2	Feb	6
Upton Warren	12	5	6	3	5	Jan	6
Langstone Harbour	2	0	13	0	12	Mar	5
North Cave Wetlands	3	15	4	3	2	Jan	5
Other sites surpassing table qualifying levels in Winter 2004/2005 in Great Britain†							
Hill Ridware Lake	4	4	3	4	7	Dec	4
Langford Lowfields Gravel Pits	0	7	1	4	7	Dec	4
Coombe Hill Canal	0	0	3	2	5	Mar	2
Henfield Brooks	0	0	1	0	5	Mar	1

† *as few sites exceed the All-Ireland threshold and no British threshold has been set a qualifying level of 5 has been chosen to select sites for presentation in this report*

Jack Snipe (Tommy Holden)

Snipe

Gallinago gallinago

International threshold: 20,000**
Great Britain threshold: ?†
All-Ireland threshold: ?†

GB max:	8,167	Nov
NI max:	135	Nov

**50 is normally used as a minimum threshold*

Despite their abundance and wide distribution, Snipe are inevitably under-recorded during WeBS, being difficult to detect. Peak counts are generally recorded when counters make a particular effort to concentrate on this (and the previous) species. The British maximum has risen a third on the previous year yet this value has fluctuated around the 8,000 mark for the past ten years. The Northern Ireland maximum has declined by two-thirds since 2003/04 although as for Britain, it is difficult to put this down to anything but variable detection of the species during counts. The Somerset Levels remained top of the list of sites holding mean peak counts of 200 or more; the January peak of 1,513 was the highest single site count during 2004/05. Annual peaks exceeding 200 were recorded at an additional nine sites, and the 2004/05 peak counts at Malltraeth RSPB, Adur Estuary and Woolmer Common Ponds were the highest ever for each site.

	00/01	01/02	02/03	03/04	04/05	Mon	Mean
Sites with mean peak counts of 200 or more birds in Great Britain†							
Somerset Levels	1,817	854	972	308	1,513	Jan	1,093
Lower Derwent Ings			997	269			633
Severn Estuary	(301)	(217)	(240)	(519)	(349)	Mar	(519)
Ouse Washes	62	1,685 [13]	126	233	302	Mar	482
Doxey Marshes SSSI	149	544	(239)	390	716	Jan	450
North Norfolk Coast	207	1,169 [11]	92	121	77	Dec	333
Maer Lake	280	510	0	403	378	Jan	314
Middle Yare Marshes	(217)	(545)	257	124	(210)	Nov	271
Cleddau Estuary	215	189	283	311	144	Jan	228
Arun Valley	335	166	242	134	153	Nov	206
Dee Estuary (England & Wales)	189	334 [11]	154	(52)	127	Dec	201
Sites with mean peak counts of 50 or more birds in Northern Ireland†							
Loughs Neagh and Beg	33	(16)	129	151	22	Nov	84
Belfast Lough	65	61	48	86 [11]	45 [11]	Nov	61
Strangford Lough	31	29	97 [11]	55	56	Mar	54
Ballysaggart Lough			51	53			52
Other sites surpassing table qualifying levels in Winter 2004/2005 in Great Britain†							
Malltraeth RSPB	10	55	131	54	570	Nov	164
Adur Estuary	144	54	148	147	273	Feb	153
Morecambe Bay	(218)	109	101	(147)	(265)	Nov	168
Woolmer Common Ponds		1	1	12	205	Jan	55

† *as no British or All-Ireland thresholds have been set qualifying levels of 200 and 50 have been chosen to select sites, in Great Britain and Northern Ireland respectively, for presentation in this report*

Woodcock

Scolopax rusticola

International threshold: 20,000**
Great Britain threshold: ?
All-Ireland threshold: ?

GB max:	41	Nov
NI max:	0	

**50 is normally used as a minimum threshold*

The preference for non-wetland habitats and its secretive nature mean that Woodcock are inevitably grossly under-recorded during WeBS. This is stressed by the peak national total of just 41 in November; a tiny fraction of the total numbers present in the country. Although most records were of single birds, site peaks of five were noted at River Cam - Kingfishers Bridge and Traeth Melynog, both in November, and Longueville Marsh in March. Two of these sites were also named as holding the highest counts during 2003/04.

Black-tailed Godwit

Limosa limosa

International threshold:	350	
Great Britain threshold:	150	
All-Ireland threshold:	90	

GB max:	35,666	Sep
NI max:	1,715	Sep

	S	M	L
GB change	+	++	++
NI change	+	+	++

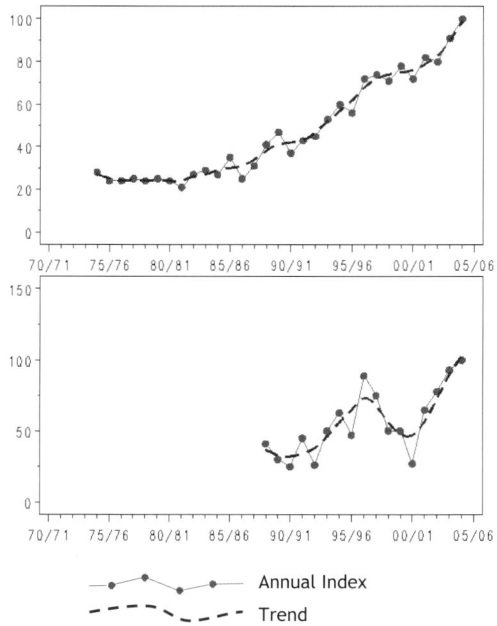

Figure 51.a, Annual indices & trend for Black-tailed Godwit for GB (above) & NI (below).

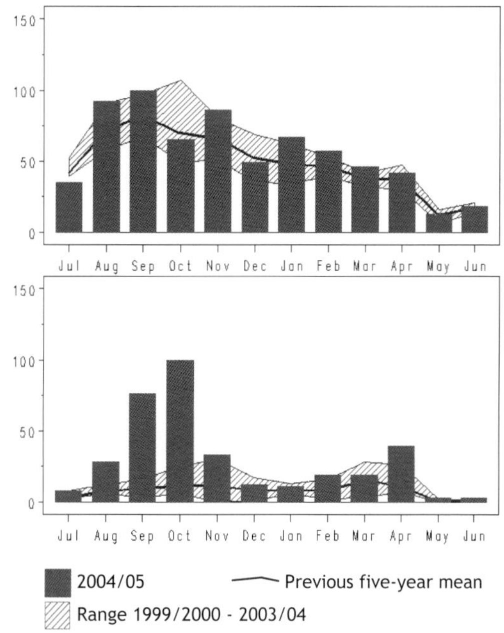

Figure 51.b, Monthly indices for Black-tailed Godwit for GB (above) & NI (below).

The increase in the Icelandic race of the Black-tailed Godwit *islandica* shows no signs of abating. The annual British and Northern Ireland indices increased by 10% and 7% respectively to their highest ever values. If the current rate of increase continues into 2005/06 then this species will out-number Bar-tailed Godwit, which itself is decreasing, as has already happened in Northern Ireland this year.

Although sites are compared with national and international thresholds, both of these figures are clearly now well out of date, with the peak British count in September 2004 exceeding the nominal international population estimate for islandica. Population estimates and threshold values are due to be updated shortly. However, based on the existing thresholds, a further six sites now qualify as supporting internationally important numbers compared to the year before, and a further three support nationally important numbers. The increase has been particularly great in Wales, on the Dee Estuary, Severn Estuary, Burry Inlet and Carmarthen Bay, whilst other sites with proportionally large increases include the Dee Estuary, North Norfolk Coast, Morecambe Bay, Forth Estuary, Eden Estuary, Portsmouth Harbour, Tamar Complex and Lough Foyle. A Low Tide Count on the Crouch-Roach Estuary located far higher numbers than have been recorded on recent Core Counts, suggesting local movement between different parts of the Essex coast here. The only sites apparently experiencing declining numbers seem to be the Stour Estuary, Mersey Estuary, Southampton Water and Hamford Water. Whilst the count was very low on the River Avon: Ringwood to Christchurch, and at Meadow Lane and Fen Drayton Gravel Pits, such fluctuations are typical for inland sites that depend upon winter flooding.

During passage, numbers exceeded the winter peak at 15 sites of national or international importance and at Mersey Estuary, Humber Estuary, Langstone Harbour, Thames Estuary, Stour Estuary and North

Norfolk Coast passage numbers were at least twice that of the peak winter count.

	00/01	01/02	02/03	03/04	04/05	Mon	Mean	
Sites of international importance in the UK								
The Wash	3,555	9,163	2,773	3,031	5,492	Nov	4,803	
Dee Estuary (England & Wales)	2,366	4,624 [11]	3,955	4,493	5,362	Dec	4,160	
Ouse Washes	268	3,273	3,468	3,137 [13]	3,424	Mar	2,714	
Blackwater Estuary	(2,094)	(926)	(2,939)	1,232	2,356	Mar	2,155	
Stour Estuary	2,846 [11]	2,593	1,927	1,607	1,151	Jan	2,025	
Ribble Estuary	3,271	1,733	975	1,385	2,629	Jan	1,999	
Poole Harbour	1,134	(2,115)	(2,691)	(2,133)	(1,732)	Jan	1,961	
Thames Estuary	2,306	1,967	1,584	1,380	1,931	Nov	1,834	
Swale Estuary	2,153	1,580 [11]	1,045	1,511	1,782	Nov	1,614	
Breydon Wtr & Berney Marshes	1,376	1,607	1,142	1,277	1,566	Mar	1,394	
R. Avon: R'wood to Christchurch	2,630 [13]	3	3,002	170	26	Feb	1,166	
Exe Estuary	880	737	890	(1,079)	1,054	Dec	928	
Humber Estuary	545	921	1,311	914 [11]	(629)	Feb	923	
Crouch-Roach Estuary	(272)	(260)	(162)	(261)	729 [11]	Nov	729	▲
Medway Estuary	(0)	(662)	(199)	(154)	(518)	Nov	(662)	
Mersey Estuary	810	313	1,002	740	241	Jan	621	
Chichester Harbour	136	552	715	1,050	545	Nov	600	
Southampton Water	1,265	(358)	196	(434)	291	Dec	584	
Pagham Harbour	248	252	826	541	664	Jan	506	
North Norfolk Coast	108	233	477	631	998	Mar	489	▲
Belfast Lough	383 [11]	492	545	367 [11]	479	Jan	453	
Hamford Water	601	366 [11]	490	414	314	Jan	437	
Morecambe Bay	219	(117)	(143)	(403)	655	Nov	437	▲
North West Solent	323	452	(261)	373	(300)	Jan	383	
Colne Estuary	450	344	(190)	253	472	Mar	380	▲
Newtown Estuary	(86)	231	510	(173)	(113)	Feb	371	▲
Beaulieu Estuary	495	725	147	116	326	Jan	362	▲
Sites of national importance in Great Britain								
Langstone Harbour	(97)	442	314	245 [11]	290	Mar	323	
Alde Complex	30	113	355	600	298	Mar	279	
Orwell Estuary	73 [11]	260 [11]	407 [11]	389 [11]	255 [11]	Dec	277	
Blyth Estuary	271	244					258	
Deben Estuary	114	260	304	258	305	Nov	248	
Forth Estuary	55	232	243	291	348	Feb	234	
Burry Inlet	7	30	60	222	845	Feb	233	▲
Eden Estuary	170	221 [11]	206	220	305	Nov	224	
Portsmouth Harbour	(70)	(84)	246 [11]	78	340	Feb	221	
Carmarthen Bay	2	(8)	(29)	(331)	307	Mar	213	▲
Meadow Lane Gravel Pits	800	0	(3)	0	0		200	
Severn Estuary	5	141	193	200	(450)	Nov	198	▲
Fen Drayton Gravel Pits	780	1	0	0	0		156	
Sites of all-Ireland importance in Northern Ireland								
Strangford Lough	83	153	189 [11]	267	176	Feb	174	
Other sites surpassing table qualifying levels in Winter 2004/2005 in Great Britain								
Nene Washes	281	39	51	185	188	Feb	149	
Fal Complex	(103)	87	163	105	176	Jan	133	
Tamar Complex	(130)	106	53 [11]	80	175	Nov	109	
Christchurch Harbour	1	18	7	174	169	Feb	74	
Other sites surpassing table qualifying levels in Winter 2004/2005 in Northern Ireland								
Lough Foyle	0	0	32	161	213	Nov	81	

Sites surpassing international passage threshold in the UK in 2004/2005

Dee Estuary (England and Wales)	6,452	Sep	Belfast Lough	857	Sep
The Wash	5,546	Sep	Ouse Washes	800	Apr
Thames Estuary	3,757	Oct	Nene Washes	770	Apr
Mersey Estuary	2,950	Sep	Langstone Harbour	758	Oct
Ribble Estuary	2,936	Sep	Morecambe Bay	722	Apr
Humber Estuary	2,435	Sep	Exe Estuary	706	Sep
Stour Estuary	1,972	Sep	Burry Inlet	700	Sep
Swale Estuary	1,705	Jul	Blackwater Estuary	645	Apr
Breydon Wtr & Berney Marshes	1,612	Aug	Chichester Harbour	533	Oct
North Norfolk Coast	1,577	Apr	Eden Estuary	374	Apr
Poole Harbour	1,210	Sep	Portsmouth Harbour	360	Sep
Lough Foyle	983	Oct			

Sites surpassing national passage threshold in Great Britain in 2004/2005

North West Solent	311	Apr	Carmarthen Bay	239	Oct
Medway Estuary	301	Oct	Crouch-Roach Estuary	219	Apr
Deben Estuary	298	Apr	Southampton Water	206	Oct
Severn Estuary	294	Oct	Dengie Flats	205	Oct
Forth Estuary	288	Sep	Colne Estuary	197	Apr
Alde Complex	286	Oct	Beaulieu Estuary	154	Apr
Orwell Estuary	277	Apr	Sandbach Flashes	151	May
Solway Estuary	275	Aug			

Sites surpassing national passage threshold in Northern Ireland in 2004/2005

Carlingford Lough	100	Aug

Bar-tailed Godwit

Limosa lapponica

International threshold:	1,200	
Great Britain threshold:	620	
All-Ireland threshold:	175	

GB max:	37,497	Jan
NI max:	1,494	Feb

	S	M	L
GB change	o	o	o
NI change	o	o	o

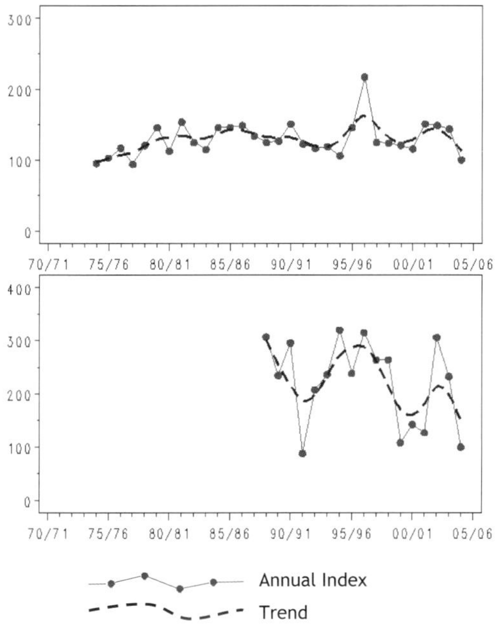

Figure 52.a, Annual indices & trend for Bar-tailed Godwit for GB (above) & NI (below).

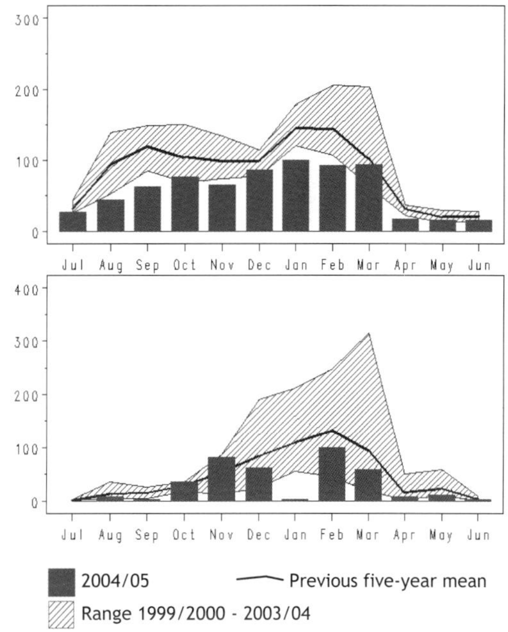

Figure 52.b, Monthly indices for Bar-tailed Godwit for GB (above) & NI (below).

The 2004/05 winter saw a substantial decline in the Bar-tailed Godwit British annual index, of about 30% since the previous year. At the same time, the peak counted number in Britain fell by over 8,000 birds. Monthly indices suggest that numbers were low through most of the winter. Over the longer-term, this species has displayed overall stability since the 1970s and it is probably too soon to make any judgements based on this one low year, but the results for 2005/06 will be of interest. Moreover, numbers across the North Sea in the Netherlands have increased sharply in recent years (van Roomen *et al.* 2006). This backs up other work showing how certain species, including Bar-tailed Godwit, are basically shifting their wintering areas in response to warmer winters (Maclean *et al.* in press).

Numbers in Northern Ireland typically fluctuate to a far greater extent than in Britain, due to the fewer birds present, but there was also a major decline in 2004/05 to the second-lowest level yet for the Northern Ireland index.

Much of the decline was due to the lowest ever numbers on the two key sites of the Wash and Ribble Estuary, whilst low peaks were also recorded from the Alt, Morecambe, Dee, Solway and Lough Foyle, all in the west. At no key sites, except for the Dornoch Firth and

Swale Estuary, were peak counts substantially higher than average.

At the majority of sites passage numbers peak during the autumn, with much lower numbers during spring. Passage numbers at The Wash were the lowest for 15 years.

	00/01	01/02	02/03	03/04	04/05	Mon	Mean
Sites of international importance in the UK							
The Wash	17,223	23,751	18,374	16,280	11,268	Jan	17,379
Ribble Estuary	(4,118)	20,950	(3,111)	11,301	4,657	Jan	12,303
Alt Estuary	6,146	12,098	7,103	8,120	3,900	Jan	7,473
Thames Estuary	(3,019)	(6,460)	3,941	8,989	6,595	Mar	6,508
Lindisfarne	4,066	5,237	(3,000)	(4,078)	2,900	Dec	4,070
Morecambe Bay	1,685	(938)	5,718	4,424	1,752	Jan	3,395
Humber Estuary	2,065	3,669	2,688	4,291 [11]	(2,460)	Jan	3,178
Dee Estuary (England & Wales)	990	12,163 [11]	127	1,209	132	Jan	2,924
Dengie Flats	1,388	4,970	3,112	1,550	1,250	Mar	2,454
Cromarty Firth	2,193	1,044	2,212	3,439	2,311	Feb	2,240
Solway Estuary	(1,434)	(2,106)	1,761	1,572	1,050	Feb	1,622
Strangford Lough	1,543	1,949 [11]	1,079	2,019	1,422	Feb	1,602
Forth Estuary	(1,542)	964	1,793	1,750 [11]	1,599	Jan	1,530
North Norfolk Coast	1,676	1,678 [11]	1,555	1,271	1,203	Jan	1,477
Lough Foyle	208	1,328	4,108	1,019	630	Mar	1,459
Tay Estuary	1,400	1,944	1,351	910	1,680	Feb	1,457
Sites of national importance in Great Britain							
Dornoch Firth	406	1,136 [11]	1,561	1,068	1,495	Dec	1,133
Inner Moray and Inverness Firth	1,510	995	997	830	901	Dec	1,047
Chichester Harbour	925	910	872	(910)	863	Feb	896
South Ford	1,042		549	950	1,040	Feb	895
Loch Bee SSSI Coast				713 [55]			713
Swale Estuary	(700)	595	606	462	922	Feb	657 ▲
Hamford Water	334	1,002	485	803	(431)	Dec	656

Sites surpassing international passage threshold in the UK in 2004/2005					
The Wash	6,165	Sep	Ribble Estuary	1,761	Oct
Thames Estuary	4,673	Oct	Forth Estuary	1,545	Sep
Alt Estuary	4,138	Sep	North Norfolk Coast	1,360	Oct
Humber Estuary	1,825	Oct	Lindisfarne	1,294	Oct
Sites surpassing national passage threshold in Great Britain in 2004/2005					
Tay Estuary	1,083	Oct	Inner Moray and Inverness Firth	760	Oct
Lindisfarne	997	Aug	Hamford Water	647	Oct
Sites surpassing national passage threshold in Northern Ireland in 2004/2005					
Strangford Lough	470	Oct			

Whimbrel

Numenius phaeopus

International threshold: 6,100
Great Britain threshold: +†
All-Ireland threshold: +†

GB max: 1,790 May
NI max: 20 Jul

As is typical for this species, national maxima were recorded during spring, when birds that have over-wintered in western Africa return to their breeding grounds in Iceland and Scandinavia. As usual, spring peaks were higher than during autumn, although this is due to autumn passage occurring over a longer time period, therefore, the numbers of birds recorded at a single site at anyone time are likely to represent only a small proportion of the total number using the area. WeBS counts actually underestimate spring passage as for most sites this passage tends to peak around the end of April and beginning of May, away from the usual recommended mid-month WeBS Core Count dates. As a result, supplementary peak spring counts from four key passage sites have also been presented; all of these supplementary peaks were recorded between 29th April and 2nd May 2005.

The British maximum was the highest for four years, and in Northern Ireland the highest for seven years. However, these figures were in line with the expected variation for the species. As usual, small numbers of birds were recorded during mid-winter months, mostly from south coast sites but with singles also on Montrose Basin (December) and the Solway Estuary (February).

Sites with more than 50 birds during passage periods in Great Britain in 2004/2005†

Barnacre Resr. & Grizedale Lea	553 [56]	Apr	Lower Derwent Ings	139 [53]	Apr
The Wash	414	Aug	Llyn Alaw	120	May
Rye Harbour & Pett Level	338 [50]	Apr	Morecambe Bay	106	May
Brockholes Quarry	289 [57]	May	Humber Estuary	82	Jul
Severn Estuary	197	May	Breydon Wtr & Berney Marshes	74	May
Burry Inlet	175	May	Langstone Harbour	69	Jul
North Norfolk Coast	166	May	Foryd Bay	67	May
Chichester Harbour	143	Aug			

† *as no British or All-Ireland thresholds have been set a qualifying level of 50 has been chosen to select sites for presentation in this report*

Curlew

Numenius arquata

International threshold: 4,200
Great Britain threshold: 1,500
All-Ireland threshold: 875

GB max: 75,140 Jan
NI max: 7,653 Jan

	S	M	L
GB change	o	o	+
NI change	o	o	o

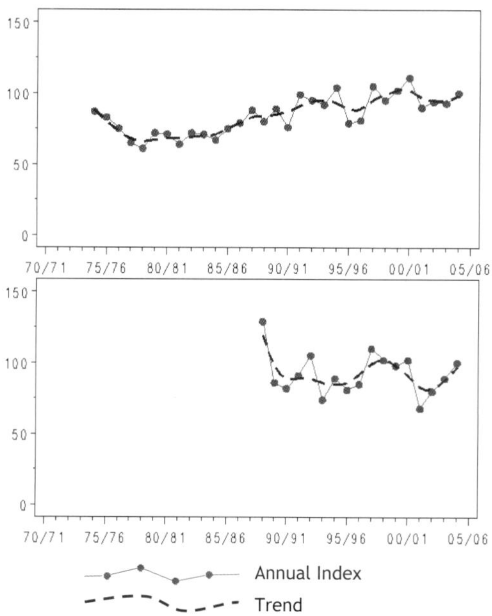

Figure 53.a, Annual indices & trend for Curlew for GB (above) & NI (below).

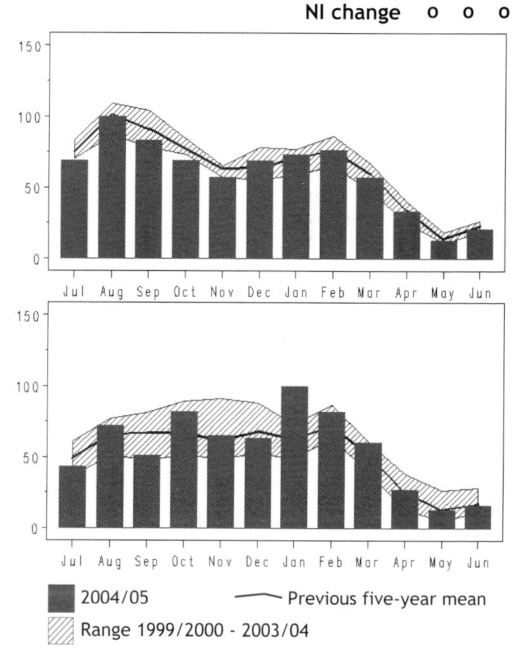

Figure 53.b, Monthly indices for Curlew for GB (above) & NI (below).

The British annual index for Curlew increased by 7% as part of a long run of steady increase in the wintering population. Similarly, in Northern Ireland the index increased for the third consecutive year. Curlew numbers in Britain peak in August as birds rapidly leave the moorland breeding grounds and head to the coast.

Two sites continue to support internationally important mean peak numbers of Curlews, but the peak at the Wash, which had been consistent for several years, dropped sharply in 2004/05 meaning that the site now only supports numbers exceeding the national importance threshold. Elsewhere, higher than expected peaks were recorded at the Inner Moray Firth, Poole Harbour and Lough Foyle, whilst the peak at Breydon Water and Berney Marshes was a new record for the site. Apart from the Wash, the lowest counts compared to recent peaks were recorded from the Duddon Estuary, Lavan Sands and Inner Firth of Clyde, whilst three sites no longer support nationally important numbers on the basis of five-year mean peak counts.

Sites from which much higher counts were recorded during passage, compared to winter, included the Wash and the Dee Estuary. At the Wash, the autumn passage peak fell sharply from that seen in autumn 2003. Conversely, at the Dee, the autumn peak was the highest for about 30 years.

	00/01	01/02	02/03	03/04	04/05	Mon	Mean
Sites of international importance in the UK							
Morecambe Bay	13,756	9,522	10,868	(10,866)	(7,338)	Jan	11,382
Solway Estuary	(4,497)	(4,311)	(3,701)	(4,561)	(3,328)	Dec	(4,561)
Sites of national importance in Great Britain							
Dee Estuary (England & Wales)	4,583	4,305 [11]	3,270	4,978	3,668	Jan	4,161
The Wash	4,058	4,339	4,774	4,036	2,937	Feb	4,029 ▼
Humber Estuary	4,044	4,277	3,941	3,530 [11]	3,751	Jan	3,909
Thames Estuary	(3,160)	(2,354)	4,093	(2,651)	2,786	Mar	3,440
Forth Estuary	(2,524)	(3,638)	3,229	(2,897)	2,669	Mar	3,179
Severn Estuary	(1,695)	(2,164)	3,615 [11]	2,528	(2,545)	Dec	3,072
Duddon Estuary	2,516	2,041	2,280	2,756	1,326	Nov	2,184
Lavan Sands	2,240	2,381	1,922 [11]	1,433	1,212	Jan	1,838
Inner Moray and Inverness Firth	1,698	1,473	1,961	1,809	2,137	Jan	1,816
Poole Harbour	1,484	1,577	1,605	1,427	(2,472)	Nov	1,713
North Norfolk Coast	1,686	2,302 [11]	1,430	1,539	1,523	Mar	1,696
Mersey Estuary	1,976	1,562	1,270	1,804	1,632	Jan	1,649
Lindisfarne	1,636	1,822 [11]	1,338 [11]	(1,072)	1,715 [11]	Dec	1,628
Inner Firth of Clyde	(1,604)	(1,294)	(1,455)	(1,485)	(1,133)	Mar	(1,604)
Sites of all-Ireland importance in Northern Ireland							
Lough Foyle	2,682	1,358	1,956	2,127	3,115	Jan	2,248
Strangford Lough	2,305	1,676	1,200	1,342 [11]	1,594	Jan	1,623
Sites no longer meeting table qualifying levels in Winter 2004/2005							
Chichester Harbour	1,501	1,511	1,414	1,670	1,262	Nov	1,472
Ribble Estuary	(1,709)	990	1,553	(1,857)	1,248	Jan	1,471
Outer Ards Shoreline	1,270		357	282	838	Jan	687
Other sites surpassing table qualifying levels in Winter 2004/2005 in Great Britain							
Breydon Wtr & Berney Marshes	1,029	1,366	981	857	1,593	Jan	1,165
Stour Estuary	1,378	1,673	(1,042)	1,118 [11]	1,511	Feb	1,420

Sites surpassing international passage threshold in the UK in 2004/2005					
Morecambe Bay	8,328	Aug	Dee Estuary (England and Wales)	6,933	Aug
The Wash	6,978	Aug			
Sites surpassing national passage threshold in Great Britain in 2004/2005					
Thames Estuary	3,352	Oct	North Norfolk Coast	1,835	Aug
Solway Estuary	3,086	Sep	Burry Inlet	1,831	Sep
Forth Estuary	2,827	Sep	Mersey Estuary	1,830	Sep
Humber Estuary	2,661	Sep	Eden Estuary	1,731	Aug
Severn Estuary	2,613	Sep	Inner Moray and Inverness Firth	1,648	Oct
Lavan Sands	2,413	Sep	Chichester Harbour	1,628	Sep
Duddon Estuary	1,883	Aug	Alt Estuary	1,601	Aug
Blackwater Estuary	1,848	Aug	Langstone Harbour	1,525	Aug
Sites surpassing national passage threshold in Northern Ireland in 2004/2005					
Lough Foyle	2,194	Oct			
Strangford Lough	1,531	Oct			

Spotted Redshank

Tringa erythropus

International threshold: 1,000
Great Britain threshold: +†
All-Ireland threshold: +†

GB max: 129 Oct
NI max: 0

Peak counts of Spotted Redshank were lower than during the previous year. The British maximum of 129 occurred in autumn as usual but was the lowest of the last 25 years; no birds were recorded from Northern Ireland at all in 2004/05. Mid-winter numbers, for example 56 in January, were also a little on the low side compared to other recent years. There was little evidence of spring passage and numbers fell to their lowest in May, but increased again in June due to early-returning birds.

Due to the nature of WeBS numbers recorded during passage at individual sites remain variable as turnover may be high and numbers can vary day-to-day. Saying this, peak numbers at The Wash, North Norfolk Coast, Humber and Blackwater Estuaries were similar to past years.

	00/01	01/02	02/03	03/04	04/05	Mon	Mean
Sites with mean peak counts of 10 or more birds in winter in Great Britain†							
Severn Estuary	(19)	15	6 [11]	6	(13)	Dec	12
Dee Estuary (England & Wales)	9	14	12	4	12	Feb	10
North Norfolk Coast	11	6	6	11	16	Nov	10
Tamar Complex	(7)	15	8 [11]	(3)	7	Dec	10
Thames Estuary	0	(10)	26	3	(3)	Feb	10

Sites with more than 10 birds during passage periods in Great Britain in 2004/2005†

Blackwater Estuary	42	Oct	North Norfolk Coast	34	Aug
The Wash	39	Aug	Humber Estuary	21	Aug

† *as no British or All-Ireland thresholds have been set a qualifying level of 10 has been chosen to select sites for presentation in this report*

Redshank

Tringa totanus

International threshold: 1,300
Great Britain threshold: 1,200
All-Ireland threshold: 245

GB max: 93,619 Oct
NI max: 10,498 Oct

	S	M	L
GB change	o	o	o
NI change	o	o	o

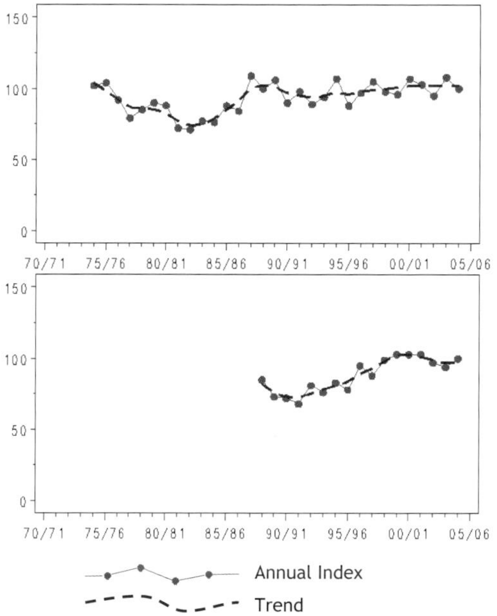

Figure 54.a, Annual indices & trend for Redshank for GB (above) & NI (below).

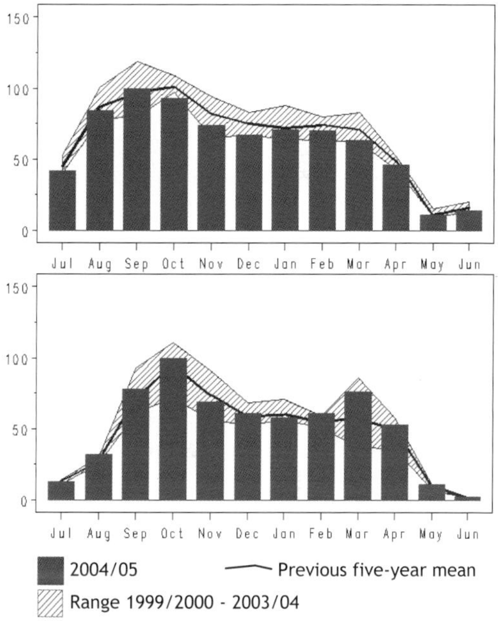

Figure 54.b, Monthly indices for Redshank for GB (above) & NI (below).

Both the British and Northern Irish maxima were in line with those of recent years. Similarly, national indices showed little variation from previous years, the British trend being broadly stable although there has been a very slight increase over the past ten or so years. The Northern Irish index has risen slightly since 2003/04, although again this is fairly trivial. Monthly indices for both regions were comparable to the past five years.

Thirty-five sites held internationally important numbers during 2004/05, two more than in the previous year. Numbers recorded during Low Tide Counts at the Crouch-Roach Estuary were well in excess of those seen during Core Counts and brought up the five-year mean to exceed the international threshold; presumably birds move in off the adjacent high tide roosts at Foulness and Dengie Sands to feed at low tide. Numbers of Redshank at both Lavan Sands and Lindisfarne have risen over the international threshold while numbers on the Outer Ards Shoreline and the Alt Estuary have just fallen below this level. Peak numbers at the Mersey Estuary were the lowest for over 10 years. Passage numbers exceeded the international threshold at 28 sites across the Britain and two in Northern Ireland. In general, passage numbers were similar to past years.

	00/01	01/02	02/03	03/04	04/05	Mon	Mean	
Sites of international importance in the UK								
Morecambe Bay	8,604	6,274	6,650	6,715	7,106	Feb	7,070	
Dee Estuary (England & Wales)	5,893	8,579 [11]	5,847	5,736	5,812	Jan	6,373	
Humber Estuary	4,990	4,526	4,787	8,229 [11]	(5,247)	Nov	5,633	
Mersey Estuary	6,045	4,690	4,143	6,050	3,290	Nov	4,844	
Forth Estuary	4,040	4,204	4,194	4,587	5,501	Dec	4,505	
Thames Estuary	(4,168)	4,479	3,763	(4,383)	3,735	Nov	4,106	
Blackwater Estuary	(4,199)	(3,539)	(2,849)	(1,818)	3,034	Mar	3,591	
The Wash	3,286	4,501	3,619	3,410	2,902	Feb	3,544	
Crouch-Roach Estuary	(742)	(1,220)	(592)	(496)	3,299 [11]	Nov	3,299	▲
Strangford Lough	2,729	3,273 [11]	2,879 [11]	3,146 [11]	2,692	Nov	2,944	
Inner Moray and Inverness Firth	2,862	2,714	2,942	2,317	2,846	Dec	2,736	
Ribble Estuary	(1,734)	1,877	3,882	(1,911)	2,211	Jan	2,657	
Severn Estuary	(1,528)	2,616	2,439 [11]	(1,865)	2,516	Dec	2,524	
Solway Estuary	(3,023)	1,668 [11]	(2,528)	(2,324)	(2,786)	Dec	2,466	
Inner Firth of Clyde	(2,324)	(2,433)	(1,589)	1,974	1,964	Dec	2,174	
Deben Estuary	2,881	1,999	2,017	1,869	1,707	Jan	2,095	
Duddon Estuary	2,816	1,596	1,849	2,508	1,658	Feb	2,085	
Chichester Harbour	1,702	2,422	1,829	2,450	1,695	Nov	2,020	
Hamford Water	1,473	2,575 [11]	2,334	1,892	1,699	Jan	1,995	
Stour Estuary	2,038 [11]	2,261 [11]	1,769 [11]	2,010 [11]	1,773 [11]	Dec	1,970	
Alde Complex	2,742	2,071 [11]	1,456	1,430	1,957	Nov	1,931	
Orwell Estuary	1,637 [11]	2,279 [11]	1,825 [11]	1,939 [11]	1,799	Jan	1,896	
North Norfolk Coast	1,412	3,915 [11]	1,299	1,416	1,180	Nov	1,844	
Montrose Basin	1,509	2,511	1,830	1,803	1,349	Feb	1,800	
Cromarty Firth	1,157	1,849	1,604	2,569	1,784	Dec	1,793	
Belfast Lough	1,677	2,261	1,540	1,452	1,547 [11]	Dec	1,695	
Colne Estuary	1,342	1,871	(97)	(868)	(797)	Jan	1,607	
Swale Estuary	1,569	2,481	959	(1,352)	974	Dec	1,496	
Lough Foyle	1,974	1,104	1,606	1,198	1,404	Mar	1,457	
Tees Estuary	1,441	1,332	1,398	1,926	1,183	Nov	1,456	
Breydon Wtr & Berney Marshes	1,456 [11]	1,207	1,497 [11]	1,630 [11]	1,406	Mar	1,439	
Lavan Sands	1,270	1,126	1,525 [11]	1,248	(1,947)	Nov	1,423	▲
Lindisfarne	975	1,825	(1,371)	1,503	1,365	Dec	1,417	▲
Blyth Estuary	1,265	1,481					1,373	
Medway Estuary	(858)	(1,537)	(972)	(814)	1,068 [11]	Dec	1,303	
Sites of all-Ireland importance in Northern Ireland								
Outer Ards Shoreline	1,428		1,351	1,228	1,121	Jan	1,282	▼
Carlingford Lough	1,325	1,525	1,211	1,027	1,324	Nov	1,282	
Dundrum Bay	1,051	(696)	(530)	(942)	(256)	Nov	1,051	
Larne Lough	379	363	427	356	462	Mar	397	
Bann Estuary	422	260	324	240	290	Feb	307	
Sites no longer meeting table qualifying levels in Winter 2004/2005								
Alt Estuary	1,470	1,090	931	945	931	Feb	1,073	

Sites surpassing international passage threshold in the UK in 2004/2005					
Dee Estuary (England and Wales)	10,208	Aug	North Norfolk Coast	1,845	Aug
Humber Estuary	8,494	Aug	Ythan Estuary	1,797	Sep
Morecambe Bay	6,846	Aug	Lindisfarne	1,737	Sep
The Wash	6,760	Sep	Alde Complex	1,727	Oct
Forth Estuary	5,136	Oct	Tees Estuary	1,723	Aug
Thames Estuary	5,081	Oct	Swale Estuary	1,715	Oct
Strangford Lough	4,505	Oct	Belfast Lough	1,667	Sep
Mersey Estuary	3,618	Oct	Montrose Basin	1,641	Sep
Solway Estuary	3,617	Oct	Chichester Harbour	1,600	Sep
Severn Estuary	2,440	Oct	Lavan Sands	1,504	Oct
Cromarty Firth	2,094	Oct	Deben Estuary	1,494	Sep
Inner Moray and Inverness Firth	2,072	Oct	Carlingford Lough	1,471	Oct
Inner Firth of Clyde	1,977	Oct	Hamford Water	1,449	Oct
Duddon Estuary	1,956	Oct	Ribble Estuary	1,388	Aug
Blackwater Estuary	1,941	Sep	Tay Estuary	1,347	Sep
Sites surpassing national passage threshold in Great Britain in 2004/2005					
Eden Estuary	1,238	Oct	Breydon Wtr & Berney Marshes	1,229	Apr
Sites surpassing national passage threshold in Northern Ireland in 2004/2005					
Lough Foyle	940	Oct	Bann Estuary	290	Apr
Outer Ards Shoreline	881	Oct	Larne Lough	255	Oct
Dundrum Bay	594	Oct			

Greenshank

Tringa nebularia

International threshold: 3,100
Great Britain threshold: 6*†
All-Ireland threshold: 9*†

GB max: 1,630 Aug
NI max: 170 Nov

**50 is normally used as a minimum threshold*

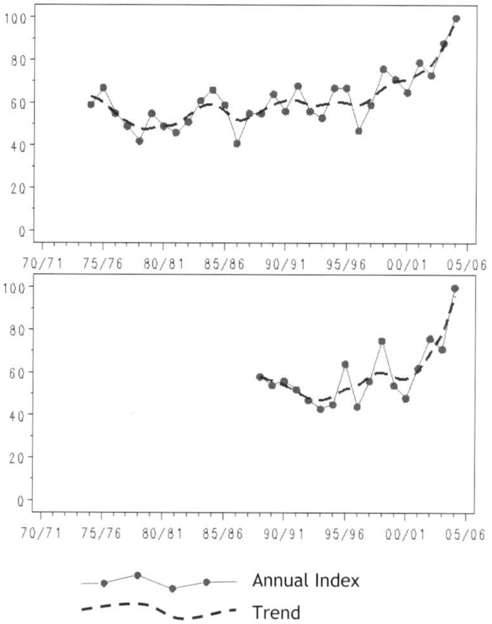

Figure 55.a, Annual indices & trend for Greenshank for GB (above) & NI (below).

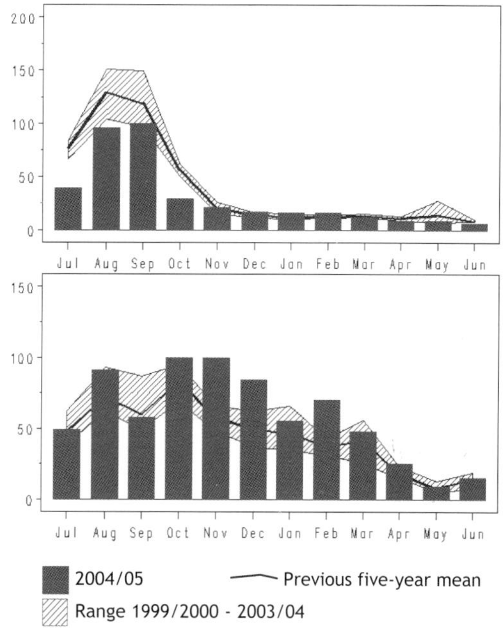

Figure 55.b, Monthly indices for Greenshank for GB (above) & NI (below).

As illustrated by the annual index the numbers of Greenshank over-wintering in Britain have reached their highest ever level. In contrast autumn passage numbers were low, the peak of 1,630 being the second lowest for over ten years. Wintering numbers in Northern Ireland also increased since the previous year and were at their highest since 1998/99. Monthly indices in Northern Ireland were above or near to average throughout the year. Mean peak winter numbers at Carmarthen Bay, Hunterston Sands and Forth and Eden Estuaries have risen above the national threshold during 2004/05, whilst the winter peak on the Tamar Complex was a new record for the site. During passage periods, the highest counts on individual count sections were 188 at Crouchside on the Thames Estuary in August and 115 at Frampton South on the Wash in September.

Greenshank (John Bowers)

	00/01	01/02	02/03	03/04	04/05	Mon	Mean	
Sites of national importance in Great Britain								
Chichester Harbour	43	44	35 [13]	42	20	Nov	37	
Tamar Complex	29	30	31 [11]	26	42	Jan	32	
Kingsbridge Estuary	14	26	41	(36)	45	Nov	32	
Fal Complex	26	26	27	32	37	Nov	30	
Cleddau Estuary	27	34	28	20	(23)	Dec	27	
Queens Valley Reservoir	17	20	19		22	Nov	20	
Grouville Marsh	15	11	31	25	15	Feb	19	
Blackwater Estuary	(17)	(12)	27 [11]	12	12	Jan	17	
Exe Estuary	14	14	18	14	22	Nov	16	
Taw-Torridge Estuary	19	14	16	19	11	Nov	16	
Camel Estuary	9	(6)	(17)	17	15	Jan	15	
Southampton Water	8	15	13	19	(14)	Dec	14	
Inner Firth of Clyde	13	(12)	14	10	9	Dec	12	
North West Solent	8	11	(5)	11	(13)	Nov	11	
Jersey Shore	7	(13)					10	
Yealm Estuary	6	7	15	8	15	Jan	10	
Morecambe Bay	(5)	7	6	15	12	Nov	10	
Foryd Bay	9	8	10	16	9	Nov	10	
Solway Estuary	11	11	8	6	(6)	Nov	9	
Carmarthen Bay	(2)	(1)	(0)	2	15	Nov	9	▲
Tyninghame Estuary	8	7	11	9	11	Nov	9	
Lavan Sands	5	7	9	9	9	Jan	8	
Loch nan Capull (South Uist)				8			8	
Poole Harbour	4	6	11	(6)	7 [11]	Jan	7	
Thames Estuary	(3)	8	(3)	6	6	Nov	7	
Hunterston Lagoon		3		9	10	Nov	7	
Broadford Bay		(8)	(7)	6	7	Nov	7	
Brading Harbour	7	5	6	8	4	Jan	6	
North Norfolk Coast	7	9	3	(3)	3	Jan	6	
The Wash	0	3	18	7	4	Nov	6	
Dee Estuary (England and Wales)	8	6	3	9	5	Nov	6	
Rough Firth	9	7	2	6	7	Nov	6	
Hunterston Sands			8	3			6	▲
Forth Estuary	(7)	2	(6)	7 [11]	9	Feb	6	▲
Eden Estuary	3	(0)	6	9	7	Jan	6	▲
Ceann a Bhaigh		4	8		5	Mar	6	
Sites of all-Ireland importance in Northern Ireland								
Strangford Lough	41	56	72	61	117	Nov	69	
Lough Foyle	16	20	22	27	37	Feb	24	
Carlingford Lough	13	18	14	16	21	Dec	16	
Dundrum Bay	11	18	15	15	11	Nov	14	
Larne Lough	9	(15)	15	11	11	Nov	12	
Sites no longer meeting table qualifying levels in Winter 2004/2005								
Helford Estuary	5	6	4	5	7	Nov	5	
Burry Inlet	2	4	4	6	4	Dec	4	
Medway Estuary	3	8	(1)	5	(0)		5	
Other sites surpassing table qualifying levels in Winter 2004/2005 in Great Britain								
Avon Estuary	5	(6)	3	6	7	Feb	5	
Helford Estuary	5	6	4	5	7	Nov	5	

Sites with more than 50 birds during passage periods in Great Britain in 2004/2005†

Thames Estuary	259	Aug	Stour Estuary	88	Aug
The Wash	204	Sep	Chichester Harbour	80	Sep
Blackwater Estuary	147	Sep	Hamford Water	69	Sep
North Norfolk Coast	118	Sep	Exe Estuary	56	Sep
Morecambe Bay	94	Aug	Fal Complex	52	Aug

Sites with more than 50 birds during passage periods in Northern Ireland in 2004/2005†

Strangford Lough	93	Oct

† *as no British or All-Ireland passage thresholds have been set a qualifying level of 50 has been chosen to select sites for presentation in this report*

Lesser Yellowlegs

Tringa flavipes

Vagrant
Native Range: N & S America

GB max:	1	Sep
NI max:	0	

A long-staying bird was present at Stiffkey on the North Norfolk Coast, being noted on six counts between September and April. This species has now been recorded by WeBS for four consecutive years.

Green Sandpiper

Tringa ochropus

International threshold: 14,500
Great Britain threshold: ?†
All-Ireland threshold: ?†

GB max:	478	Aug
NI max:	1	Jul

As usual, Green Sandpiper numbers peaked in Britain in August, the 478 birds recorded being about average compared to the past five years. Wintering numbers remained just below 100 between December and February before rising to an early spring peak of 109 in March.

As usual, site peaks were all in the early autumn. In the winter, between November and March, birds were noted from 124 sites, average for the past five years, with five or more at five sites. Spring passage is typically much less obvious than autumn passage, birds moving through more rapidly towards the breeding grounds. Birds were recorded from 45 sites between April and June; a peak of eight was at Tophill Low Reservoirs in June may well have involved early returning birds however.

The one in Northern Ireland, at Lough Foyle, was the first recorded here since 1999/2000.

Sites with more than 15 birds during passage periods in Great Britain in 2004/2005

King George V Reservoirs	28	Aug	Tophill Low Reservoirs	22	Aug
North Norfolk Coast	26	Aug	Rutland Water	21	Aug
Swale Estuary	23	Aug	Abberton Reservoir	19	Aug
Thames Estuary	23	Aug	Blackwater Estuary	19	Aug

† *as no British or All-Ireland thresholds have been set a qualifying level of 15 has been chosen to select sites for presentation in this report*

Wood Sandpiper

Tringa glareola

International threshold: 10,400†

GB max:	72	Aug
NI max:	0	

Whilst there were very few birds recorded during July 2004 compared to 2003, the August peak of Wood Sandpipers was the highest ever recorded by WeBS and was made up of counts from 25 sites. Following this glut of records, there were then just two in September and no more until a handful of spring birds in May and June. All of the sites holding three or more birds were in south and eastern England.

Obviously, counts from a single date per month during passage periods do not fully reflect the true numbers of individuals passing through.

Sites with more than 2 birds during passage periods in Great Britain in 2004/2005†

Breydon Wtr & Berney Marshes	19	Aug	Thames Estuary	7	Aug
Rye Harbour and Pett Level	8	Aug	Staines Reservoirs	6	Aug
Blackwater Estuary	7	Aug	North Norfolk Coast	3	Aug

† *as no British or All-Ireland thresholds have been set a qualifying level of 2 has been chosen to select sites for presentation in this report*

Common Sandpiper

Actitis hypoleucos

International threshold: 17,000
Great Britain threshold: ?†
All-Ireland threshold: ?†

GB max:	1,058	Aug
NI max:	4	Aug

As usual, peak numbers of Common Sandpipers were recorded during late summer passage. Numbers in Britain were a little higher than in 2003/04 and around average for the past five years. The peak of four in Northern Ireland was also about average for this species. Most of the sites with the largest numbers were in the southeast corner of Britain.

Small numbers remained to overwinter, with birds recorded from 105 sites between November and March; the summed maxima for these sites totalled 154. This was slightly below the numbers recorded during the same period of the previous winter, which totalled 170 at 110 sites. The highest numbers of birds in the winter period of 2004/05 were seven recorded at Camas Rudh a` Mhurain in March, although these could have been early spring arrivals. There were also counts of four at Walthamstow Reservoirs in January and Loch Leven in December. Numbers then increased in April and to a peak in May that was only about 25% of the autumn peak. The highest count in May was of 14 at Abberton Reservoir.

Sites with more than 40 birds during passage periods in Great Britain in 2004/05†

The Wash	48	Aug	Morecambe Bay	45	Jul
North Norfolk Coast	46	Aug	Thames Estuary	45	Aug
Rye Harbour and Pett Level	46	Aug	Abberton Reservoir	43	Aug

† *as no British or All-Ireland thresholds have been set a qualifying level of 40 has been chosen to select sites for presentation in this report*

Spotted Sandpiper

Actitis macularius

Vagrant
Native Range: America

GB max: 1 Sep
NI max: 0

A single Spotted Sandpiper was reported during the September WeBS count at Darwell Reservoir, East Sussex. This was the first during a WeBS count since December 2001.

Turnstone

Arenaria interpres

International threshold:	**1,000**
Great Britain threshold:	**500**
All-Ireland threshold:	**225**

GB max: 13,287 Oct
NI max: 1,940 Oct

	S	M	L
GB change	o	-	o
NI change	o	o	-

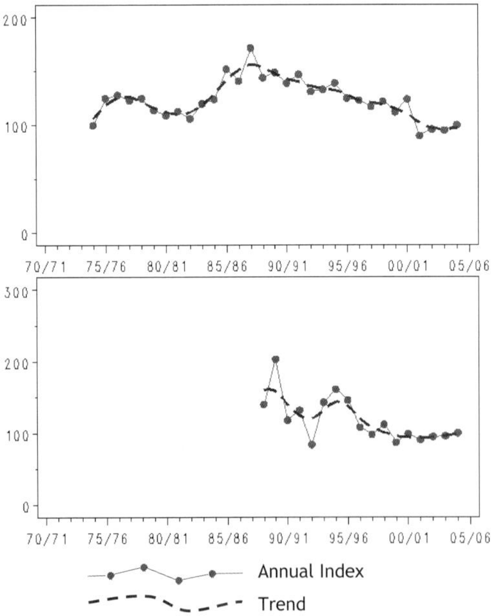

Figure 56.a, Annual indices & trend for Turnstone for GB (above) & NI (below).

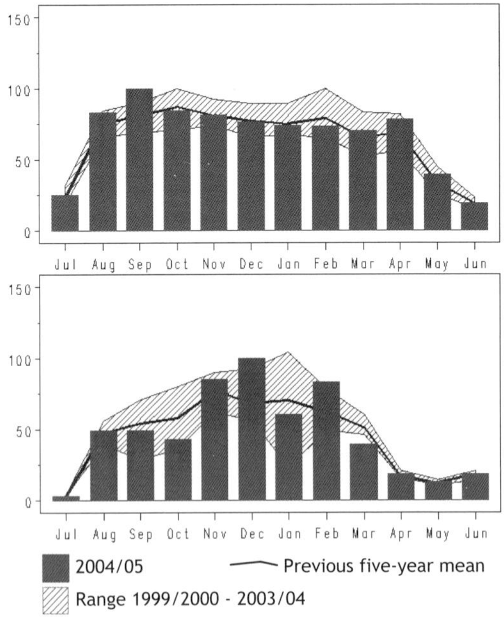

Figure 56.b, Monthly indices for Turnstone for GB (above) & NI (below).

After a long period of decline in wintering Turnstone numbers in Great Britain, there was an indication that trends were levelling off, with the annual index at its highest point for the last four years. Likewise, Turnstones numbers also appear to be level in Northern Ireland. However, this is predominantly a species of rocky coasts, a habitat not well covered by WeBS, and so the relationship of the trends recorded here to the wider picture is not readily discernible. The forthcoming Non-estuarine Coastal Waterbird Survey (NEWS) in January 2007 should help to provide further detail on the status of this wader.

Reflecting the slight up-turn in national fortunes, peak numbers were above average at the Thanet Coast, Swale Estuary, Stour Estuary, Solway Estuary and, particularly, at Carlingford Lough where the March peak was over three times the previous highest count here. Peak numbers were relatively low, however, on the Forth Estuary, Dee Estuary and Belfast Lough, whilst the five-year mean at the Outer Ards means that the numbers at this site no longer exceed the international threshold.

Although the monthly indices indicate that numbers tend to be very stable throughout the whole non-breeding season from August to April, passage numbers at some sites exceed those seen in the winter. In autumn 2004 this was particularly the case at the Wash, Morecambe Bay and the North Norfolk Coast.

	00/01	01/02	02/03	03/04	04/05	Mon	Mean	
Sites of national importance in Great Britain								
Thanet Coast	827	964	694	1,192	1,130	Dec	961	
Morecambe Bay	1,175	825	588	766	691	Jan	809	
Forth Estuary	989	610	940	701	619	Nov	772	
Thames Estuary	632	(879)	488	(465)	702	Dec	675	
Stour Estuary	708	614	(640)	537	705	Feb	641	
North Norfolk Coast	461	744	833	473	655	Nov	633	
Humber Estuary	659	499	(529)	723 [11]	(570)	Jan	627	
Dee Estuary (England & Wales)	791	(405)	726	415	421	Jan	588	
Solway Estuary	(165)	(308)	(300)	(259)	(509)	Jan	(509)	▲
Sites of all-Ireland importance in Northern Ireland								
Outer Ards Shoreline	879		1,086	1,081	923	Mar	992	▼
Belfast Lough	(524)	432	401	485	305	Jan	429	
Carlingford Lough	125	140	157	181	624	Mar	245	▲
Strangford Lough	248	227	206	225	(235)	Dec	228	
Sites no longer meeting table qualifying levels in Winter 2004/2005								
The Wash	515	(270)	579	354	488	Nov	484	
Other sites surpassing table qualifying levels in Winter 2004/2005 in Great Britain								
Swale Estuary	(387)	389 [11]	476	(244)	(515)	Feb	460	

Sites surpassing international passage threshold in the UK in 2004/2005					
The Wash	1,244	Aug	North Norfolk Coast	1,028	Aug
Morecambe Bay	1,054	Oct	Thanet Coast	1,025	Oct
Outer Ards Shoreline	1,035	Oct			
Sites surpassing national passage threshold in Great Britain in 2004/2005					
Forth Estuary	760	Oct	Thames Estuary	711	Oct
Sites surpassing national passage threshold in Northern Ireland in 2004/2005					
Belfast Lough	508	Oct			
Carlingford Lough	368	Apr			

Wilson's Phalarope

Vagrant

Phalaropus tricolor

Native Range: America

GB max: 1 Jan
NI max: 0

The first Wilson's Phalarope reported during a WeBS count for over five years was a single bird in January at Seaton Common in the Tees Estuary. This was a very unusual time of year for this transatlantic vagrant, all previous WeBS records falling between August and October.

Red-necked Phalarope

Scarce

Phalaropus lobatus

GB max: 1 Jul
NI max: 0

All three records of Red-necked Phalarope referred to passage birds; early autumn singles at Loch Ken in July and Blithfield Reservoir in August, and a late spring (or very early autumn bird) at Norton Marsh on the North Norfolk Coast in the following June.

Grey Phalarope

Phalaropus fulicarius

Scarce

GB max:	2	Jan
NI max:	0	

Three were recorded during WeBS counts in 2004/05. A single bird was present during October at Frensham Great Pond, Surrey, while two were recorded at Filey Bay, North Yorkshire during January.

Mediterranean Gull

Larus melanocephalus

International threshold: 8,400†

GB max:	136	Aug
NI max:	3	Sep

Mediterranean Gulls have increased in number and range over the past ten years. With the exception of 2003/04, maximum counts from WeBS Core Counts in Britain have surpassed 100 in every year since 2000/01. The current British maximum was slightly higher than average for the previous five years and is the second highest to date. This total is likely to include post-breeding as well as juvenile birds. Peak numbers in Northern Ireland equaled the previous record in January 2004 and were maintained from September through to November, most were recorded at Lough Foyle.

Birds were recorded at 83 sites in Great Britain with peak counts of five or more birds recorded at a record 17 sites, the majority of which were along the south coast. The largest count was at Folkestone and was part of the Winter Gull Roost Survey. The Brading Harbour total was the second-highest single site Core Count total, being second only to the 2002/03 count at the same site.

Away from the south coast numbers remained high at sites in Wales and along the east coast. The Ribble remained the only key site in northern Britain.

	00/01	01/02	02/03	03/04	04/05	Mon	Mean
Sites with mean peak counts of 5 or more birds in Great Britain†							
Folkestone: Copt Pt/E. Wear Bay					157 [45]	Jan	157
Brading Harbour	35	28	126	57	92	Aug	68
Newtown Estuary	2	65	80	(15)	(42)	Apr	49
Thames Estuary	13	(13)	20	27	27	Dec	22
Camel Estuary	(3)	(1)	8	25	26	Oct	20
Ryde Pier to Puckpool Point	16	8	45	9	23	Jul	20
Tamar Complex	28	14	30	0	(26)	Aug	20
Swansea Bay	11	20	16	19	12 [45]	Jan	16
Chichester Harbour	36 [13]	4 [13]	(16)	(14)	4	Jul	15
North Norfolk Coast	(4)	(6)	(13)	8	(3)	Nov	11
Ribble Estuary	(0)	9	7	8	(2)	Mar	8
The Wash	(0)	1	2	8	15	Jul	7
Poole Harbour	2	(2)	3	(7)	(12)	Mar	6
Llanon and Llansantffraid		6	3				5
Medway Estuary	(0)	(6)	(10)	(2)	2	Nov	5
Pagham Harbour	1	0	16		1	Jan	5
Southampton Water	12 [11]	2	0	(1)	(0)		5
Other sites surpassing table qualifying levels in WeBS-Year 2004/2005 in Great Britain†							
Fleet and Wey	3	(2)	2	4	8	Jan	4
Aberarth			0	0	6	Feb	2
Portsmouth Harbour	(0)	2	5 [11]	1	6	Mar	4
Hamford Water	0	4	2	1	5	Nov	2
Kenfig Pool	0	2	3	3	5	Dec	3
Taw-Torridge Estuary	2	3	7	(3)	5	Jul	4

† *as no British or All-Ireland thresholds have been set a qualifying level of 5 has been chosen to select sites for presentation in this report*

Little Gull

Larus minutus

International threshold: 840
Great Britain threshold: ?†
All-Ireland threshold: ?†

GB max: 7,038 Sep
NI max: 1 Jan

The British maximum was the highest ever recorded, this was largely due to the large count at Hornsea Mere, which was the highest ever Core Count from a single site. This particular count, however, was carried out at dusk so will refer to roosting birds; still an impressive number. With the Hornsea peak removed from the British maximum numbers are on a par with the past five years. Whilst the Forth Estuary count was also a site record, numbers at the majority of key sites were unremarkable.

	00/01	01/02	02/03	03/04	04/05	Mon	Mean
Sites of international importance in the UK							
Hornsea Mere	163	3,150 [12]	1,350 [12]	(940) [13]	7,000	Sep	2,916
Sites with mean peak counts of 5 or more birds in Great Britain							
Alt Estuary	2	80	218	432	24	Sep	151
Forth Estuary	1	22	41	75 [31]	321	Oct	92
Tophill Low Reservoirs	2	0	10	110 [13]	75	Aug	39
Tay Estuary	(0)	(22)	50	36	28	Jul	38
North Norfolk Coast	(17)	70	9	38	4	Jul	30
Monikie Reservoirs	0		60	0	38	Aug	25
Eden Estuary	44	(1)			3	Oct	24
Minsmere	1	2	(15)	73	1	Mar	19
East Chevington Pools		4	29	12	18	Jul	16
Morecambe Bay	3	31	1	36	7	Jan	16
Benacre Broad	30		0				15
Lindisfarne	(0)	0	(0)	26	(0)		13
Humber Estuary	(10)	(0)	(2)	(12)	(0)		(12)
Outer Tay & St Andrews Bay				13 [31]	4 [31]	Dec	9
Moray Firth					8 [1]	Nov	8
St Andrews Bay		8			7	Sep	8
Tring Reservoirs	0	0	1	40	0		8
King George VI Reservoir		1	1	20			7
Thames Estuary	1	(3)	17	(2)	3	Apr	7
Dee Estuary (England & Wales)	(0)	(0)	3	(12)	3	Feb	6
Rescobie Loch		0	(11)				6
Staines Reservoirs	1	0	18	11	1	Sep	6
Dungeness Gravel Pits		(7)	(0)		2	Oct	5
Loch Fleet Complex	4	4	5	8	5	Oct	5
Walthamstow Reservoirs	22	1	0	0	0		5
Other sites surpassing table qualifying levels in WeBS-Year 2004/2005 in Great Britain							
Lunan Bay		0	0	4	5	Oct	2

† *as no British or All-Ireland thresholds have been set a qualifying level of 5 has been chosen to select sites for presentation in this report*

Bonaparte's Gull

Larus philadelphia

Vagrant
Native Range: N America

GB max: 1 Sep
NI max: 0

Two sites hosted Bonaparte's Gull between July 2004 and June 2005. These were Thurso Bay in September and January, and Traigh Luskentyre (Harris) in March.

Black-headed Gull

Larus ridibundus

International threshold: 20,000**
Great Britain threshold: 19,000†
All-Ireland threshold: ?†

GB max:	208,015	Jan
NI max:	14,218	Feb

British maxima have surpassed 200,000 in every year since gulls were first regularly counted for WeBS in 1993/94. However, the current maximum is the lowest recorded, being almost 20% lower than the ten-year average. As with the rest of the gulls their inclusion in WeBS counts is optional so the number of sites at which they have been counted will influence these totals. The Northern Ireland total was in line with recent figures and was only slightly below the five-year average.

Many of the key sites listed below warrant their place due to Winter Gull Roost Survey figures, either from 2003/04 or 2004/05 counts. These figures provide additional information for key sites where gulls are not normally recorded as part of standard WeBS counts, such as Chew Valley Lake, Grafham Water and Queen Mary, Church Wilne, Southfield and Eyebrook Reservoirs. The roost at Bewl Water was particularly large, as indeed it was for Common Gull. Mean numbers at The Wash have fallen below the international qualifying threshold; however, the current peak is similar to the past five-year mean of the Core Counts.

Numbers at Lower Lough Erne were the highest recorded at the site; however, this is only the second year that gulls have been widely covered.

	00/01	01/02	02/03	03/04	04/05	Mon	Mean	
Sites of international importance in the UK								
Thames Estuary	(35,410)	(22,911)	(30,275)	43,601 [45]	40,048	Feb	41,825	
Bewl Water	800	33,000 [12]	63,000 [45]	31,000 [45]	69,000 [45]	Jan	39,360	
Chew Valley Lake				29,800 [45]			29,800	
Lower Derwent Ings			25,300	28,000			26,650	
Humber Estuary	(3,264)	(2,217)	(363)	21,450 [45]	(1,028)	Sep	21,450	▲
Sites with mean peak counts of 10,000 or more birds in Great Britain†								
Queen Mary Reservoir				16,836 [45]			16,836	
The Wash	(9,008)	(16,136)	(15,999)	17,582 [45]	11,093	Aug	15,203	▼
Church Wilne Reservoir				15,000 [45]			15,000	
Morecambe Bay	(17,610)	7,795	(17,772)	12,574	16,757	Aug	14,502	
Grafham Water				14,470 [45]			14,470	
Forth Estuary	(16,521)	(2,343)	(2,195)	11,554 [45]	(3,039)	Nov	14,038	
Poole Harbour	10,162	(7,386)	(12,461)	17,707 [45]	(11,811)	Mar	13,935	
Ribble Estuary	(6,793)	(24,460)	(821)	7,419 [45]	9,750 [45]	Jan	13,876	
Tophill Low Reservoirs	12,500	25,000	11,900	8,900	8,385	Sep	13,337	
Southampton Water	9,751 [11]	(1,788)	(826)	14,822 [45]	(2,280)	Dec	12,287	
Southfield Reservoir				12,000 [45]			12,000	
Exe Estuary					11,577 [12]	Jan	11,577	
Eyebrook Reservoir				11,300 [45]			11,300	
Severn Estuary	(4,478)	(5,725)	9,209 [11]	13,139 [45]	(3,017)	Jul	11,174	
Portsmouth Harbour	(14,247)	4,881	15,311 [11]	14,836 [45]	6,192	Sep	11,093	
Hurleston Reservoir		1,500	6,500	14,500 [45]	20,000 [45]	Dec	10,625	
Pitsford Reservoir	10,000 [12]	10,000 [12]	12,000 [12]	10,000 [45]			10,500	
Inner Moray and Inverness Firth	(375)		7,452 [11]	12,760 [45]	(210)	Feb	10,106	
Sites with mean peak counts of 1,000 or more birds in Northern Ireland†								
Belfast Lough	7,496 [11]	8,986 [13]	5,503 [11]	7,095 [11]	7,515 [11]	Feb	7,319	
Loughs Neagh and Beg	(3,543)	(2,787)	(4,036)	(1,593)	(2,267)	Mar	(4,036)	
Outer Ards Shoreline	3,290		4,945	5,113	2,419	Oct	3,942	
Strangford Lough	3,588	3,503 [11]	3,518 [11]	3,388	3,111	Dec	3,422	
Lough Foyle	1,214	1,627	2,780	1,300 [45]	1,057	Oct	1,596	
Larne Lough	942	2,060	733	831	1,396	Mar	1,192	
Other sites surpassing table qualifying levels in WeBS-Year 2004/2005 in Great Britain†								
Blithfield Reservoir		0			16,500 [45]	Jan	8,250	
Cropston Reservoir	195		50	282	15,000 [45]	Jan	3,882	
Chasewater				3,500 [45]	14,000 [45]	Jan	8,750	
Doddington Pool	300	0	600	150	11,000 [45]	Jan	2,410	
Other sites surpassing table qualifying levels in WeBS-Year 2004/2005 in Northern Ireland†								
Lower Lough Erne			(39)	10	1,615	Feb	813	

† *as few sites exceed the British threshold and no All-Ireland threshold has been set qualifying levels of 10,000 and 1,000 have been chosen to select sites, in Great Britain and Northern Ireland respectively, for presentation in this report*

Ring-billed Gull

Larus delawarensis

Vagrant
Native Range: N America

GB max:	**4**	**Jan**
NI max:	**2**	**Nov**

Singles were recorded from ten sites, of which four held birds during more than one month. Both the Taw-Torridge and Thames Estuaries hosted birds in five months between August and September to March respectively. Another single was at Gerrans Bay in November and January, while in Northern Ireland one was present at Belfast Lough in November, December and February. Other birds were at Cleddau Estuary, Lough Foyle, Par Sands Pools and St Andrews Road, Drift Reservoir, Hayle Estuary and Alt Estuary.

Common Gull

Larus canus

International threshold: 17,000
Great Britain threshold: 9,000†
All-Ireland threshold: ?†

GB max:	**57,817**	**Feb**
NI max:	**4,853**	**Dec**

The British maximum was the lowest for ten years and was almost exactly 10,000 less than the previous year's peak. In Northern Ireland, however, the summed total was similar to the ten-year mean.

Several sites have merited inclusion in the list of key sites on the basis of Winter Gull Roost Survey counts; numbers at Tees Estuary and Willen Lake were particularly higher than the Core Count figures. The large numbers of Common Gulls recorded roosting at Bewl Water were sustained for the second year running. The peak Core Count at Loch of Lintrathen was the highest for the site.

Numbers at Tophill Low Reservoirs were the lowest recorded by Core Counts and were 75% lower than the ten-year average. Peaks at Moray Firth, Dee Estuary and Eccup Reservoir were also lower than average.

	00/01	01/02	02/03	03/04	04/05	Mon	Mean
Sites of international importance in the UK							
Bewl Water	50	63,000 [12]	52,000 [45]	75,000 [45]	75,000 [45]	Mar	53,010
Haweswater Reservoir	26,480 [12]	16,566 [12]	13,674 [12]	27,986 [13]			21,177
Tophill Low Reservoirs	24,500	33,000	23,100	16,530	6,500	Oct	20,726
Hallington Reservoir	19,000 [12]	4	24,000 [12]	25,000 [12]			17,001
Sites of national importance in Great Britain							
Eyebrook Reservoir				16,100 [45]			16,100
Humber Estuary	(502)	(366)	(2,077)	29,000 [45]	2,005	Feb	15,503 ▼
Derwent Reservoir	41,000 [12]	6,500	11,800 [12]	(6,500) [12]	1,714	Jan	15,254 ▼
Ullswater	(0)	(0)		11,470 [13]	(0)		11,470
West Water Reservoir				10,050 [45]			10,050
Colt Crag Reservoir	16,000 [12]		8,200 [12]	4,700 [45]			9,633 ▲
Ribble Estuary	(3,077)	8,653	(146)	(6,036)	(9,817)	Feb	9,235 ▲
Chew Valley Lake	2	(0)		18,200 [45]			9,101 ▲
Sites with mean peak counts of 3,000 or more birds in Great Britain†							
Lower Derwent Ings			14,200	3,720 [45]			8,960
Rye Harbour and Pett Level				8,600 [45]			8,600
Solway Estuary	(2,247)	(1,398)	7,193	9,564 [45]	(2,275)	Oct	8,379
Blyth Estuary	2,750	(1,337)		12,000 [45]			7,375
Rutland Water	4,000	50 [12]	100	12,080 [45]	14,500	Mar	6,146
Southwold Sole Bay				5,000 [45]			5,000
Tees Estuary	3,268	8,130	2,970	4,033	6,193 [45]	Jan	4,919
Forth Estuary	(2,197)	(1,658)	(1,356)	6,321 [45]	2,500 [45]	Jan	4,411
Moray Firth	4,494 [1]	5,961 [1]	5,037 [1]	5,208 [1]	809 [1]	Dec	4,302
Wigtown Bay	4,503	(4,277)	1,427	3,251	7,269	Jan	4,145
Loch of Skene	1,390	570	433	17,284 [45]	361	Dec	4,008
Morecambe Bay	4,860	3,632	3,194	4,358	3,633	Oct	3,935
Eccup Reservoir	3,500	9,000	5,000	579 [45]	1,200	Sep	3,856
Hule Moss	1,600 [13]	2,200 [13]	6,300 [13]	5,600 [13]	3,550	Oct	3,850
Lindisfarne	(630)	(2,920)	(370)	(3,644)	(580)	Mar	(3,644)
Thames Estuary	6,848	3,135	2,041	2,319 [45]	(3,669)	Feb	3,602

	00/01	01/02	02/03	03/04	04/05	Mon	Mean
Dee Estuary (England &Wales)	(572)	(1,519)	4,182	5,311	692	Sep	3,395
Hamilton Low & Strathclyde Pks				3,200 [45]			3,200
The Wash	3,681	(1,784)	2,482	4,912 [45]	1,239	Jun	3,079
North Norfolk Coast	460	(1,420)	(1,283)	5,600 [45]	(2,163)	Mar	3,030
Pitsford Reservoir	2,000 [12]	3,000 [12]	4,000 [12]	3,000 [45]			3,000
Sites with mean peak counts of 1,000 or more birds in Northern Ireland[†]							
Lough Foyle	6,095	3,300	4,606	(5,930)	2,322	Sep	4,451
Belfast Lough	1,416	2,103 [13]	2,718	2,644 [11]	1,937 [11]	Feb	2,164
Outer Ards Shoreline	706		772	2,543	1,171	Mar	1,298
Other sites surpassing table qualifying levels in WeBS-Year 2004/2005 in Great Britain[†]							
Loch of Lintrathen	1	1,450		0	10,000	Oct	2,863
Cameron Reservoir	650	5,600	400	240	6,500	Oct	2,678
Chichester Harbour	618	2,062	4,142	3,389	3,778	Feb	2,798
Willen Lake	15	11	7	7	3,500 [45]	Jan	708
Loch of Boardhouse	650	430	1,080	1,550	3,000	Jan	1,342

[†] *as few sites exceed the British threshold and no All-Ireland threshold has been set qualifying levels of 3,000 and 1,000 have been chosen to select sites, in Great Britain and Northern Ireland respectively, for presentation in this report*

Lesser Black-backed Gull

Larus fuscus

International threshold: 5,300
Great Britain threshold: 500
All-Ireland threshold: ?[†]

GB max: 45,841 Jul
NI max: 600 Oct

The British maximum was similar to the average of the past ten years. In total, average numbers at thirty-four sites surpassed the national qualifying level, six less than in the previous year. In addition, five sites surpassed the international qualifying level; the same sites as in as 2003/04. Three sites, Blithfield and Haweswater Reservoirs and Cotswold Water Park (East) attained nationally importance status; counts at the former were attained entirely though Winter Gull Roost Survey data. Core Count totals were particularly high at Great Pool Westwood Park, Hule Moss, Thames Estuary and Longnewton Reservoir. Britain's top site for this species remains South Walney Island, where 97% of the birds in the Morecambe Bay site total were recorded.

The summed total across sites in Northern Ireland was half that of recent years, largely due to poor coverage at the region's main site Loughs Neagh & Beg.

	00/01	01/02	02/03	03/04	04/05	Mon	Mean
Sites of international importance in the UK							
Morecambe Bay	40,590	31,620	36,461	31,479	33,004	Jul	34,631
Theale Gravel Pits		(0)	(3)	20,000 [45]	1,152 [45]	Jan	10,576
Chew Valley Lake	(0)	(0)		7,015 [45]			7,015
Queen Mary Reservoir				6,656 [45]			6,656
Cotswold Water Park (West)	(203)	(687)	(25)	5,800 [45]	(44)	Mar	5,800
Sites of national importance in Great Britain							
Severn Estuary	669	945	(3,072)	(8,073)	(8,064)	Jul	4,165
R. Avon: F'bridge-Ringwood	960	3,478	2,309	6,550 [45]	3,500	Oct	3,359
Llys-y-fran Reservoir	11,000 [12]	6	2,000	90	650	Dec	2,749
Great Pool Westwood Park	2,000	1,350	2,000	3,800 [45]	2,500	Jan	2,330
Hule Moss	3,300 [13]	3,090 [13]	2,100 [13]	250 [13]	2,400	Sep	2,228
Rutland Water	600	2,000 [12]	5,000	2,500	140	Mar	2,048
Thames Estuary	1,783	1,560	1,507	1,898 [45]	2,966	Oct	1,943
Alt Estuary	1,122	1,619	4,341	(945)	556	Sep	1,910
Alde Complex	(36)	767 [11]	4,474	388 [45]	1,833	Mar	1,866
Longnewton Reservoir	340	970	2,680	1,890	2,930	Sep	1,762
Lower Windrush Valley GPs	2,424	3,166	871	484	1,343	Jan	1,658
Roadford Reservoir		52	70	6,031 [45]	110	Dec	1,566
Belvide Reservoir				3,000 [45]	0		1,500
Hurleston Reservoir		65	700	1,500 [45]	3,500 [45]	Oct	1,441
Chelmarsh Reservoir	(300)	500	(34)	3,500 [45]	83	Dec	1,361
Blithfield Reservoir		0			2,620 [45]	Jan	1,310 ▲
Bartley Reservoir				1,200 [45]			1,200
Wellington Gravel Pits	(2,400)		(1,400)	750	100	Aug	1,163

	00/01	01/02	02/03	03/04	04/05	Mon	Mean	
Pitsford Reservoir	1,000 [12]	550 [12]	1,000 [12]	2,000 [13]			1,138	
The Wash	1,139	(582)	855	898	1,039	Aug	983	
Solway Estuary	(195)	(243)	(673)	(971)	(154)	Jun	(971)	
Portworthy Mica Dam	750	(2,000)	419	700	960 [45]	Jan	966	
Llangorse Lake	1,050	1,170 [13]	1,110 [12]	1,140 [13]	16	Sep	897	
Cleddau Estuary	625	825	659	723	1,537	Dec	874	
Frampton Pools	120	1,500	(250)				810	
NE Glamorgan Moorland Pools	732						732	
Hayle Estuary	852	(340)	130	940	980	Feb	726	
Burghfield Gravel Pits				618 [45]			618	
Cotswold Water Park (East)	(104)	299	(156)	(133)	821	Nov	560	▲
Blyth Estuary	886	(93)		200 [45]			543	
Inner Firth of Clyde	393	(557)	544	705	509	Jul	542	
Haweswater Reservoir	15 [12]	231 [12]	400 [12]	1,450 [13]			524	▲
Hollowell Reservoir				500 [45]			500	
Farne Islands	500			(0)			500	
Sites no longer meeting table qualifying levels in WeBS-Year 2004/2005								
Colliford Reservoir	52	140	144	92	455	Jan	177	
Crowdy Reservoir	1,000	60	850	34	300	Jan	449	
Dee Estuary (England & Wales)	(130)	(170)	(342)	648 [45]	259	Oct	454	
Poole Harbour	565	(237)	285	997	116	Sep	491	
Swithland Reservoir	2	2	36	2,050 [45]	70 [45]	Jan	432	
Middle Tame Valley Gravel Pits	425	(170)	(384)	(80)	(50)	Nov	425	
Nosterfield Gravel Pits	70	94	1,560	218	32	Jul	395	
Sites with mean peak counts of 500 or more birds in Northern Ireland†								
Loughs Neagh and Beg	1,459	(228)	1,218	1,115	(434)	Oct	1,264	
Other sites surpassing table qualifying levels in WeBS-Year 2004/2005 in Great Britain†								
Heaton Park Reservoir	21	350 [13]	920 [12]	200 [45]	870 [45]	Oct	472	
Fernworthy Reservoir	33	4	61	(139)	663	Oct	190	
Willen Lake	2	12	3	4	650 [45]	Jan	134	
Blucks Pool to Freshwater West	375	240	630	235	625	Sep	421	
Ditchford Gravel Pits	240	146		276	534	Nov	299	

† *as no All-Ireland threshold has been set a qualifying level of 500 has been chosen to select sites for presentation in this report*

Yellow-legged Gull / Caspian Gull

International threshold: 7,000†

Larus michahellis michahellis / Larus (argentatus) cachinnans

GB max:	**165**	**Aug**
NI max:	**0**	

Following a recent taxonomic review by the British Ornithologists' Union Yellow-legged Gull has been promoted to specific rank after many years of classification as a subspecies of Herring Gull (Sangster et al. 2005). The status of 'Caspian Gull' remains unresolved and it is still officially considered a race of Herring Gull. However, as reports of 'Yellow-legged Gull' undoubtedly refer to individuals of both Yellow-legged Gull and Caspian Gull these are presented together for the purpose of this report.

Yellow-legged Gulls were reported from 38 sites, with Caspian Gulls specified as being at five of these. This was similar to during 2003/04, as was the British maximum, which was average for the past five years and consisted almost entirely of a count of Yellow-legged Gulls at Southampton Water (only one Caspian Gull was recorded in August). Caspian Gulls were reported every month between August and February, all singles except for two in the Thames Estuary in February. Other sites at which Caspian Gulls were recorded include Aston On Trent Gravel Pits, Clifford Hill Gravel Pits, Pegwell Bay and Welbeck Estate.

Yellow-legged Gulls were specified as being at four sites, the highest counts away from Southampton Water being at Thames Estuary (16) and Kingsbridge Estuary (10).

Between July and October monthly counts of 'unspecified' Yellow-legged Gulls exceeded 130 birds; this fell to double figures with a low of 11 in February to April. None were recorded in May but June counts totalled 56. However, as the recording of gulls is optional numbers, particularly outside the main winter period will largely be dependent on the amount of coverage.

Southampton Water remains the top WeBS site for Yellow-legged Gulls supporting the highest ever site total in August. Peak counts at River Avon: Fordingbridge to Ringwood, Clifford Hill Gravel Pits, Hamford Water, King George VI Reservoir, Kinsham Pool, Netherfield Gravel Pits, Queen Elizabeth II Reservoir, Staines Reservoirs and Thorpe Malsor Reservoir were also new records for the respective sites.

Sites with 5 or more birds in Great Britain†

Southampton Water	164	Aug	Kingsbridge Estuary	10	Apr
River Avon - Fordingbridge to Ringwood	72	Oct	Staines Reservoirs	7	Jul
Thames Estuary	36	Oct	Lower Windrush Valley Gravel Pits	6	Nov
Poole Harbour	16	Sep	Great Pool Westwood Park	6	Dec

† *as no sites excced the international threshold and no British or Northern Ireland thresholds have been set, a qualifying level of 5 has been chosen to select sites for presentation in this report*

Herring Gull

Larus argentatus

International threshold: 13,000
Great Britain threshold: 4,500†
All-Ireland threshold: ?†

GB max: 88,754 Feb
NI max: 8,499 Nov

The British maximum was the highest recorded and was almost 70% higher than that of 2003/04. The Northern Ireland figure was more typical being only slightly down on the past three years' peaks. As with all gulls, however, the optional recording of this group means that numbers are largely dependent on the amount of coverage. The number of British sites at which five-year mean values exceeded 2,500 remained the same as in the previous year.

Numbers at Morecambe Bay have fallen over the past few years and are currently the lowest since gulls were first regularly included as part of WeBS in 1993/94. As with Lesser Black-backed Gull the majority of birds at the site are at South Walney Island. In contrast, numbers at the Alt Estuary were the highest recorded at the site for over five years.

Mean peak numbers at the North Norfolk Coast have fallen below the national qualifying level. The highest single-site count was at Ribble Estuary and was one-and-a-half times the previous five-year mean of the site.

	00/01	01/02	02/03	03/04	04/05	Mon	Mean
Sites of international importance in the UK							
Ribble Estuary	(9,032)	(9,767)	(209)	14,859 [45]	(31,090)	Feb	22,975 ▲
Inner Moray and Inverness Firth	650		27,956 [11]	(2,341)	(2,003)	Feb	14,303 ▲
Morecambe Bay	20,530	12,170	14,373	10,551	8,311	Apr	13,187
Sites of national importance in Great Britain							
Forth Estuary	(3,828)	(1,868)	(1,925)	7,376 [45]	(15,434)	Dec	11,405
Queen Mary Reservoir				8,279 [45]			8,279
The Wash	10,003	(7,603)	7,640	10,703 [45]	3,258	Aug	7,901
Moray Firth	10,429 [1]	9,564 [1]	10,335 [1]	6,468 [1]	2,349 [1]	Dec	7,829
Rye Harbour and Pett Level				5,850 [45]			5,850
Hastings to Bexhill				5,700 [45]			5,700
Thames Estuary	4,180	2,867	3,330	(4,349)	8,504	Feb	4,720 ▲
Llandegfedd Reservoir					4,710 [45]	Jan	4,710 ▲
Sites with mean peak counts of 2,500 or more birds in Great Britain†							
Dee Estuary (England & Wales)	(2,000)	(778)	3,602	4,052 [45]	4,244	Nov	3,966
Alt Estuary	3,967	1,440	3,153	3,825 [45]	7,155	Feb	3,908
North Norfolk Coast	3,895	5,062	3,964	3,047	2,684	Aug	3,730 ▼
Chew Valley Lake				3,400 [45]			3,400
Roughrigg Reservoir	133	47	1,121	15,144 [45]	416	Jan	3,372
Troon Meikle Craigs				3,174 [45]			3,174
Carmarthen Bay	(2,600)	(1,494)	(2,111)	(371)	(3,066)	Sep	(3,066)
Broadwater Lake: S. Harefield				3,000 [45]			3,000
Guernsey Shore	(3,525)	1,972	2,127	(2,759)	(3,744)	Aug	2,825
Caldey Island				2,800 [45]			2,800
Solway Estuary	(2,165)	(2,719)	3,281	2,189 [45]	(1,051)	Feb	2,735
Pegwell Bay	3,890	3,000	1,897 [11]	1,569 [45]	(106)	Sep	2,589
Sites with mean peak counts of 1,000 or more birds in Northern Ireland†							
Belfast Lough	6,749	9,157	7,046	7,536 [11]	7,903 [11]	Nov	7,678
Outer Ards Shoreline	898		1,001	(1,351)	1,179	Mar	1,107

	00/01	01/02	02/03	03/04	04/05	Mon	Mean
Other sites surpassing table qualifying levels in WeBS-Year 2004/2005 in Great Britain†							
Heaton Park Reservoir	2,300	700 [13]	2,050 [12]	1,755 [45]	3,400 [45]	Jan	2,041
Hayle Estuary	712	864	2,206	2,325	2,822	Aug	1,786

† *as few sites exceed the British threshold and no All-Ireland threshold has been set qualifying levels of 2,500 and 1,000 have been chosen to select sites, in Great Britain and Northern Ireland respectively, for presentation in this report*

Iceland Gull

Larus glaucoides

International threshold: 2,000

GB max:	**25**	**Feb**
NI max:	**11**	**Jan**

Numbers of Iceland Gulls recorded during WeBS Core Counts rose following the low totals of the previous year. British maxima reached double figures during January to March with a peak of 25 in February; the Northern Irish peak was a month earlier. Birds were present at 36 sites including 18 in Scotland, 14 in England and four in Northern Ireland.

The majority of records were of single birds although eight were at Belfast Lough (Jan) and Scalloway Islands (Feb), seven at Thurso Bay (Jan), four at Larne Lough (Feb) and Burra and Trondra (Feb), three at Loch Innis Na Ba Buidhe (Feb) and two at Camel Estuary, Loch A Chairn Bhain, Loch Inver, Loch Shieldaig, Lough Foyle, Moray Firth, Purfleet Chalk Pit, Thurso Bay and Traigh Luskentyre.

Glaucous Gull

Larus hyperboreus

International threshold: 10,000

GB max:	**21**	**Feb**
NI max:	**15**	**Feb**

Numbers of Glaucous Gulls were the highest ever recorded during WeBS. An influx during February saw the highest ever British maximum, as was the Northern Irish peak of 15, 11 of which were at Belfast Lough. Birds were reported from a record 29 sites across Britain and six in Northern Ireland. Most reports were of single birds, except for two at Island of Papa Westray and Thurso Bay in January and The Ouse and Lairo Water in March.

Great Black-backed Gull

Larus marinus

International threshold: 4,700
Great Britain threshold: 400
All-Ireland threshold: ?†

GB max:	**7,539**	**Oct**
NI max:	**1,348**	**Mar**

The British maximum was the lowest since gulls were regularly included in WeBS in 1993/94 and was over a third down on the ten-year average. Conversely, the Northern Ireland maximum was the highest yet recorded. Twenty-six British sites surpassed national qualifying levels, one less than in 2003/04.

The Belfast Lough total was the highest Core Count there for over five years. Counts at the Tees and Thames Estuaries were the highest since 1994/95 and 1993/94 (when gulls were first included in WeBS) respectively, whereas counts at Tophill Low Reservoirs, Don Mouth to Ythan Estuary and North Norfolk Coast were well below average.

	00/01	01/02	02/03	03/04	04/05	Mon	Mean
Sites of national importance in Great Britain							
The Wash	1,303	4,515	1,959	4,628	(1,480)	Oct	3,101
Humber Estuary	(313)	(83)	(113)	2,200 [45]	(226)	Oct	2,200
Tees Estuary	1,564	(1,038)	702	1,523	1,657	Dec	1,362
Thames Estuary	1,530	(412)	1,236 [11]	857 [45]	1,648	Jan	1,318
Tophill Low Reservoirs	1,880	900	3,030	223 [45]	120	Dec	1,231
Lynemouth Ash Lagoons				1,074			1,074
Grafham Water				1,050 [45]			1,050

	00/01	01/02	02/03	03/04	04/05	Mon	Mean	
Dungeness Gravel Pits		(0)		1,000 [45]	(0)		1,000	
Coquet Island				980 [45]			980	
Pegwell Bay	1,050	1,000	1,305 [11]	305	(350)	Sep	915	
Lower Derwent Ings			777	1,041 [45]			909	
Ogston Reservoir				900 [45]			900	
Don Mouth to Ythan Mouth	(1,225)	(67)	(55)	(200)	495	Jul	860	
Durham Coast	(106)	(16)	(21)	(41)	(684)	Oct	(684)	▲
Loch of Strathbeg	1,280 [13]	(129)	569	(606)	191	Feb	680	
North Norfolk Coast	567	748	617	1,051	327	Oct	662	
Inner Moray and Inverness Firth	330		1,432 [11]	70	(93)	Dec	611	
Hastings to Bexhill				520 [45]			520	
Hanningfield Reservoir	(0)	(0)	1,098 [45]	437 [45]	0		512	
Eyebrook Reservoir				500 [45]			500	
Morecambe Bay	(716)	331	353	(322)	(296)	Nov	467	
Guernsey Shore	(273)	(205)	353	(560)	404	Nov	439	
Forth Estuary	575	(108)	(211)	286 [11]	(239)	Sep	431	
Southfield Reservoir				408 [45]			408	
Hoveringham & Bleasby GPs	0	0		1,600 [45]	2	Dec	401	
Romney Sands				400 [45]			400	
Sites no longer meeting table qualifying levels in WeBS-Year 2004/2005								
Portsmouth Harbour	1,028	54	304	206 [45]	130	Sep	344	
Sites with mean peak counts of 500 or more birds in Northern Ireland†								
Belfast Lough	398	458	397	436 [11]	1,008	Mar	539	
Other sites surpassing table qualifying levels in WeBS-Year 2004/2005 in Great Britain								
Alt Estuary	213	294	(375)	(275)	589	Feb	368	

† *as no All-Ireland threshold has been set a qualifying level of 500 has been chosen to select sites for presentation in this report*

Kittiwake

Rissa tridactyla

International threshold: 20,000**
Great Britain threshold: ?†
All-Ireland threshold: ?†

GB max: 1,117 Sep
NI max: 7 Sep

Kittiwakes are predominantly coastal, and this is reflected in the list of sites attaining a five-year mean peak of 200 or more. The highest numbers are typically recorded at Scottish sites with many of the summer peaks being recorded at sites near to breeding colonies.

Despite the numbers recorded during WeBS increasing over the previous few years the 2004/05 maximum was 80% down on the previous year and was the lowest for over ten years. Additionally, no four-figure counts made in 2004/05.

	00/01	01/02	02/03	03/04	04/05	Mon	Mean
Sites with mean peak counts of 200 or more birds in Great Britain†							
Loch of Strathbeg	200	0	940	6,300	41	Jul	1,496
Lunan Bay		0	400	3,400	100	Oct	975
Farne Islands		920					920
Arran	225	1,700	185	290	340	Oct	548
Loch Linnhe: Camas Shallachain		500					500
Tweed Estuary		340	470	860	114	Aug	446
Tay Estuary	8	266	1,100	133	690	Sep	439
Tees Estuary	153	20	30	1,492	(56)	Jul	424
Loch a` Phuill (Tiree)	36	406	1,128	276	104	Jul	390
Don Mouth to Ythan Mouth	595	(18)	0	153	534	Jul	321
Forth Estuary	(254)	(274)	(453)	(426)	170	Oct	315
Dee Estuary (Scotland)	774	36	162	248	161	Aug	276
Inner Moray and Inverness Firth	250						250
Solway Estuary	120	(574)	(300)	(47)	0		249
Broadford Bay		(0)	200	(20)	(50)	Sep	200
Loch Nan Gabhar			200				200

† *as no British or All-Ireland thresholds have been set a qualifying level of 200 has been chosen to select sites for presentation in this report*

Little Tern

Sternula albifrons

International threshold: 340
Great Britain threshold: ?†
All-Ireland threshold: ?†

GB max: 760 Jul
NI max: 0

Southern and eastern England remains the stronghold for breeding Little Terns in the UK with the majority of key sites occurring between The Wash and Dorset coasts. Numbers on the North Norfolk Coast were the lowest for some years, resulting in the five-year mean falling below the international threshold. Numbers at the Thames Estuary also remained low for a second year. As with all terns, however, the optional recording of the group means that any changes in site peaks should be treated with caution. Moreover, this species is well known for shifting its breeding sites from year to year.

Outwith the southeast, numbers at both the Dee Estuary and the Duddon Estuary have increased, the Dee Estuary actually supporting the highest peak in summer 2004. Another high site total was recorded at the Tees Estuary and was the highest at this site for ten years.

	2000	2001	2002	2003	2004	Mon	Mean
Sites with mean peak counts of 50 or more birds in Great Britain†							
North Norfolk Coast	(241)	(265)	(280)	(405)	233	Jun	285 ▼
Dee Estuary (England & Wales)	111	(0)	242	(256)	300	Jul	227
Thames Estuary	161	(1)	(100)	(28)	33	Aug	98
The Wash	(56)	(103)	(36)	68	(108)	Aug	93
Chichester Harbour	15	200 [13]	42	28	36	May	64
Fleet and Wey	125	0	59	62	69	May	63
Langstone Harbour			(140)	50	0		63
Blackwater Estuary	(101)	(50)	(3)	(20)	28	Jun	60
Duddon Estuary	44		28	42	84	Jul	50
Other sites surpassing table qualifying levels in Summer 2004 in Great Britain†							
Tees Estuary	64	0	26	16	(110)	Jul	43

† *as no British or All-Ireland thresholds have been set a qualifying level of 50 has been chosen to select sites for presentation in this report*

Black Tern

Chlidonias niger

International threshold: 4,000
Great Britain threshold: ?†
All-Ireland threshold: ?†

GB max: 86 Aug
NI max: 0

Black Terns were noted at 40 sites between May and October 2004, with single records only in May to July; these were from North West Solent (May), Seaton and Murton Ponds (Jun) and Severn Estuary (Jul). Most records were during August and September with the latest birds at Knight and Bessborough Reservoirs, Southampton Water (both singles) and Chichester Gravel Pits (three) in October. The British maximum was slightly higher than the previous year and was around average for the past five years.

Sites with 5 or more birds in 2004†

Taw-Torridge Estuary	30	Aug	Killington Reservoir	5	Sep
Dungeness Gravel Pits	22	Aug	Langstone Harbour	5	Aug

† *as no British or All-Ireland thresholds have been set a qualifying level of 5 has been chosen to select sites for presentation in this report*

Sandwich Tern

Sterna sandvicensis

International threshold: 1,700
Great Britain threshold: ?†
All-Ireland threshold: ?†

GB max: 11,877 Jul
NI max: 289 Sep

Sandwich Terns were recorded in every month during 2004, being were noted at 101 sites in Britain and six in Northern Ireland. The British peak was, as usual, recorded during July and was similar to the past few years.

The North Norfolk Coast remains the top site for Sandwich Tern with a record total this year. A second year of high reported summer counts saw Cemlyn Bay and Lagoon becoming the third site to support internationally important numbers; this is largely due to the inclusion of summer counts that have only been submitted during the past two years. Numbers recorded at the Duddon Estuary were the highest since 1999. Counts at the top two Northern Irish sites were unexceptional; numbers at Belfast Lough were the lowest recorded at the site.

As with all terns the optional coverage of this group during WeBS counts means that numbers recorded will largely be dependent on coverage at each site.

	2000	2001	2002	2003	2004	Mon	Mean	
Sites of international importance in the UK								
North Norfolk Coast	(5,015)	(3,365)	(4,600)	4,170	5,533	Jun	4,906	
Forth Estuary	3,424	(994)	(2,317)	2,802	(1,526)	Aug	3,113	
Cemlyn Bay and Lagoon		0		2,455	2,700	May	1,718	▲
Sites with mean peak counts of 200 or more birds in Great Britain†								
Humber Estuary	1,329	(124)	(396)	(303)	(324)	Sep	1,329	
Tees Estuary	897	35	974	2,601	(333)	Jul	1,127	
Dee Estuary (England & Wales)	(672)	(11)	1,632	716	759	Jul	1,036	
Duddon Estuary	994	(0)	704	955	1,144	Jul	949	
Pegwell Bay	320	660	360	(930)	(680)	Jul	590	
Solway Estuary	(78)	(235)	(206)	(548)	(282)	Jul	(548)	
Ythan Estuary			930	150			540	
Morecambe Bay	110	(0)	220	531	(500)	Apr	340	
The Wash	310	512	150	223	208	Aug	281	
Tay Estuary	132	167	461	310	(96)	Aug	268	
Lindisfarne	(260)	(100)		(4)	(80)	May	(260)	
Sites with mean peak counts of 200 or more birds in Northern Ireland†								
Dundrum Bay	166	296	722	264	173	Aug	324	
Belfast Lough	195	409	357	136	99	Sep	239	
Other sites surpassing table qualifying levels in Summer 2004 in Great Britain†								
Lavan Sands	147	(7)	25	170	250	Aug	148	
Exe Estuary	194	(116)	81	110	237	Jul	156	
Lade Sands				85	232	Apr	159	
Alt Estuary	47	19	41	178	219	Aug	101	
Tyninghame Estuary	26	180	110	23	203	Aug	108	

† *as no British or All-Ireland thresholds have been set a qualifying level of 200 has been chosen to select sites for presentation in this report*

Common Tern

Sterna hirundo

International threshold: 1,900
Great Britain threshold: ?†
All-Ireland threshold: ?†

GB max:	5,222	Aug
NI max:	10	Sep

The British peak was below that of the previous year by around 15% and in Northern Ireland this figure fell by around 75%. As with all tern species, however, the national totals are dependent on coverage, as the recording of terns is optional. As usual peak numbers were recorded in July and August, the Northern Irish peak was later than usual. The species was recorded at 264 British sites and two in Northern Ireland, compared to 325 and one in the previous year.

The top site remained the Humber Estuary, although mean peak numbers here have fallen below the threshold of international importance. Most sites' five year means remained similar to previous years, although increases were witnessed at the Tees and Thames Estuaries.

	2000	2001	2002	2003	2004	Mon	Mean
Sites with mean peak counts of 200 or more birds in Great Britain†							
Humber Estuary	2,165	(6)	(291)	280	(160)	Aug	1,223
Tees Estuary	877	(12)	696	1,678	1,251	Aug	1,126
Alt Estuary	1,292	129	868	1,664	1,135	Aug	1,018
Thames Estuary	(284)	(190)	(158)	(224)	(553)	Aug	(553)

	2000	2001	2002	2003	2004	Mon	Mean
North Norfolk Coast	(611)	(213)	(321)	419	476	Jul	502
Dee Estuary (England & Wales)	(246)	(3)	422	(384)	(180)	Aug	422
Forth Estuary	356	(40)	(691)	193	(183)	Aug	413
Southampton Water	(238)	(300)	(50)	(7)	(63)	Aug	(300)
Chichester Harbour	209	500 [13]	(167)	(314)	44	Jul	267
The Wash	262	(435)	(102)	122	199	Aug	255
Ythan Estuary			18	415			217
Other sites surpassing table qualifying levels in Summer 2004 in Great Britain†							
Blackwater Estuary	81	(55)	(126)	43	(223)	Jun	118

† *as no British or All-Ireland thresholds have been set a qualifying level of 200 has been chosen to select sites for presentation in this report*

Roseate Tern

Sterna dougallii

Scarce

GB max: 39 Aug
NI max: 0

The peak for 2004 fell in August and was chiefly made up of totals from sites in northeast England, with 32 from St Mary's Island, three at the Tees Estuary, and one each at the Tyne Estuary and East Chevington Pools. Other sites at which this species were recorded include the Forth (July and August), the Alt Estuary (August) and two at the Exe Estuary (May).

Arctic Tern

Sterna paradisaea

International threshold: ?
Great Britain threshold: ?†
All-Ireland threshold: ?†

GB max: 922 Jul
NI max: 0

In Britain Arctic Terns breed mostly in the north. Peak numbers are often recorded in Scotland, indeed only two sites outside of this region, the Farne Islands and Morecambe Bay, hold mean peak numbers above table qualifying levels. The British maximum in 2004 was the second lowest recorded during WeBS. This was reflected in the site totals, the highest single site total being the lowest since 2000. However, as with all terns recording is optional so peak numbers will largely be dependent on the sites at which the group is counted.

	2000	2001	2002	2003	2004	Mon	Mean
Sites with mean peak counts of 50 or more birds in Great Britain†							
Farne Islands		(600)	(0)		(0)		(600)
Ythan Estuary			106	860			483
Forth Estuary	76	2	(1,214)	197	(186)	Aug	335
Loch a` Phuill (Tiree)	0	473	477	150	120	May	244
Tay Estuary	80	32	660	290	0		212
Eden Estuary	220	(53)	125	320	4	Jun	167
Morecambe Bay	(103)		94	(178)	(59)	May	125
The Houb (Whalsay)	100	0	120	82	300	Jul	120
St Andrews Bay	170	44	29	(0)	192	Jul	109
Loch of Clumlie	250	150	0	0			100
Loch of Brow	200	100	0	0			75
Loch of Tankerness	75						75
Braewick Loch	45	70	170	50	30	Jul	73
Cambois to Newbiggin	0	5	246	0			63
Loch Indaal	61		51	76			63
Loch of Beith	5		150 [13]	31			62
Other sites surpassing table qualifying levels in Summer 2004 in Great Britain†							
Don Mouth to Ythan Mouth	34	0	0	0	146	Jul	36
Balta Sound	26	7	9	0	55	May	19

† *as no British or All-Ireland thresholds have been set a qualifying level of 50 has been chosen to select sites for presentation in this report*

Kingfisher

Alcedo atthis

International threshold: 2,000
Great Britain threshold: ?†
All-Ireland threshold: ?†

GB max:	**456**	**Nov**
NI max:	**5**	**Oct**

With the majority of the UK population occurring away from surveyed WeBS sites, Kingfishers are not well monitored by the scheme. The Great Britain peak fell slightly from the previous year's figure and is similar to that of 2002/03. Thirty sites held five-year mean peaks of five or more, three less than in the previous year. However, it should be noted that the table qualifying level has been raised from five to seven.

Despite the often-elusive nature of Kingfishers, the key sites do tend to remain the same from year to year. The Somerset Levels remained at the top of the list, with the count of 20 being the highest for a single site ever recorded by WeBS. High counts at Ditchford Gravel Pits during the past two years have helped to elevate the mean peak of the site. Many of the sites with the highest numbers of Kingfishers are large gravel pit complexes, allowing a large area of often tree-fringed shorelines along which the species can forage.

	00/01	01/02	02/03	03/04	04/05	Mon	Mean
Sites with mean peak counts of 7 or more birds in Great Britain†							
Somerset Levels	(10)	16	(14)	(12)	20	Nov	18
Wraysbury Gravel Pits	8	14	19	12	18	Dec	14
Ditchford Gravel Pits	8	7		13	12	Aug	10
Pitsford Reservoir	5	11	9	11	(3)	Sep	9
Colne Valley Gravel Pits	(11)	(3)	(4)	(4)	5	Sep	8
Eversley Cross & Yateley GPs	6	11	8	10	6	Oct	8
Lee Valley Gravel Pits	(6)	12	4	10	6	Sep	8
Middle Tame Valley Gravel Pits	(8)	(3)	(5)	(4)	(3)	Oct	(8)
Deben Estuary	7	5	10	(7)	4	Nov	7
Lower Derwent Ings			11	3			7
Old Moor	8	4	(8)	7			7
Southampton Water	(8)	5	(5)	6	9	Dec	7
Stour Estuary	4	6	(11)	6	(6)	Oct	7
Thames Estuary	(6)	3	9	(7)	7	Dec	7
Other sites surpassing table qualifying levels in WeBS-Year 2004/2005 in Great Britain†							
Chichester Gravel Pits	5	3	4	7	9	Dec	6
North Norfolk Coast	3	4 [11]	7	(6)	8	Oct	6

† *as no British or All-Ireland thresholds have been set a qualifying level of 7 has been chosen to select sites for presentation in this report*

Kingfisher (Dawn Balmer)

PRINCIPAL SITES

Table 6 below lists the principal sites for non-breeding waterbirds in the UK as monitored by WeBS. All sites supporting more than 10,000 waterbirds are listed, as are all sites supporting internationally important numbers of one or more waterbird species. Naturalised species (*e.g.* Canada Goose and Ruddy Duck) and non-native species presumed to have escaped from captive collections have been excluded from the totals, as have gulls and terns since the recording of these species is optional (see *Analysis*). Table 7 lists other sites holding internationally important numbers of waterbirds, which are not routinely monitored by standard WeBS surveys but rather by the Icelandic Goose Census and aerial surveys.

A total of 220 sites are listed in tables 6 and 7. Of these 208 supported one or more species in internationally important numbers and 84 held a five-year mean peak of 10,000 or more birds. Typically there are few changes to the top twenty sites listed in the principal sites table, with the order of the top ten rarely changing. The Wash remains as the key waterbird site with regard to numbers and in 2004/05 held the highest numbers of the preceding five years. Numbers on the Ribble Estuary again exceeded those on Morecambe Bay and for the first time since 1999/2000 the five-year mean surpassed that of Morecambe Bay to become the second most important site in the UK in terms of numbers of waterbirds. Numbers on the Thames Estuary rose on the previous year, and combined with lower than average figures from the Humber Estuary, the Thames Estuary climbs back into the top five sites. Total numbers on the Mersey Estuary continue to fall and are now the lowest for over ten years. Following a low total during 2003/04 current numbers at the Somerset Levels were similar to the average of the past five-years. The recent decline in the numbers of diving ducks at Loughs Neagh and Beg that have contributed to the fall in numbers at the site seem not to be as severe as first thought. Scaup were again present at the site in internationally important numbers and total numbers at the site remained similar to those of the past few years.

Five-year averages of sites holding 100,000 or more waterbirds were relatively similar compared to the previous year, with 74 of the 84 sites undergoing changes of less than 10%. Six sites witnessed an increase and four sites a decrease of 10% or more. The greatest increase was experienced at Loch Spynie (35%). The greatest decreases were seen at Hule Moss (17%) and the Medway Estuary (15%).

Table 6. Total number of waterbirds at principal sites in the UK, 2000/2001 to 2004/05 (includes data from all available sources) and species occurring in internationally important numbers at each. (Species codes for those listed are provided in Table 8.)

Site	00/01	01/02	02/03	03/04	04/05	Average	Int.Imp.species
The Wash	291,311	332,637	343,462	338,379	369,518	335,061	PG DB SU PT OC RP GP GV L. KN SS DN BW BA CU RK TT
Ribble Estuary	220,066	210,305	255,011	252,372	245,093	236,569	WS PG SU WN T. PT OC RP GV L. KN SS DN BW BA RK
Morecambe Bay	246,841	211,417	250,768	249,247	203,855	232,426	PG SU PT OC KN DN BW BA CU RK
North Norfolk Coast	182,596	185,125	212,438	284,591	234,310	219,812	PG DB WN T. PT RP KN BW BA RK
Thames Estuary	182,103	169,953	197,458	160,179	172,352	176,409	DB GA T. SV OC AV RP GV KN DN BW BA RK
Humber Estuary	163,061	159,693	174,920	217,792	163,055	175,704	PG SU RP GP GV L. KN DN BW BA RK
Dee Estuary (England and Wales)	135,078	195,680	127,007	171,906	115,259	148,986	SU T. PT OC KN DN BW BA CU RK
Solway Estuary	150,642	124,127	153,364	145,052	140,961	142,829	WS YS SU PT KN BA
Somerset Levels	114,207	115,001	102,741	85,154	103,894	104,199	MS WN GA T. PT SV L.
Mersey Estuary	109,877	102,666	108,736	97,784	85,554	100,923	SU T. DN BW RK
Forth Estuary	95,622	88,204	109,401	91,992	83,304	93,705	PG SU SZ KN BA RK
Swale Estuary	81,239	88,524	86,101	86,965	73,828	83,331	WN T. PT SV RP GP BW RK
Blackwater Estuary	122,134	66,473	81,684	64,533	78,182	82,601	DB GP GV KN DN BW RK
Breydon Wtr/Berney Mshs	70,962	78,423	64,804	75,823	110,614	80,125	PG WN T. SV AV GP BW RK
Strangford Lough	66,164	80,345	79,352	88,429	78,610	78,580	MS QN SU GP KN BA RK
Ouse Washes	48,585	76,165	66,409	85,698	87,945	72,960	MS BS WS WN GA T. PT SV PO BW
Alt Estuary	58,651	90,066	63,501	72,792	50,073	67,017	GV KN SS BA

Site	00/01	01/02	02/03	03/04	04/05	Average	Int.Imp.species
Loch of Strathbeg	55,470	68,134	49,356	79,231	81,614	66,761	WS PG YS
Loughs Neagh and Beg	101,142	63,479	53,401	59,026	56,261	66,662	MS WS PO TU SP GN CA
Inner Moray / Inverness Firth	56,580	59,615	60,386	80,199	65,877	64,531	PG JI RK
Severn Estuary	62,768	60,723	68,649	64,668	63,861	64,134	SU T. PT DN RK
Lindisfarne	61,271	63,463	64,138	56,842	53,325	59,808	PG YS QS WN BA RK
Stour Estuary	59,566	59,772	50,227	48,642	46,852	53,012	RP GV KN BW RK
Montrose Basin	49,767	63,742	37,019	35,461	50,147	47,227	PG RK
Chichester Harbour	47,412	51,567	44,824	43,732	43,369	46,181	DB DN BW RK
Burry Inlet	39,105	41,432	43,830	52,852	49,185	45,281	PT OC
Dengie Flats	48,665	55,927	39,325	23,895	40,229	41,608	GV KN BA
Hamford Water	42,671	44,224	40,135	37,990	39,938	40,992	DB RP GV BW RK
Langstone Harbour	42,703	33,738	37,453	43,773	45,637	40,661	DB DN BW
Carmarthen Bay	34,175	31,358	36,917	47,069	42,392	38,382	GP
Cromarty Firth	39,151	36,643	26,278	41,307	37,956	36,267	JI BA RK
Loch Leven	31,865	38,128	39,588	37,334	33,773	36,138	MS PG T. SV
Dornoch Firth	35,954	35,227	37,151	35,851	35,087	35,854	WN
WWT Martin Mere	22,408	37,479	39,104	30,880	45,272	35,029	WS PG T.
Lough Foyle	37,207	29,448	34,153	37,291	33,078	34,235	WS QN BA RK
West Water Reservoir	26,500	23,276	40,000	34,210	.	30,997	PG
Nene Washes	32,179	19,344	52,880	20,910	29,269	30,916	BS PT
Alde Complex	26,534	29,355	29,649	22,907	31,838	28,057	AV BW RK
Medway Estuary	23,452	29,829	27,329	26,174	27,066	26,770	PT BW
Duddon Estuary	23,559	25,065	22,185	32,592	29,117	26,504	PT RK
Poole Harbour	24,474	24,637	25,953	24,857	26,362	25,257	AV BW
Abberton Reservoir	25,920	20,620	20,675	31,375	24,021	24,522	SV PO
Rutland Water	19,182	22,140	26,169	28,216	26,165	24,374	MS GA SV
Tees Estuary	23,194	22,192	25,350	30,079	20,859	24,335	RK
Crouch-Roach Estuary	24,729	21,024	21,376	18,218	34,219	23,913	DB BW RK
Inner Firth of Clyde	23,697	22,904	23,200	23,740	19,838	22,676	RK
Orwell Estuary	20,044	21,530	24,842	25,284	19,981	22,336	BW RK
Exe Estuary	21,162	24,754	20,589	22,893	20,093	21,898	BW
Tay Estuary	21,432	25,699	20,724	21,303	19,847	21,801	PG JI BA RK
Colne Estuary	36,322	25,745	4,183	19,334	18,288	20,774	DB BW
Lavan Sands	19,060	17,282	21,779	21,062	22,025	20,242	RK
Cleddau Estuary	15,548	18,585	17,879	20,487	27,752	20,050	
Belfast Lough	21,042	18,841	18,789	19,436	20,080	19,638	BW RK
Loch Spynie	15,998	15,467	19,524	15,748	30,736	19,495	PG JI
Pegwell Bay	17,453	21,463	28,815	25,509	3,720	19,392	
Deben Estuary	18,877	16,970	17,082	17,873	19,047	17,970	RK
Carsebreck / Rhynd Lochs	23,454	19,941	15,252	16,529	12,262	17,488	PG
Loch of Skene	25,163	17,136	11,004	13,693	18,833	17,166	PG JI
Southampton Water	21,490	16,875	16,684	15,440	15,037	17,105	BW
Wigtown Bay	9,583	18,028	14,473	21,604	19,074	16,552	PG YS
Fleet and Wey	14,126	19,703	14,552	16,289	17,241	16,382	MS
Blyth Estuary	18,876	13,713	.	.	.	16,295	BW RK
Pagham Harbour	16,057	14,312	13,217	14,565	20,506	15,731	DB BW
Eden Estuary	15,441	15,071	15,124	15,366	14,110	15,022	
Mersehead RSPB	14,508	13,464	15,666	11,117	15,251	14,001	YS PT
WWT Caerlaverock	7,952	11,815	19,416	14,144	16,132	13,892	WS YS
Walland Marsh	9,713	21,016	5,111	5,951	23,639	13,086	
Taw-Torridge Estuary	12,240	14,195	11,847	10,314	16,374	12,994	
North West Solent	10,847	12,306	10,071	15,136	16,145	12,901	BW
Arun Valley	17,399	10,719	15,686	11,911	8,653	12,874	PT
Middle Yare Marshes	10,575	13,853	10,686	9,741	17,520	12,475	
Dungeness Gravel Pits	14,750	11,072	13,884	11,665	10,974	12,469	
Old Moor	13,251	12,091	12,437	11,855	.	12,409	
Rye Harbour / Pett Level	10,156	11,452	10,846	16,911	9,710	11,815	
Loch of Harray	17,058	9,129	10,010	12,330	8,501	11,406	MS JI
Hule Moss	15,245	9,727	7,110	15,858	8,994	11,387	PG
Beaulieu Estuary	16,975	11,774	11,147	6,140	10,775	11,362	BW
Dyfi Estuary	10,952	9,419	11,608	11,646	12,405	11,206	
Portsmouth Harbour	6,036	8,190	15,004	16,420	9,651	11,060	
Cotswold Water Park (W)	11,544	10,032	9,173	12,476	10,178	10,681	
Thanet Coast	9,956	9,640	16,146	7,964	8,412	10,424	TT
Loch of Lintrathen	4,537	10,476	9,241	16,418	11,068	10,348	PG
R. Avon: R'wood/ C'church	13,817	5,035	24,594	4,766	2,795	10,201	PT BW
Carlingford Lough	8,431	9,763	10,762	10,516	11,235	10,141	QN RK
Lower Derwent Ings	113	92	24,458	23,427	139	9,646	PT
Outer Ards Shoreline	11,647	210	12,694	12,748	9,552	9,370	QN TT
Cameron Reservoir	6,454	18,249	4,495	11,225	4,825	9,050	PG
Chew Valley Lake	8,189	8,775	8,369	10,109	9,261	8,941	SV

Site	00/01	01/02	02/03	03/04	04/05	Average	Int.Imp.species
Lee Valley Gravel Pits	8,118	9,579	9,312	8,989	8,394	8,878	GA
Upper Lough Erne	7,800	7,717	8,777	9,239	9,369	8,580	MS WS
Dundrum Bay	10,016	9,257	6,859	6,959	8,311	8,280	QN
Loch Fleet Complex	8,986	11,677	8,275	6,489	5,177	8,121	JI
R. Clyde: Carstairs/ Thankertn	8,043	14,501	4,683	6,623	4,510	7,672	PG
R. Nith: Keltonbnk/Nunholm	7,883	7,487	6,665	10,005	5,973	7,603	PG YS
R. Avon: F'bridge- R'wood	7,416	7,420	10,022	6,781	6,094	7,547	GA
Hornsea Mere	6,047	5,737	8,926	7,347	6,906	6,993	MS
Loch of Stenness	5,553	9,130	7,099	5,636	5,072	6,498	JI
Holburn Moss	6,066	1,774	6,375	10,501	5,398	6,023	PG
Loch of Boardhouse	5,620	5,096	5,257	7,111	5,907	5,798	JI
Horsey Mere	3,620	5,000	4,000	8,465	7,231	5,663	PG
Orchardton/Auchencairn Bays	3,107	.	5,563	8,146	3,070	4,972	YS
Larne Lough	4,674	4,462	5,043	5,299	4,989	4,893	QN
R. Eden: Warcop/Little Salkeld	4,760	.	.	.	.	4,760	JI
Slains Lochs	2,093	2,832	360	575	17,215	4,615	PG
Wraysbury Gravel Pits	4,256	4,970	6,262	3,361	3,716	4,513	GA
Loch Eye	2,205	5,211	1,926	4,474	8,354	4,434	JI
Lake of Menteith	5,327	732	4,958	4,639	6,462	4,424	PG
Kilconquhar Loch	3,777	3,285	3,941	5,870	2,891	3,953	JI
Tweed Estuary	3,926	3,806	3,516	4,180	3,501	3,786	MS
Loch Ken	3,905	2,887	4,216	4,030	2,301	3,468	JI
Dee Flood Meadows	4,130	4,603	3,888	2,859	1,551	3,406	PT
Lochs Davan and Kinord	5,327	5,568	3,124	1,336	881	3,247	JI
Loch Watten	1,596	5,356	3,023	2,199	3,102	3,055	JI
Loch Bee (South Uist)	3,944	1,416	3,048	2,672	4,037	3,023	JH
Lower Lough Erne	.	.	2,931	2,485	3,342	2,919	MS
Dornoch Firth	2,050	6,571	1,955	1,965	1,236	2,755	WN
Loch of Wester	1,129	5,776	.	.	1,129	2,678	WS JI
Lower Teviot Valley	4,828	1,134	2,780	.	1,621	2,591	JI
Heigham Holmes	.	2,500	.	.	.	2,500	PG
Balranald (RSPB Reserve) (North Uist)	1,328	3,445	2,792	2,136	2,739	2,488	JH
Loch a` Phuill (Tiree)	1,551	2,170	2,643	3,075	1,271	2,142	JH
Killough Harbour	.	1,832	2,731	1,158	2,736	2,114	QN
Loch Garten	2,700	2,804	1,000	1,133	2,417	2,011	JI
Loch Paible (North Uist)	1,626	1,169	2,253	2,609	2,287	1,989	JH
Haddo House Lakes	1,104	1,639	1,675	1,816	.	1,559	JI
Melbost Sands, Broad Bay & Tong Saltings (Lewis)	1,542	1,729	.	394	1,804	1,367	JH
Loch of Tankerness	5,680	4	1	1	1	1,137	JI
Loch Mor Baleshare	792	686	789	1,354	1,014	927	JH
Loch Bhasapoll (Tiree)	1,090	1,229	777	1,136	190	884	JH
Loch Sandary (North Uist)	659	1,188	767	1,071	652	867	JH
Loch Riaghain (Tiree)	1,016	408	647	604	306	596	JH
Rispond Bay	.	.	.	.	581	581	YN
Loch Ashie	481	.	.	.	.	481	SZ
Loch An Eilein (Tiree)	708	321	570	492	224	463	JH
Branahuie Saltings	.	324	.	.	.	324	JH
Kentra Moss & Lower Loch Shiel (Moidart)	252	207	216	277	340	258	JH
Loch Broom	367	.	.	.	42	205	JH
Loch Urrahag	203	27	.	.	87	106	JH
Balranald Nature Reserve	18	44	23	31	112	46	JH

Table 7. Other sites in the UK holding internationally important numbers of waterbirds in 2004/05, which are not routinely monitored by standard WeBS surveys. (Species codes for those listed are provided in Table 8.)

Site	Int.Imp.species
Aberlady Bay Roost	PG
Beauly Firth Roost	JI
Benbecula	JH
Bridge of Crathies	WS
Bute	JI
Caithness Lochs	JI
Coll and Tiree offshore	ND
Dingwall Bay	JI
Dupplin Lochs	PG
East Mainland	JI
East Mains Flood	JI
Fala Flow	PG
Findhorn Bay Roost	PG
Hule Moss (West)	PG
Island of Islay	NW YN
Islands of Shapinsay	JI
Isle Of Oronsay	YN
Isle of Coll	NW JH YN
Isle of Colonsay	NW JH
Isle of South Ronaldsay	JI
Keills Peninsula and Isle of Danna	NW
Loans of Tullich	WS
Loch Eye and Cromarty Firth	WS PG JI
Loch Slapin	ND
Loch Tullybelton	PG
Lochhill	PG
Machrihanish	NW
Martin Mere and Ribble Estuary	WS
Moray Firth	ND SZ
Munlochy Bay Roost	JI
North Norfolk Coast & The Wash	PG
Holbeach St Matthew Roost	PG
Holkham Bay Roost	PG
Scolt Head Roost	PG
Snettisham Roost	PG
Thornham Roost	PG
Wells-next-the-Sea	PG
North Sutherland	YN
North Uist	JH YN
Baleshare and Carinish (Grimsay)	JH
Balmartin To Vallay	JH
Balranald Clettraval and Tigharry	JH
Berneray	JH YN
Boreray and Lingay	JH YN
Clachan Na Luib to Bayhead	JH
Malaclate To Grenitote	JH
Oronsay	JH
Paible	JH
Trumisgarry Clachan and Newton	JH
Trumisgarry to Newton	JH
Orkney	JI YN ND
Deer and Shapinsay Sounds	ND
South Walls (Hoy)	YN
Pilling to Cockerham	PG
Read`s Island Flats	PG
Rhunahaorine	NW
Sanday	JI
Scapa Flow	ND
Scarp to Vatersay offshore	ND
Severn Hams	PT
Skinflats Roost	PG
Sound of Barra (Barra)	YN ND
Sound of Gigha	ND
Sound of Harris (NW) (Harris)	YN
Sound of Taransay (Harris)	SZ
South Uist	JH YN ND
Askernish To Smerclate	JH
Bornish To Askernish	JH
Drimore To Howmore	JH
Howbeg To Bornish	JH
Lochdar, Gerinish To Drimore	JH
West Coast	ND
Southwest Lancashire	PG
Stranraer Lochs	NW
Strathearn (West)	PG JI
Stronsay (Whole Island)	JI
Tay and Isla Valley	PG JI
Tiree	NW JH YN TT
West Mainland	JI
Whiteness to Skelda Ness	SZ
Wyre Estuary	PG

Table 8. Species codes for species listed in tables 6, 7 and 10.

AV	Avocet
BA	Bar-tailed Godwit
BS	Bewick's Swan
BW	Black-tailed Godwit
CA	Cormorant
CU	Curlew
DB	Dark-bellied Brent Goose
DN	Dunlin
GA	Gadwall
GN	Goldeneye
GP	Golden Plover
GV	Grey Plover
JH	Greylag Goose (Northwest Scotland population)
JI	Greylag Goose (Icelandic population)
KN	Knot
L.	Lapwing
MS	Mute Swan
ND	Great Northern Diver
NW	Greenland White-fronted Goose
OC	Oystercatcher
PG	Pink-footed Goose
PO	Pochard
PT	Pintail
QN	Light-bellied Brent Goose (Nearctic population)
QS	Light-bellied Brent Goose (Svalbard population)
RK	Redshank
RP	Ringed Plover
SP	Scaup
SS	Sanderling
SU	Shelduck
SV	Shoveler
SZ	Slavonian Grebe
T.	Teal
TT	Turnstone
TU	Tufted Duck
WN	Wigeon
WS	Whooper Swan
YN	Barnacle Goose (Nearctic population)
YS	Barnacle Goose (Svalbard population)

WeBS Low Tide Counts

AIMS

Estuarine sites in the UK provide the most important habitat for non-breeding waterbirds, acting as wintering grounds for many migrants but also as stopover feeding locations for other waterbirds passing along the East Atlantic Flyway. Core Counts on estuaries tend to quantify birds present at high tide roosts. Although important, knowledge of roost sites provides only part of the picture, and does not elucidate the use that waterbirds make of a site for feeding.

The WeBS Low Tide Counts scheme has flourished since its inception in the winter of 1992/93, with most of the major estuaries covered. The scheme aims principally to monitor, assess and regularly update information on the relative importance of intertidal feeding areas of UK estuaries for wintering waterbirds and thus to complement the information gathered by WeBS Core Counts.

The data gathered contribute greatly to the conservation of waterbirds by providing supporting information for the establishment and management of UK Ramsar sites and Special Protection Areas (SPAs), other site designations and whole estuary conservation plans. In addition, WeBS Low Tide Counts enhance our knowledge of the low water distribution of waterbirds and provide data that highlight regional variations in habitat use, whilst also informing protection of the important foraging areas identified. WeBS Low Tide Counts provide valuable information needed to gauge the potential effects on waterbirds of a variety of human activities which affect the extent or value of intertidal habitats, such as proposals for dock developments, recreational activities, tidal power barrages, marinas and housing schemes. Designing mitigation or compensation for such activities can be assisted using data collected under the scheme. Furthermore, the effects on bird distributions of climate change and sea level rise can be assessed.

METHODS

The scheme provides information on the numbers of waterbirds feeding on subdivisions of the intertidal habitat within estuaries. Given the extra work that Low Tide Counts entail, often by the same counters that carry out the Core Counts, WeBS aims to cover most individual estuaries about once every six years, although on some sites more frequent counts are made. Coordinated counts of waterbirds are made by volunteers each month between November and February on pre-established subdivisions of the intertidal habitat in the period two hours either side of low tide.

DATA PRESENTATION

Tabulated statistics

Table 10 presents three statistics for 18 of the more numerous waterbird species present on 18 estuaries covered during the 2004/05 winter: the peak number of a species over the whole site counted in any one month (with checks for count synchronicity made from assessing proximity of count dates and consultation with Local Organisers); an estimate of the mean number present over the winter for the whole site (obtained by summing the mean counts of each species for each count section) and the mean density over the site (in birds per hectare), which is the mean number divided by the total area surveyed (in hectares). The area value used for these calculations is the sum of the intertidal and nontidal components of each count section but omits the subtidal areas (*i.e.* those parts of the count section which are under water on a mean low tide).

Dot density maps

WeBS Low Tide Count data are presented as dot density maps, with subdivision of count sections into basic habitat elements. The reason for such a subdivision is to ensure species are plotted on appropriate habitat areas and to improve the accuracy of density estimates. Each section for which a count has been made is divided into a maximum of three different habitat components:

Intertidal: Areas that lie between mean high water and mean low water.

Subtidal: Areas that lie below mean low water. In more 'open-coast'-type situations, a subtidal zone reaching 500 m out from the intertidal sections has been created arbitrarily, to indicate the approximate extent of visibility offshore from land-based counts.

Nontidal: Areas that lie above mean high water (usually saltmarsh although some grazing marshes are also covered).

The mean count for the sector is then divided amongst a varying number of the different components, dependent on the usual habitat preferences of the species involved. For example, Dunlin dots are plotted exclusively on intertidal sections whereas Wigeon dots are spread across intertidal, subtidal and nontidal areas (in proportion to the relative areas of these three components).

Currently, throughout all WeBS Low Tide Count analyses, mean low tide and mean high tide are taken from the most recent Ordnance Survey 1:25000 maps (in Scotland, the lines on the OS maps are mean low water springs and mean high water springs instead). It is recognised, unfortunately, that these maps represent the current real shape of the mudflats, water channels and saltmarshes to varying degrees of accuracy. However, in the interests of uniformity across the UK, the Ordnance Survey outlines are adhered to throughout the analyses.

The maps display the average number of birds in each count section as dots spread randomly across habitat components of count sections, thus providing an indication of both numbers and density. **It is important to note that individual dots do not represent the precise position of individual birds; dots have been assigned to habitat components proportionally and are then randomly placed within those areas. No information about the distribution of birds at a finer scale than the count sector level should be inferred from the dot density maps.** For all maps in the present report, one dot is equivalent to one bird, except where stated. The size of individual dots has no relevance other than for clarity.

As most estuaries have now been covered more than once at low tide, density maps show the relative distributions of species in the winter of 2004/05 compared to an earlier winter of survey. It is hoped that comparative dot density distributions will lead to an easier and fuller appreciation of low tide estuarine waterbird distribution, and changes therein. The following colour conventions apply to density maps: red dots = 2004/05 winter; blue dots = earlier winter; pale blue = water; yellow = intertidal habitat (e.g. mudflat, sandflat); pale green = nontidal habitat (e.g. saltmarsh, reedbed); grey or brown = not covered in one survey winter; dark blue = sector never covered. More detailed information concerning analysis and presentation of WeBS Low Tide Counts can be obtained from Alex Banks, the National Organiser (WeBS Low Tide Counts), or from the publication *Estuarine Waterbirds at Low Tide* (Musgrove *et al.* 2003).

ESTUARY ACCOUNTS

The main estuaries counted at low tide in the winter of 2004/05 are discussed. WeBS Low Tide Counts were carried out on 21 different sites, with site accounts for 17 of these. Other counts on limited numbers of sectors in the winter of 2004/05 were made on Burry Inlet, Langstone Harbour, Swansea Bay and Morecambe Bay (the latter a mid-tide count). These sites are not included in site accounts, but data can be obtained from the WeBS Low Tide Count National Organiser upon request.

For the main site accounts, data were collected during the period November to February. Assessment of national and international importance is based on five-year peak mean counts from the main species accounts in this volume of *Waterbirds in the UK* (Table 9). Figure 57 shows the location of the sites discussed, and a site description is presented for each estuary. Distribution maps are presented for selected species, which are those of international or national importance, or are known to be undergoing site-level changes, where possible. General bird distribution is described for the winter of 2004/05, focusing on species held in important numbers at the site in question.

Table 9. Sites with Estuary Accounts and important bird numbers held. Numbers in parentheses refer to the location in Figure 57. For species codes see Table 8.

	International Importance	National Importance
Adur Estuary *(7)*	*None*	*None*
Belfast Lough *(19)*	BW, RK	SU, SP, E., GN, RM, RH, BV, GG, OC, RP, PS, DN, TT
Breydon Water *(2)*	PG, WN, T., SV, GP, BW, RK	BS, EW, PT, AV, RU
Camel Estuary *(13)*	*None*	GK
Crouch & Roach Estuaries *(5)*	DB, BW, RK	SU, PT, AV, GP, RU
Duddon Estuary *(17)*	PT, RK	RM, OC, SS, DN, CU
Fal Complex *(12)*	*None*	GK
Fowey Estuary *(11)*	*None*	*None*
Lindisfarne *(1)*	PG, YS, QS, WN, BA, RK	WS, SU, PT, E., RX, SZ, GP, GV, KN, SS, DN, CU
Medway Estuary *(6)*	PT, BW, RK	DB, SU, CA, AV, RP, GV, DN
North West Solent *(9)*	BW	DB, GK
Orwell Estuary *(3)*	RK	DB, GA, PT, BW
Poole Harbour *(10)*	AV, BW	DB, SU, T., PT, RM, GG, CA, DN, CU, GK
Rough Firth *(20)*	*None*	SP, GK
Solway Firth *(21)*	WS, PG, YS, SU, PT, OC, KN, DN, BA, CU, RK	WN, T., SP, CA, RP, GP, GV, SS, GK, TT
Stour Estuary *(4)*	GV, KN, DN, BW, RK	DB, SU, PT, GN, GG, GP, TT
Strangford Lough *(18)*	MS, QN, SU, GP, KN, BA, RK	WS, WN, T., MA, PT, SV, E., GN, RM, RH, GG, CO, OC, RP, GV, L., DN, BW, CU, GK

Figure 57. Map showing estuaries counted at low tide in the winter of 2004/05. 1: Lindisfarne; 2: Breydon Water; 3: Orwell Estuary; 4: Stour Estuary; 5: Crouch & Roach Estuaries; 6: Medway Estuary; 7: Adur Estuary; 8: Langstone Harbour; 9: North West Solent; 10: Poole Harbour; 11: Fowey Estuary; 12: Fal Complex; 13: Camel Estuary; 14: Swansea Bay; 15: Burry Inlet; 16: Morecambe Bay (west, mid-tide counts); 17: Duddon Estuary; 18: Strangford Lough; 19: Belfast Lough; 20: Rough Firth; 21: Solway Firth.

Table 10. Peak and mean counts, and mean density (birds per ha) of 18 waterbird species across 18 estuaries covered by the 2004/05 WeBS Low Tide Counts. Orwell Estuary displayed by Stour Estuary. '+' indicates non-zero densities of <0.01 birds per ha.

	Belfast Lough			**Breydon Water**			**Camel Estuary**		
	Peak	*Mean*	*Mean*	*Peak*	*Mean*	*Mean*	*Peak*	*Mean*	*Mean*
Species	*No.*	*No.*	*Dns.*	*No.*	*No.*	*Dns.*	*No.*	*No.*	*Dns.*
Brent Goose	61	39	0.08	0	0	0	5	2	+
Shelduck	489	361	0.73	89	47	0.12	161	86	0.16
Wigeon	220	132	0.27	14,325	7,753	19.29	253	191	0.37
Teal	436	312	0.63	150	76	0.19	11	4	0.01
Mallard	371	268	0.54	292	117	0.29	42	30	0.06
Pintail	0	0	0	115	76	0.19	1	0	+
Oystercatcher	3,909	3,565	7.25	62	31	0.08	260	195	0.37
Ringed Plover	109	94	0.19	30	10	0.02	6	2	+
Golden Plover	28	13	0.03	14,990	4,299	10.69	4,750	1,591	3.04
Grey Plover	0	0	0	150	43	0.11	2	1	+
Lapwing	1,439	1,005	2.04	12,732	4,330	10.77	5,511	2,527	4.83
Knot	155	52	0.11	295	96	0.24	0	0	0
Dunlin	1,136	919	1.87	6,190	3,628	9.02	473	309	0.59
Black-tailed Godwit	392	337	0.68	1,134	625	1.55	4	1	+
Bar-tailed Godwit	85	42	0.09	55	16	0.04	2	1	+
Curlew	536	462	0.94	710	343	0.85	220	187	0.36
Redshank	1,547	1,319	2.68	2,181	1,257	3.13	74	65	0.12
Turnstone	272	214	0.43	2	1	+	58	38	0.07

	Crouch & Roach Estuaries			**Duddon Estuary**			**Fal Complex**		
	Peak	*Mean*	*Mean*	*Peak*	*Mean*	*Mean*	*Peak*	*Mean*	*Mean*
Species	*No.*	*No.*	*Dns.*	*No.*	*No.*	*Dns.*	*No.*	*No.*	*Dns.*
Brent Goose	4,636	2,140	1.23	0	0	0	0	0	0
Shelduck	1,661	1,190	0.68	599	714	0.21	154	129	0.41
Wigeon	2,715	2,064	1.18	1,212	838	0.24	55	20	0.06
Teal	2,981	2,122	1.22	28	28	0.01	30	17	0.05
Mallard	181	101	0.06	183	236	0.07	85	80	0.26
Pintail	130	44	0.03	1,626	1,096	0.32	0	0	0
Oystercatcher	379	209	0.12	2,289	2,199	0.64	61	47	0.15
Ringed Plover	193	102	0.06	350	209	0.06	2	1	+
Golden Plover	4,771	2,724	1.56	0	0	0	0	0	0
Grey Plover	306	239	0.14	6	4	+	0	0	0
Lapwing	11,288	8,249	4.73	1,196	980	0.28	133	66	0.21
Knot	330	143	0.08	732	376	0.11	0	0	0
Dunlin	3,364	2,926	1.68	6,970	6,627	1.93	231	141	0.45
Black-tailed Godwit	729	589	0.34	69	29	0.01	114	106	0.34
Bar-tailed Godwit	129	61	0.03	7	2	+	0	0	0
Curlew	498	411	0.24	512	679	0.2	305	245	0.78
Redshank	3,299	2,590	1.48	728	877	0.26	213	180	0.58
Turnstone	39	26	0.01	34	27	0.01	8	3	0.01

	Fowey Estuary			**Langstone Harbour**			**Lindisfarne**		
	Peak	*Mean*	*Mean*	*Peak*	*Mean*	*Mean*	*Peak*	*Mean*	*Mean*
Species	*No.*	*No.*	*Dns.*	*No.*	*No.*	*Dns.*	*No.*	*No.*	*Dns.*
Brent Goose	0	0	0	1,637	1,445	3.58	2,623	2,525	0.85
Shelduck	36	31	0.3	201	142	0.35	1,773	1,773	0.6
Wigeon	0	0	0	1,253	619	1.53	2,412	2,412	0.81
Teal	0	0	0	50	32	0.08	814	814	0.27
Mallard	137	101	0.98	15	12	0.03	658	658	0.22
Pintail	0	0	0	27	19	0.05	132	132	0.04
Oystercatcher	2	1	0.01	814	678	1.68	941	941	0.32
Ringed Plover	0	0	0	112	80	0.2	67	67	0.02
Golden Plover	0	0	0	12	3	0.01	3,384	3,384	1.14
Grey Plover	0	0	0	242	214	0.53	518	518	0.17
Lapwing	0	0	0	299	253	0.63	1,795	1,795	0.61
Knot	0	0	0	32	18	0.04	1,519	1,519	0.51
Dunlin	0	0	0	4,185	3,503	8.67	3,245	3,245	1.09
Black-tailed Godwit	0	0	0	61	57	0.14	0	0	0
Bar-tailed Godwit	0	0	0	175	75	0.19	859	859	0.29
Curlew	64	54	0.52	325	255	0.63	1,715	1,715	0.58
Redshank	10	8	0.08	330	272	0.67	968	968	0.33
Turnstone	0	0	0	138	121	0.3	157	157	0.05

	Medway Estuary			Morecambe Bay			North West Solent		
Species	*Peak No.*	*Mean No.*	*Mean Dns.*	*Peak No.*	*Mean No.*	*Mean Dns.*	*Peak No.*	*Mean No.*	*Mean Dns.*
Brent Goose	1,834	1,059	0.41	77	46	0.02	1,291	965	1.28
Shelduck	2,360	1,555	0.61	2,030	1,428	0.7	136	97	0.13
Wigeon	2,250	1,737	0.68	884	743	0.36	571	451	0.6
Teal	1,146	847	0.33	36	12	0.01	258	186	0.25
Mallard	152	112	0.04	103	45	0.02	30	22	0.03
Pintail	812	456	0.18	743	428	0.21	204	88	0.12
Oystercatcher	854	713	0.28	31,394	21,527	10.55	193	135	0.18
Ringed Plover	332	157	0.06	37	25	0.01	165	91	0.12
Golden Plover	6	2	+	952	573	0.28	914	229	0.3
Grey Plover	453	302	0.12	82	33	0.02	208	166	0.22
Lapwing	3,442	1,721	0.67	709	488	0.24	204	109	0.14
Knot	3,024	1,362	0.53	37,300	24,127	11.83	282	95	0.13
Dunlin	9,373	7,374	2.89	17,497	8,484	4.16	5,082	4,211	5.59
Black-tailed Godwit	277	174	0.07	53	14	0.01	204	88	0.12
Bar-tailed Godwit	306	119	0.05	285	74	0.04	14	5	0.01
Curlew	408	320	0.13	2,095	1,501	0.74	206	160	0.21
Redshank	1,068	917	0.36	1,122	859	0.42	224	172	0.23
Turnstone	73	67	0.03	79	44	0.02	138	72	0.1

	Poole Harbour			Rough Firth			Solway Firth		
Species	*Peak No.*	*Mean No.*	*Mean Dns.*	*Peak No.*	*Mean No.*	*Mean Dns.*	*Peak No.*	*Mean No.*	*Mean Dns.*
Brent Goose	720	505	0.32	0	0	0	1	0	+
Shelduck	1,439	1,065	0.67	122	89	0.17	420	298	0.05
Wigeon	1,517	1,224	0.78	852	516	0.96	1,736	1,196	0.22
Teal	1,793	1,299	0.82	0	0	0	594	273	0.05
Mallard	71	62	0.04	90	53	0.1	577	512	0.09
Pintail	123	108	0.07	2	1	+	354	212	0.04
Oystercatcher	1,360	1,303	0.83	548	386	0.72	2,192	2,016	0.37
Ringed Plover	16	6	+	29	12	0.02	63	38	0.01
Golden Plover	2	1	+	0	0	0	6,145	4,062	0.74
Grey Plover	152	71	0.04	0	0	0	602	409	0.07
Lapwing	1,505	772	0.49	308	111	0.21	3,780	2,592	0.47
Knot	4	1	+	0	0	0	5,123	3,017	0.55
Dunlin	6,551	3,449	2.19	0	0	0	9,629	7,507	1.37
Black-tailed Godwit	1,684	1,385	0.88	0	0	0	121	40	0.01
Bar-tailed Godwit	78	52	0.03	0	0	0	112	53	0.01
Curlew	2,374	1,968	1.25	238	136	0.25	2,391	1,705	0.31
Redshank	592	493	0.31	201	142	0.26	856	681	0.12
Turnstone	26	17	0.01	0	0	0	50	41	0.01

	Stour Estuary			Orwell Estuary			Strangford Lough		
Species	*Peak No.*	*Mean No.*	*Mean Dns.*	*Peak No.*	*Mean No.*	*Mean Dns.*	*Peak No.*	*Mean No.*	*Mean Dns.*
Brent Goose	1,533	863	0.53	3,097	1,407	1.13	3,231	2,501	0.58
Shelduck	1,604	1,552	0.95	575	415	0.33	5,709	4,349	1
Wigeon	3,229	2,983	1.82	1,920	1,540	1.23	1,439	665	0.15
Teal	1,523	1,288	0.79	665	559	0.45	808	862	0.2
Mallard	408	296	0.18	438	367	0.29	593	450	0.1
Pintail	441	337	0.21	165	128	0.1	540	484	0.11
Oystercatcher	1,500	1,284	0.78	1,722	1,529	1.22	5,629	5,445	1.25
Ringed Plover	211	175	0.11	160	125	0.1	270	164	0.04
Golden Plover	2,104	1,858	1.13	200	155	0.12	12,310	7,641	1.76
Grey Plover	1,750	1,634	1	350	271	0.22	61	48	0.01
Lapwing	8,929	4,404	2.69	2,719	2,258	1.81	6,395	4,425	1.02
Knot	7,764	5,402	3.3	2,115	1,522	1.22	12,306	5,617	1.29
Dunlin	12,203	11,558	7.05	3,856	3,136	2.51	5,659	4,442	1.02
Black-tailed Godwit	1,756	1,048	0.64	255	191	0.15	151	101	0.02
Bar-tailed Godwit	103	69	0.04	14	10	0.01	904	632	0.15
Curlew	901	827	0.5	831	747	0.6	1,344	1,178	0.27
Redshank	1,889	1,745	1.06	1,562	1,473	1.18	2,651	3,208	0.74
Turnstone	589	495	0.3	240	212	0.17	128	83	0.02

ADUR ESTUARY

Site description

The Adur Estuary is a small and narrow embanked estuary, roughly in the middle of the conurbation from Brighton to Littlehampton that stretches along the south coast of England. The estuary is sheltered, flowing into a natural harbour at its mouth, and is bordered by human developments on the south slopes of the South Downs, including the town of Shoreham-by-Sea along the north, south and east banks, and Shoreham Airport to the west. The estuary is designated as an SSSI, and includes some small areas of saltmarsh; intertidal mudflats are exposed at low tide. Although colonisation by cord grass has reduced areas of intertidal mudflats in the south of the estuary since the mid nineties, it is assessed as being in favourable condition. The area is intensively farmed and agricultural land drainage has removed much of the tidal plain and created extensive lowland wet grasslands. The main leisure activities of the estuary are water-based; some shooting occurs in the upper estuary. A commercial mussel fishery is in the mouth of the estuary, and some bait digging occurs. Shoreham has a large port, for which plans to extend handling facilities and onshore oil-holding tanks exist. A number of manufacturing industries and boat-building yards are also based in Shoreham.

General bird distribution 2004/05

Area covered 74 ha; Mean total birds 1,171; Mean bird density 15.8 birds per ha.

Of 19 species counted on the Adur in 2004/05, ten occurred in single figure maximum numbers for the winter. Of these, Little Egret and Ruff were notable records. Of the remaining nine species, two wildfowl, Mallard and Teal, were most heavily concentrated on the count sector nearest the river mouth, an area containing subtidal habitat, the fringes of which may be used by dabbling ducks for foraging. This sector contains a mixture of intertidal and subtidal habitat also favoured by Grey Plover, Redshank, Snipe and Turnstone. The former two species were also found in similar densities north of the Railway Bridge. Lapwing was the most abundant species on the Adur, an average of 729 birds being counted. Density of the species was high on all sectors except the middle reaches of the estuary, and as in previous winters, large flocks were counted on the fields of Shoreham Airport. In February, peak numbers of Lapwing counted at the site exceeded the 1,000 mark.

Comparative bird distribution

The Adur is notable for concentrations of Ringed Plover, Dunlin and Redshank. WeBS Low Tide Counts have taken place regularly on the Adur for the past three winters, and here distribution maps of Ringed Plover and Dunlin are shown for the winters of 2004/05 and 2002/03 (Figure 58).

Densities of Ringed Plover toward the mouth of Shoreham Harbour were very similar in the two winters examined, with 1.65 birds per ha and 1.54 birds per ha in 2002/03 and 2004/05 respectively. This area includes a RSPB reserve and is thus managed appropriately for birds. In both winters, the stretch of intertidal habitat south of the Railway Bridge has held few individuals of this species. Further north, between the Railway Bridge and south of Old Shoreham Bridge, Ringed Plover were thinly and widely spread. The later winter returned greater densities of the species in this area: on average, 2.40 birds per ha were recorded, compared with an average of 0.98 per ha in 2002/03. Across the site as a whole, Ringed Plover density was higher in the later of the winters discussed, with an average of 19 birds more for the winter. It is clear that most of the intertidal habitat available on the Adur Estuary is used by Ringed Plover at low tide. At present, there appear to be no changes in distribution of the species that give rise for concern.

The density distribution of Dunlin is in some respects very similar to that of Ringed Plover, perhaps unsurprising in that both species will often forage on the same mudflats. In the winter of 2002/03, relative density of Dunlin was higher on the southernmost count section of the river than in 2004/05. However, overall mean site density for the later winter was higher for the species (7.34 compared with 6.31 birds per ha), largely attributable to greater concentrations upriver. Like Ringed Plover, the middle section of river held very few Dunlin, but north of the Railway Bridge flocks were much larger. In 2004/05, a peak of just over 30 birds per ha was recorded in this area, with an average of 16.33 birds per ha for the section across the whole winter. Although peak densities in 2002/03 were virtually identical two winters later, average densities were lower. It may be that on one or more months in 2002/03 birds were recorded further downriver, or that a factor such as disturbance produced lower counts in some count months. As one bird less per ha was recorded in 2002/03 than 2004/05, there would appear to be no distributional shifts that suggest the site is undergoing changes detrimental to Dunlin.

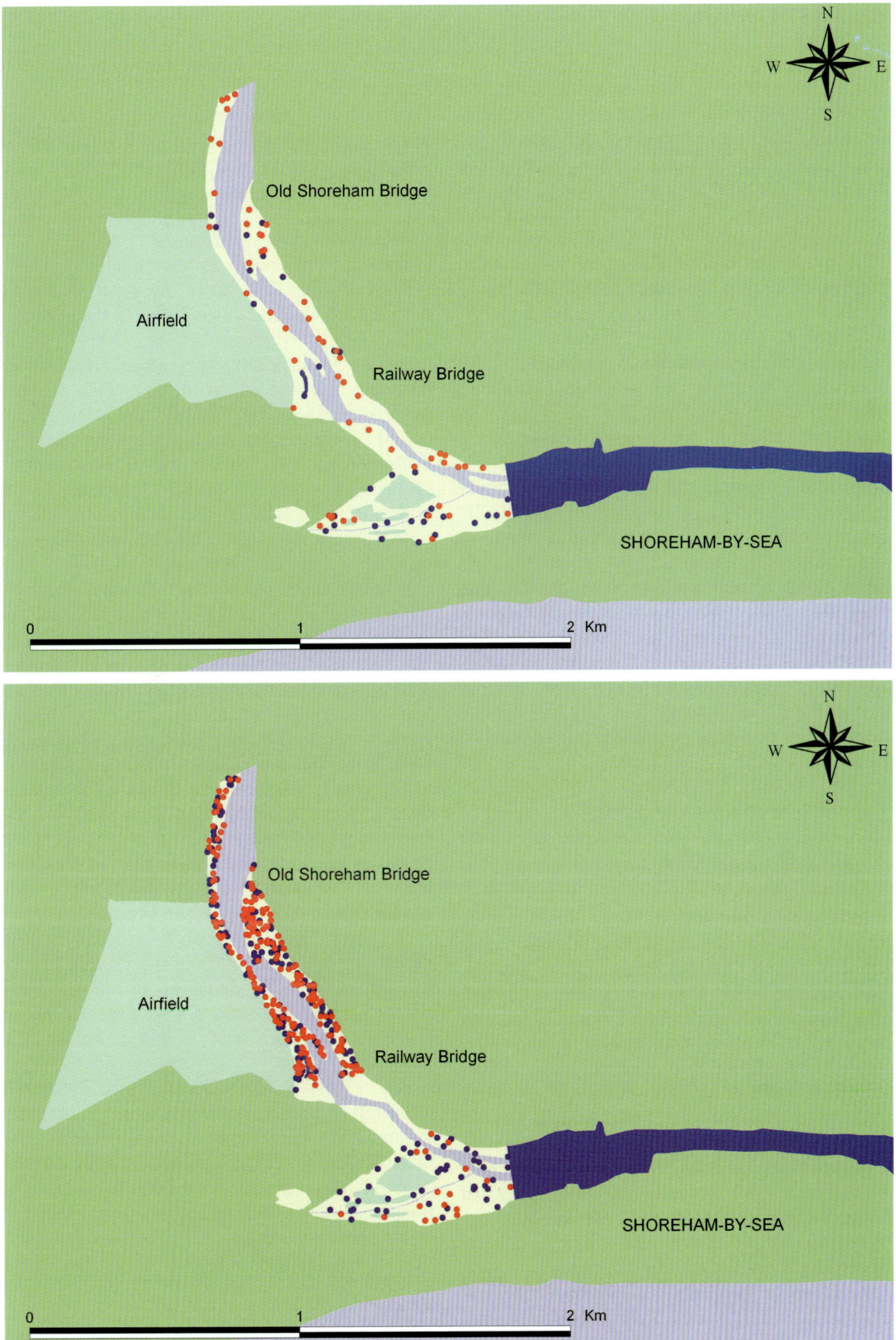

Figure 58. Low Tide distribution of Ringed Plover (above) and Dunlin (below) for the winters of 2002/03 (blue) and 2004/05 (red). Yellow = intertidal; pale green = nontidal; pale blue = subtidal. Dark blue area not covered in either winter.

BELFAST LOUGH

Site description

Belfast Lough is a large sea lough in the northeast of Ireland, with the city of Belfast at its head. The area surveyed comprised the coast from Carrickfergus on the north shore around to the eastern end of Bangor on the south shore. Much of the site is afforded SPA and Ramsar status, with a further proposed SPA over open water. The outer parts of the lough's shore are generally rocky with some sandy bays, although more extensive areas of intertidal mud are found toward Belfast. Industrial land claim has reduced the area of the mudflats over the last 150 years, and Belfast has become the main port in Northern Ireland for heavy cargo. More recently, two of the Belfast Harbour Pools were filled, leaving one permanent waterbody (part of the important Belfast Harbour RSPB reserve) and some occasional standing water to the south. The area is now protected and undergoing habitat management, though this may not fully compensate for the habitat loss. Extensive areas of the lough support commercial shellfisheries. There are problems of refuse disposal, pollution and general disturbance, but notably bait diggers on the north shore can pose potentially high levels of disturbance.

General bird distribution 2004/05

Area covered 456 ha; Mean total birds 11,808; Mean bird density 25.9 birds per ha.

The combination of large waterbodies, intertidal mud and rocky shores attract a wide diversity of species to Belfast Lough. Most wildfowl and allies were most heavily concentrated in three areas. Grebes, mergansers and seaducks (Eider and Scaup) were most densely distributed on the west shore, from Belfast to Carrickfergus. Other wildfowl, such as Mallard and Shelduck, were present in relatively high numbers in Belfast Harbour, whilst the three nontidal waterbodies - Belfast Harbour Pool, Victoria Park Lake and Whitehouse Lake typically held large flocks of Mute Swan, Coot, Mallard, Teal and Wigeon. Waders were distributed more widely according to feeding ecology, with Oystercatcher ubiquitous throughout. Black-tailed Godwit favoured Belfast Harbour Pool and Belfast Harbour, the latter preferred along with Whitehouse Lake by Dunlin. Especially high densities of Curlew (58 birds per ha) were recorded at Green Island, between Macedon Point and Carrickfergus. Ringed Plover occurred at highest density both on the outer north shore and at Victoria Park Lake, whilst Turnstone density was highest on the rocky shore at Carrickfergus.

Comparative bird distribution

Goldeneye are known to have declined within Belfast Lough SPA over the past 25 years, with a 58% decline since 1994/95 triggering a High Alert (Maclean *et al.* 2005). It is therefore unsurprising that mean site density has declined according to Low Tide Counts made in 1994/95 and 2004/05, the latest figure of 0.07 birds per ha much lower than the 0.39 birds per ha in the earlier winter. Densities of Goldeneye were greatly reduced on approximately half of the sectors counted in both winters. Greatest changes were observed on the coast between Belfast and Carrickfergus, especially at Macedon Point. A winter mean of 90 Goldeneye was recorded on the count sector here in 1994/95, decreasing to just four ten years later. Density of Goldeneye also declined along adjacent sections of coast. Various nontidal waterbodies such as Belfast Harbour Pool, Victoria Park and Whitehouse Lake now hold few if any Goldeneye, whereas in the earlier winter small flocks were present on all. Some redistribution is evident on the east coast of the Lough, where relative densities are generally greater than in 1994/95, but this is insufficient to counteract birds absent from other parts of the site.

Although Redshank were recorded at low density on the outer parts of the Lough, Figure 59 shows relative densities on those areas where Redshank densities were greatest. Mean numbers of Redshank counted across the whole site in the two winters considered were similar (1,484 in 1994/95; 1,318 in 2004/05), although a greater area was covered in the later winter, thus producing a lower mean site density. Many sectors counted in both winters showed lower densities in 2004/05. This was notable between Belfast Harbour and Macedon Point, where the main area of intertidal habitat occurs. In Belfast Harbour itself, Redshank density declined from 27 to 22 birds per ha between the winters. Interestingly, density was higher on Whitehouse Lake in the later winter; this may reflect a movement of birds feeding, for example, on periodically exposed muddy fringes of the lake.

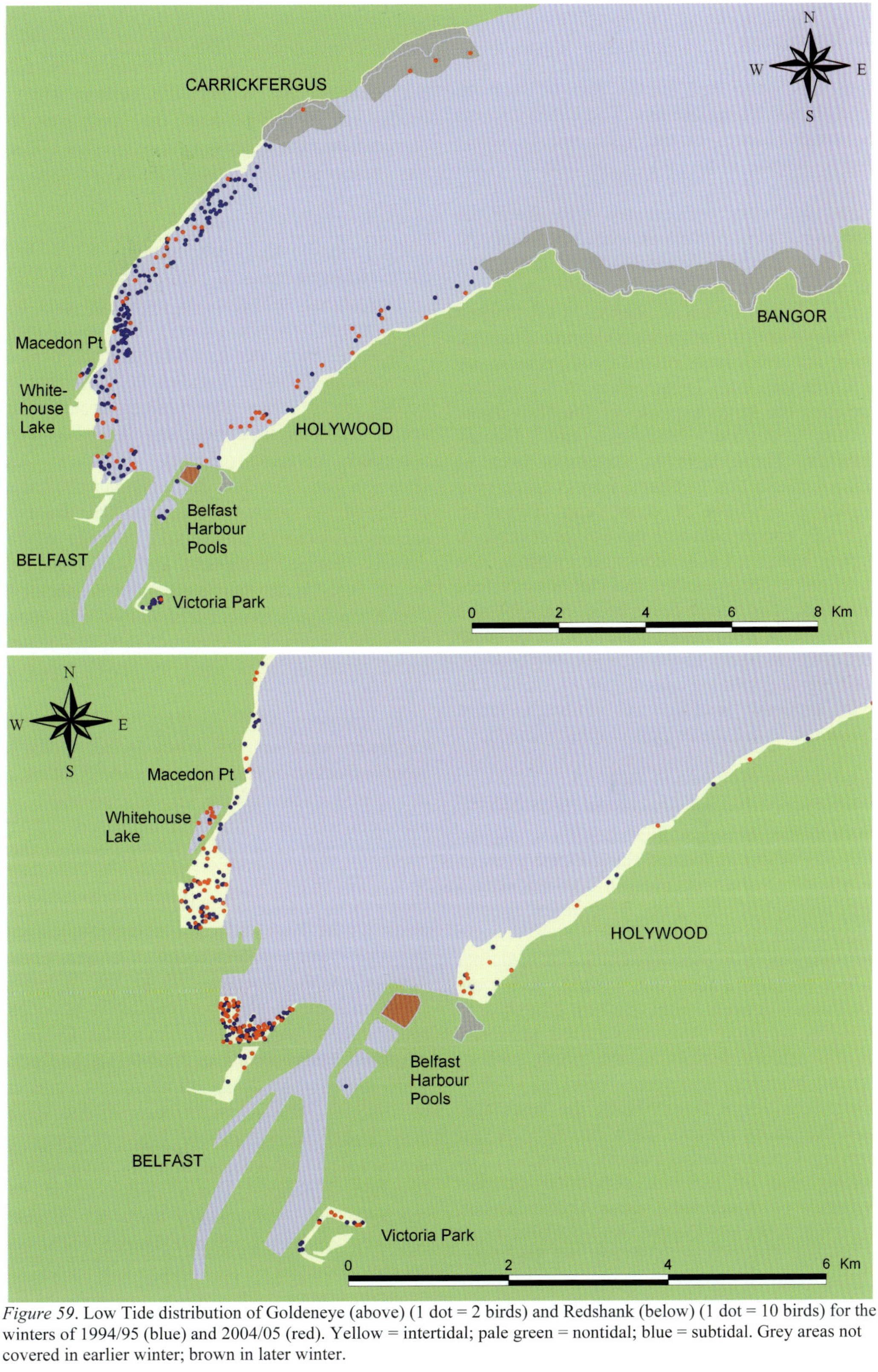

Figure 59. Low Tide distribution of Goldeneye (above) (1 dot = 2 birds) and Redshank (below) (1 dot = 10 birds) for the winters of 1994/95 (blue) and 2004/05 (red). Yellow = intertidal; pale green = nontidal; blue = subtidal. Grey areas not covered in earlier winter; brown in later winter.

BREYDON WATER

Site description

Breydon Water is a bar-built estuary separated from the North Sea by the spit of land on which Great Yarmouth sits. The estuary forms the lower reaches of the Yare and Waveney rivers, which drain much of central East Anglia. The rivers are tidal for many miles inland but only the estuary area from the confluence of the rivers is considered here. At high tide, Breydon Water forms a large lake but as the tide recedes, the only water that remains forms a narrow channel, well marked by buoys for the numerous leisure cruisers. There are small areas of saltmarsh, principally at the eastern end. To the north of the estuary stretches the huge expanse of the Halvergate Levels, Breydon Marshes and Berney Marshes. These form an extensive area of grazing marsh that has been subject to varying degrees of drainage in recent years. The main high tide roosts occur at the RSPB reserve at Berney Marshes (only accessible by boat, train or a very long walk) and in the eastern saltmarsh. The site is designated as a SPA and is judged in favourable condition. The main conservation issues in the area involve boating, shooting and grazing marsh management. The river channel leading out through Great Yarmouth to the sea is highly industrialized.

General bird distribution 2004/05

Area covered 402 ha; Mean total birds 22,995; Mean bird density 57.2 birds per ha.

Breydon Water supported the greatest average density of all sites surveyed in 2004/05, with a large number of birds from 27 different species packed into a relatively small area. Wigeon were present in highest densities (16 birds per ha on average across the winter), and were distributed densely from Berney Marshes to Breydon Junction. Avocet occurred primarily on the estuary sectors to the north east of Berney Marshes, and were found in slightly lower densities at Burgh Flats; the latter area was the most notable part of the site for Greylag Geese and held many Lapwing. The eastern end of the site supported greatest densities of waterbirds; most Oystercatcher, Lapwing, Knot and Dunlin were contained east of Pump House, in addition to Shelduck, whilst east of Breydon Junction, Mallard, Golden Plover and Teal densities were higher than elsewhere at the site. Curlew and Redshank were both comparatively thinly but widely spread around the sectors surveyed.

Comparative bird distribution

Regular counts of Breydon Water at low tide are received by WeBS, the earliest survey made in 1998/99. For the purposes of comparison, the distribution from this winter has been selected. Patterns of change in distributions of Shoveler and Black-tailed Godwit, both of which are present in internationally important numbers at the site, are considered here.

Density of Shoveler has increased over the six winters between surveys discussed, from 0.04 to 0.2 birds per ha and mean numbers at low tide in 2004/05 stood at 96. Distribution across the site has not changed greatly, but sector densities have increased. All birds were recorded between Berney Marshes and Breydon Junction, but the Acle Marshes area exhibited the greatest changes. On the two sectors covered in this region, average winter density changed from 0.04 and 0.06 birds per ha to 0.58 birds per ha (on both sectors). Therefore it is apparent that at low tide, more Shoveler now use Breydon Water, but in the same areas as were used by fewer birds in 1998/99.

In common with many sites in eastern England, numbers of Black-tailed Godwits have increased over the past decade (Collier *et al.* 2005). The effect of this increase in numbers has been for both increased density on previously used sectors and expansion into new feeding areas. In 1998/99, Black-tailed Godwits were largely restricted to the intertidal north of the Yare between Acle Marshes and Breydon Junction. Although average winter density of the species fell in the Breydon Junction area, on all other previously occupied sectors there were notable increases in mean numbers. Furthermore, in 2004/05, large flocks were recorded on areas not apparently utilised in 1998/99. The southern bank of the River now holds comparatively high densities of Black-tailed Godwits, as does the area east of Breydon Junction. The distribution of the species has expanded towards Berney Marshes, and in particular Burgh Flats; the species occurred at an average 6.67 birds per ha here, amongst the highest densities on the site. It will be interesting to monitor future distributions in response to any further changes in numbers of Black-tailed Godwits.

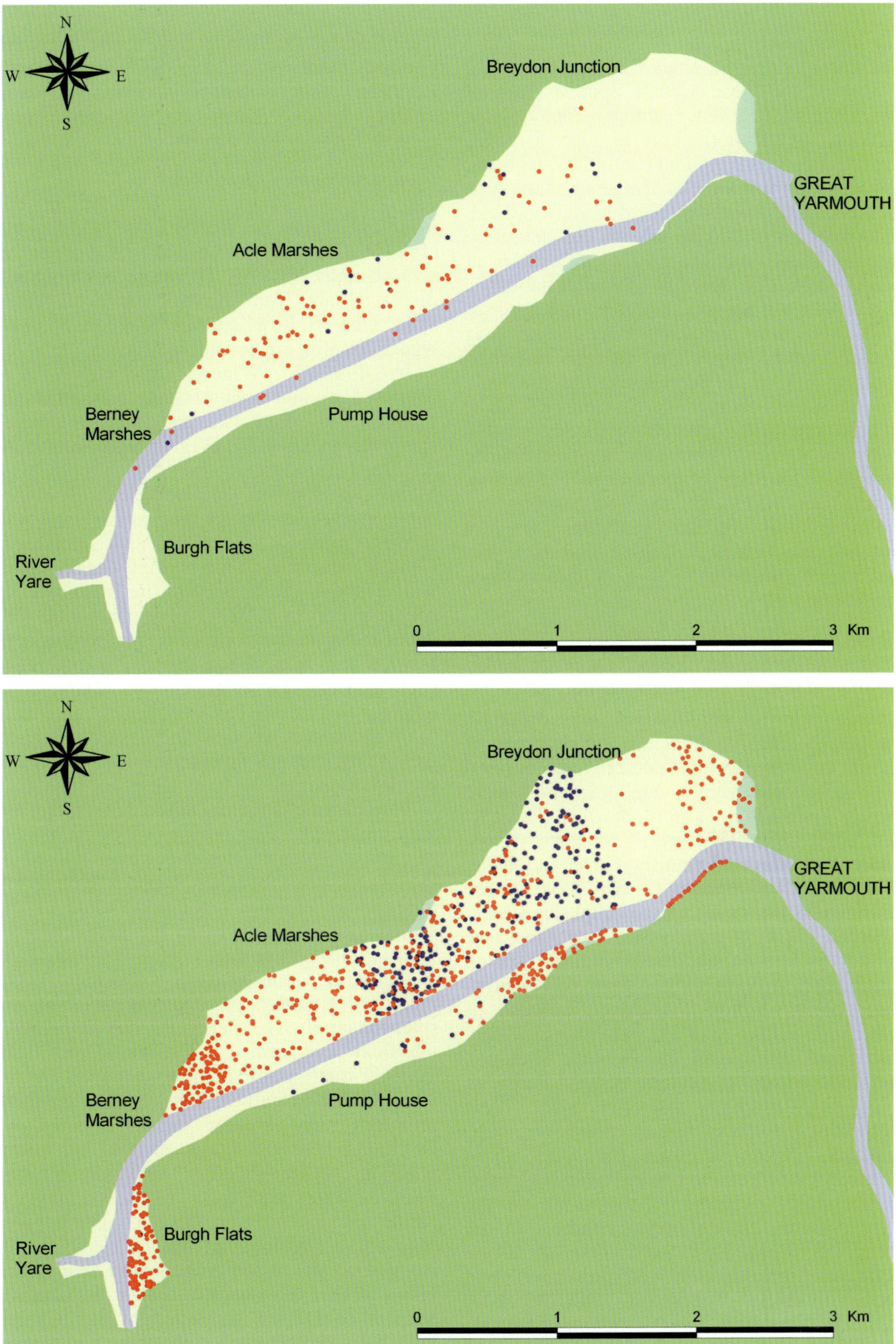

Figure 60. Low Tide distribution of Shoveler (above) and Black-tailed Godwit (below) for the winters of 1998/99 (blue) and 2004/05 (red). Yellow = intertidal; pale green = nontidal; blue = subtidal.

CAMEL ESTUARY

Site description

The gently sloping Camel Estuary, Cornwall, is part of the River Camel Special Area of Conservation (SAC), which includes a diversity of habitats, and is also a SSSI. Amongst these are mud and sandflats, lagoons, salt marshes and various fringe habitats such as scrub and especially deciduous woodland. The river rises 280 m above sea level on Bodmin Moor, and widens into an estuary at Wadebridge. The majority of marsh habitat is in this area, the narrowest stretch, with a series of muddy creeks and sandy coves developing further downstream. The fishing town of Padstow lies on the west bank, and the wide river mouth is flanked by St George's Cove and Daymer Bay. Apart from fishing vessels using the estuary, other pressures on birdlife include tourism, recreational sailing, windsurfing and jetskiing. The surrounding environment is typified by quarries (mostly disused) and farmland. English Nature report that over 50% of the SSSI is in unfavourable condition.

General bird distribution 2004/05

Area covered 523 ha; Mean total birds 5,339; Mean bird density 10.2 birds per ha.

A wide variety of waterbirds were recorded on the Camel Estuary, reflecting the diversity of habitat contained. Some species were confined to specific areas, such as Mallard, which were largely restricted to a pool between Padstow and Little Petherick Creek, Turnstone on the rocky area around Porthilly Cove and Mute Swan near Wadebridge. Golden Plover were confined to the shore east of Pinkson Creek, whereas Lapwing shared some of the same habitat preference, their distribution ranging further east to Wadebridge. North of Padstow, few count sectors returned counts of waterbirds. Other species were more widely distributed throughout the estuary. Shelduck were spread across sectors east of Padstow, with only mudflats off the town itself supporting notably large mean numbers, whereas Wigeon favoured the south shore between Little Petherick and Pinkson Creeks. Oystercatcher occurred mostly south of the river channel, greatest mean numbers in the areas off Cant Cove and east of Little Petherick Creek. Curlew were present on most sectors upstream of Padstow, with the shore west of Pinkson Creek holding highest densities. Little Petherick Creek provided suitable habitat for Redshank, which were also thinly spread along the south bank of the estuary. Other species widely but thinly spread were Cormorant, Grey Heron, Little Egret (in average numbers of 37) and Greenshank. Eighteen other waterbird species, including grebes, geese, ducks and waders, were recorded at very low density.

Comparative bird distribution

The Camel Estuary was included in the inaugural winter of WeBS Low Tide Counts (1992/93), but had not been revisited under the scheme until 2004/05. Therefore, distributions displayed are from surveys made 12 years apart. One wildfowl species, Wigeon, and one wader, Dunlin, are considered further.

In 1992/93, Wigeon distribution was restricted to one area of high concentration, between Little Petherick and Pinkson Creeks. However, in 2004/05, Wigeon appeared more dispersed throughout the site (Figure 61). Mean site numbers increased by 130 birds between the two winters, and as coverage was similar, mean site density also increased. General expansion of the species was apparent, with average site density at the main area of concentration in 1992/93 very consistent. The count sector to the north of this important area showed high relative site density, as did creeks such as Little Petherick Creek. It is unclear whether the Camel has experienced an influx of Wigeon, or whether these birds foraged previously in areas outside the low tide count sector boundaries.

Patterns of change in Dunlin distribution are very different than those of Wigeon, as average site density for the winter halved in the twelve years between surveys. Examining this change at finer resolution, some of the sectors used by Dunlin in the earlier winter were still important in 2004/05. At Wadebridge, similar site densities were recorded, although on opposite sides of the river channel. To the west, on the muddy intertidal upstream of Pinkson Creek, the 2004/05 distribution is more scattered than in 1992/93, but overall the sum of mean numbers produces very similar mean densities of Dunlin. Where the major change has taken place is on the sandier substrate off the Padstow coast known as Town Bar. On the earlier survey, a large flock of Dunlin was recorded in this location, yet by 2004/05 only isolated counts of individuals were made. The effect of this is to greatly reduce the average number of Dunlin supported by the Camel. Possible explanations of the contraction include a general reduction in Dunlin on the west coast of Britain (Austin & Rehfisch 2005); if the intertidal flats at Wadebridge are preferred by Dunlin, it may be that these areas are filled first, and that any additional birds forage elsewhere on the estuary. Should there be fewer birds present than in previous years then such secondary feeding areas may be avoided. Alternatively, local factors, such as disturbance or habitat alteration, may have influenced the change.

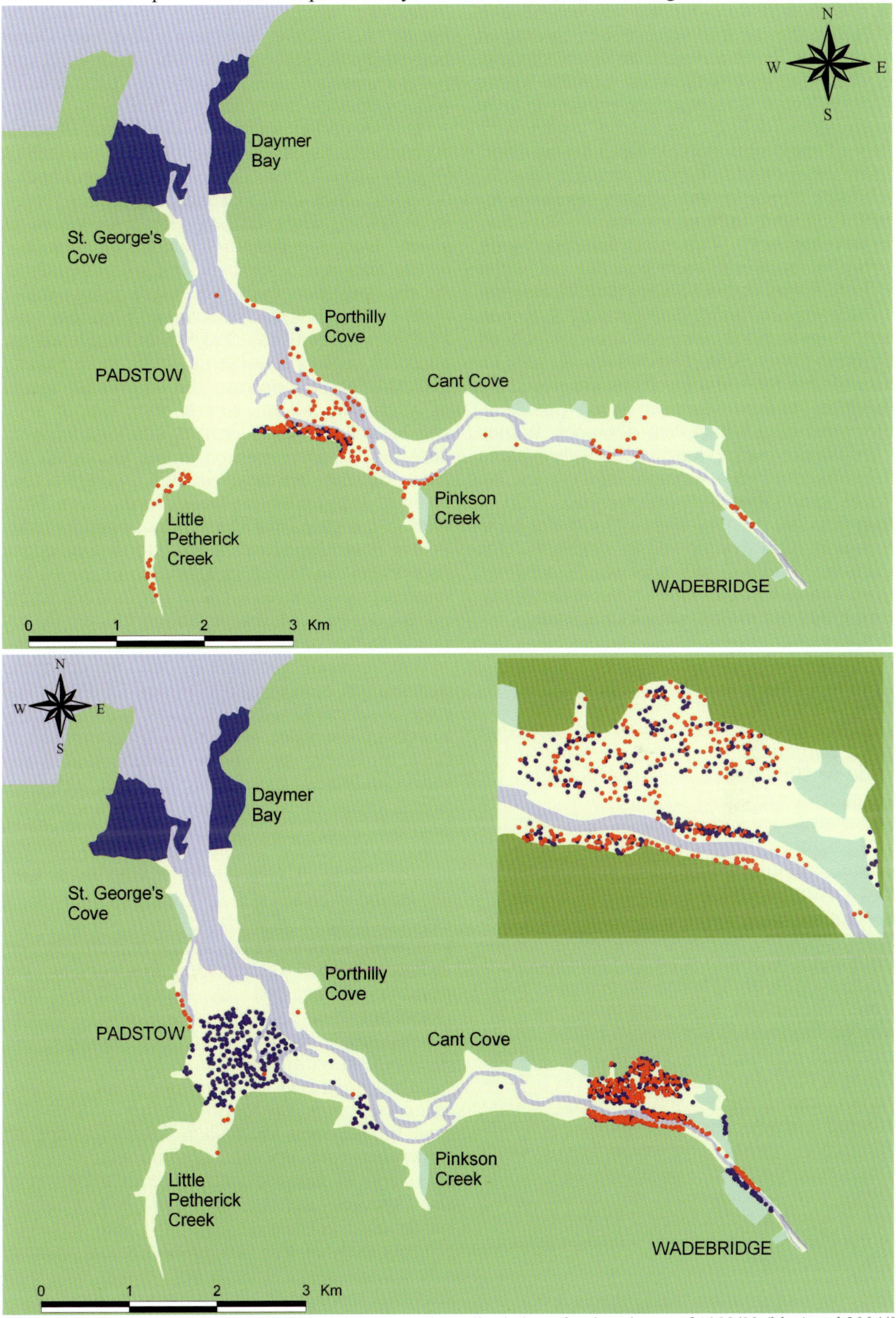

Figure 61. Low Tide distribution of Wigeon (above) and Dunlin (below) for the winters of 1992/93 (blue) and 2004/05 (red). Inset shows Wadebridge area in detail. Yellow = intertidal; pale green = nontidal; pale blue = subtidal. Grey area at Wadebridge not covered in earlier winter; dark blue areas never covered.

CROUCH & ROACH ESTUARIES

Site description

The Crouch Estuary is traditionally considered alongside its tributary, the Roach, as the two converge on the coast of south Essex in eastern England. The River Crouch carves a shallow valley between two ridges of London Clay, whilst the River Roach is set predominantly between areas of brick earth and loams with patches of sand and gravel. Both rivers form dendritic creeks and low-lying riverine islands. Surrounding habitat is almost exclusively lowland farmland and grazing marsh, with few urban developments. The intertidal zone along the rivers is 'squeezed' between the sea walls along both banks and the river channel, leaving a thin strip of tidal mud. The Crouch & Roach Estuary is an integral component of the phased Mid-Essex Coast SPA (Stroud et al. 2001). Threats to this SPA and environs are posed by disturbance caused by air activities, the development of a wharf, sea defences, homes and shops, car parks, marinas, holiday parks and an airport, and saltmarsh loss caused by sea-level rise (BirdLife International 2003). Threats specific to this site include drainage and reclamation of wetland habitats for agriculture. 75% of the site is deemed by English Nature to be in unfavourable declining condition due mainly to saltmarsh erosion.

General bird distribution 2004/05

Area covered 1,745 ha; Mean total birds 26,549; Mean bird density 15.2 birds per ha.

The large and diverse nature of the estuarine complex harboured 45 species of waterbird at low tide in 2004/05. Highest site densities for the winter were of Lapwing and Golden Plover, distributed widely but densely; the site is ideal for these species owing to the extent of bordering coastal pasture. The muddy creeks around Potton Island and inland to Rochford provide wide and shallow mud fringes, and were favoured by tall waders including Avocet and Bar-tailed Godwit, though Shelduck were also observed, with Wigeon present on marsh fringes. The latter also appeared at high density on marsh at Bridgemarsh Island, where locally high numbers of Pintail also grazed, and Stow Creek. The outer estuary was notable for Knot and Curlew, with Ringed Plover distributed from Burnham-on-Crouch to Bridgemarsh Island. The majority of other species recorded in comparatively high densities were widespread and either thinly spread (Oystercatcher, Grey Plover, Curelw) or more densely aggregated (Teal, Lapwing, Dunlin, Redshank).

Comparative bird distribution 2004/05

Dark-bellied Brent Geese are found in internationally important numbers on the Crouch & Roach Estuaries, but according to WeBS Alerts (Maclean *et al.* 2005), numbers have declined over a series of time scales triggering Medium Alerts for the species. Figure 62 shows the distribution of Dark-bellied Brent Geese at low tide in 2004/05, and for the winter of 1995/96, the last time the site was counted at low water for WeBS.

In the earlier of the two winters, notable concentrations of the species were located at Brandy Hole Creek, Bridgemarsh Island and at the confluence of the Crouch and Roach rivers. In 2004/05, Dark-bellied Brent Geese were absent from the Brandy Hole Creek area, and to the area north west of Bridgemarsh Island. At the confluence of the rivers, many more birds were counted between Foulness and Wallasea Islands in 1995/96, although on the adjacent sectors relative densities were greater in 2004/05. By contrast, the south east corner of Wallasea Island, where accretion of mud and saltmarsh has occurred, supported higher densities of Dark-bellied Brent Goose in the later of the winters. Inland, winter crops provide supplementary food. Even more birds were aggregated on the creeks around Potton Island extending toward the North Sea coast, an area not surveyed in 1995/96. In the earlier winter, many more of these geese were found on the stretch of the Roach north of Potton Island to Rochford.

The overall picture is one of changing distribution between 1995/96 and 2004/05, and overall average site densities were slightly lower in the later winter (0.73 compared to 0.81 birds per ha). It is thought that many individuals may have switched to foraging on winter crops on agricultural land, thus feeding beyond count sector boundaries. Such a behavioral switch could partially explain the absence of birds at previously used areas such as Brandy Hole Creek, though here local experts suggest disturbance may be an issue following seawall realignment, which has also induced habitat differences. Declines in site numbers suggested by Core Count data (Maclean *et al.* 2005) may explain reduced mean site density, but not changes in distribution within the site.

Black-tailed Godwit numbers have shown large increases throughout much of the UK, and the Crouch & Roach Estuaries are no exception (see species account elsewhere in this report). Consequently, the distribution of the species across the site has changed markedly between 1995/96 and 2004/05 (Figure 62). In the former winter, an average site density of 0.04 birds per ha was recorded; this had risen to 0.33 by 2004/05, though the change may be over-estimated because some birds may have initially fed in coastal pools outside the count sectors. In 1995/96, the species was largely restricted in its distribution to Rochford, with a few individuals west of Bridgemarsh Island. However, distribution in the later winter shows evidence of increasing density, in the same areas, with scattered individuals also elsewhere around Wallasea Island; the future effects of a managed breach at Wallasea (designed to mitigate for lost habitat at Lappel Bank, Medway and Fagbury Flats, Orwell) will be interesting to monitor. The muddy creeks around Potton Island are clearly suitable feeding grounds for Black-tailed Godwits, with high densities found in 2004/05.

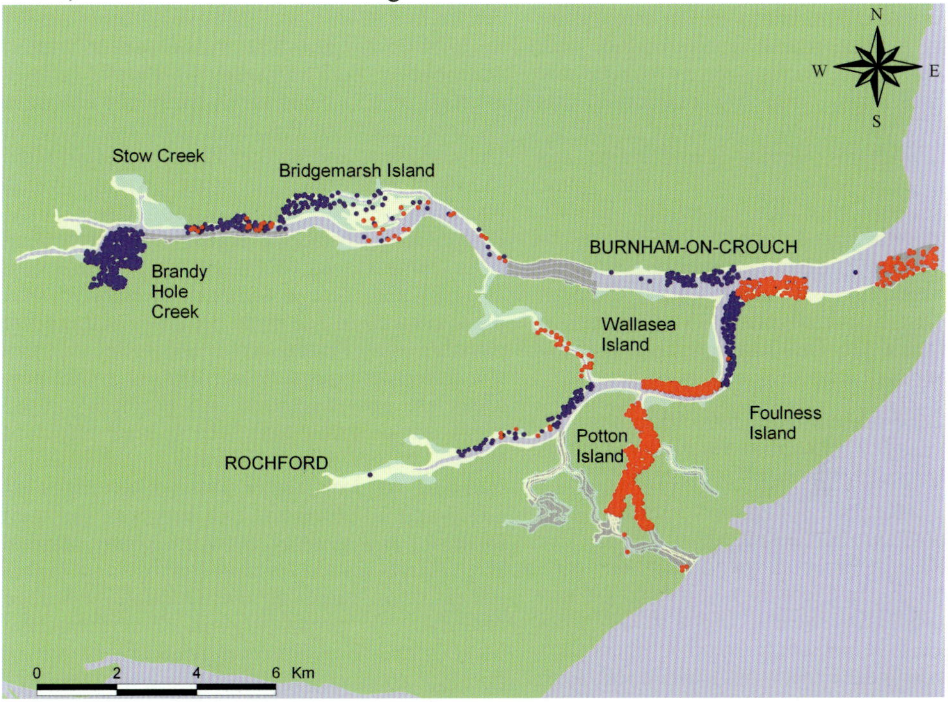

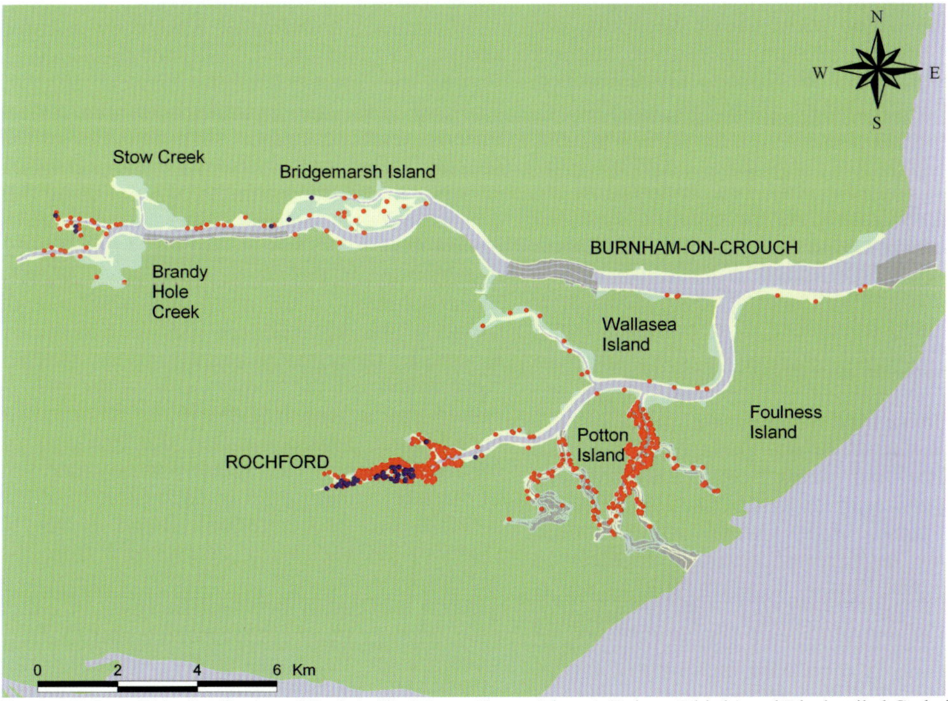

Figure 62. Low Tide distribution of Dark-bellied Brent Goose (above) (1 dot = 2 birds) and Black-tailed Godwit (below) for the winters of 1995/96 (blue) and 2004/05 (red). Yellow = intertidal; pale green = nontidal; pale blue = subtidal. Grey areas not covered in earlier winter.

DUDDON ESTUARY

Site description

The Duddon Estuary is a dynamic estuary, fed by the river Duddon that drains part of the Lake District in Cumbria. A system of large sand banks has developed, the estuary being 6 km at its mouth. The banks of the estuary are flanked by grazing marsh and farmland, with small urban developments such as Millom and Askam in Furness. Iron ore extraction was previously the main industrial activity associated with the estuary. Land claim, sea level rise and coastal defence exert some coastal squeeze on the intertidal habitat, though intertidal and saltmarsh habitats on the SPA are judged to be in favourable condition. Other pressures exist from bait digging, development and land claim. Movements of birds between the Duddon and neighbouring Morecambe Bay are likely.

General bird distribution 2004/05

Area covered 3,439 ha; Mean total birds 15,544; Mean bird density 4.5 birds per ha.

Areas of most concentrated bird distribution on the Duddon can be loosely defined into five areas, and the species occurring on each depend on feeding ecology. Firstly, Walney Channel, including Scarth Bight at the north east corner, is important for mud-foragers such as Oystercatcher, Curlew and Shelduck, with Wigeon also aggregated on the fringing marsh and Turnstone on rocky areas. Secondly, the mudflats south of Askam in Furness support high densities of Shelduck, Oystercatcher, Ringed Plover, Dunlin, Redshank and Curlew. Further upriver, the area of Whelpshead Crag was notable for Shelduck, Dunlin and Redshank, plus the only Teal recorded at the site. Millom Marsh, on the west bank of the estuary, was favoured by geese, including Greylag and Pink-footed. Finally, the nature reserve south of Millom, including Hodbarrow Lagoon, held a number of species including Lapwing and the only Black-tailed Godwits present.

Comparative bird distribution

Low Tide Counts on the Duddon Estuary had taken place four times before the winter of 2004/05, and here distributions from 1994/95 are considered. WeBS Alerts over a similar time frame have been identified (Maclean *et al.* 2005) and thus it is worthwhile to see if declines are reflected by changing distributions. Two species issued with 'High' Alerts (-50% or more) are highlighted; Pintail and Knot: the former are still found in internationally important numbers at the site, the latter are not.

In both winters, the most important area of the site for Pintail was that between Dunnerholme and Soutergate. Average densities on the count sectors here were higher than elsewhere on the Duddon, up to 11 birds per ha in 1994/95 and just over 7 birds per ha in 2004/05 (Figure 63). In the earlier winter, additionally high densities of Pintail (winter average of 125 birds at 1.1 birds per ha) were recorded south of this main location, between Dunnerholme and Askam Pier. No Pintail were found in this area in the later winter, which could indicate a change in habitat suitability. However, between Soutergate and Whelpshead Crag, large flocks of Pintail were recorded in 2004/05 that were not present in 1994/95, at a mean density of 3.2 birds per ha. It is therefore possible that Pintail have undergone within-site movement between surveys, such that new areas are now favoured and previously used areas are now avoided. The overall effect of these movements indicates stability of numbers at low water (winter averages of 1,049 and 1,096 on the two surveys), but with localised movements between count sectors. High Alerts issued for this species suggest that either birds roost at the Duddon and feed elsewhere, possibly in neighbouring Morecambe Bay, or that some birds are in nontidal areas at low tide, leading to stable low water, but not roosting, numbers.

In common with Pintail, Knot show a strong association with specific areas of the Duddon. Firstly, the intertidal habitat between Walney Channel and the mainland at Barrow-in-Furness tends to support high densities of the species. Between 1994/95 and 2004/05 surveys, the average number of Knot using this area at low tide declined from just over 2 birds per ha to only 0.25 birds per ha. Although the distribution in the later winter was slightly more scattered to the north and west, overall usage of the general area was still lower by Knot. The other major concentrations of Knot were found off the shore south of the pier at Askam-in-Furness. In 1994/95, all Knot in this area were recorded between the branches of the river channel. This sector was not surveyed in the later winter, so it is impossible to know whether the concentrations found closer to shore in 2004/05 were an overspill from the previously used area or whether there was a general movement between surveys. In either case, average numbers recorded closer to shore were greater than those on the adjacent sector in 1994/95. In conclusion, differing coverage between the surveys allows limited speculation on the extent of change in Knot distribution;

however, there are certainly less Knot in the Walney Channel. Whether other, un-surveyed, areas are now favoured, or whether declining numbers of birds at the site have led to a thinner distribution, is unclear.

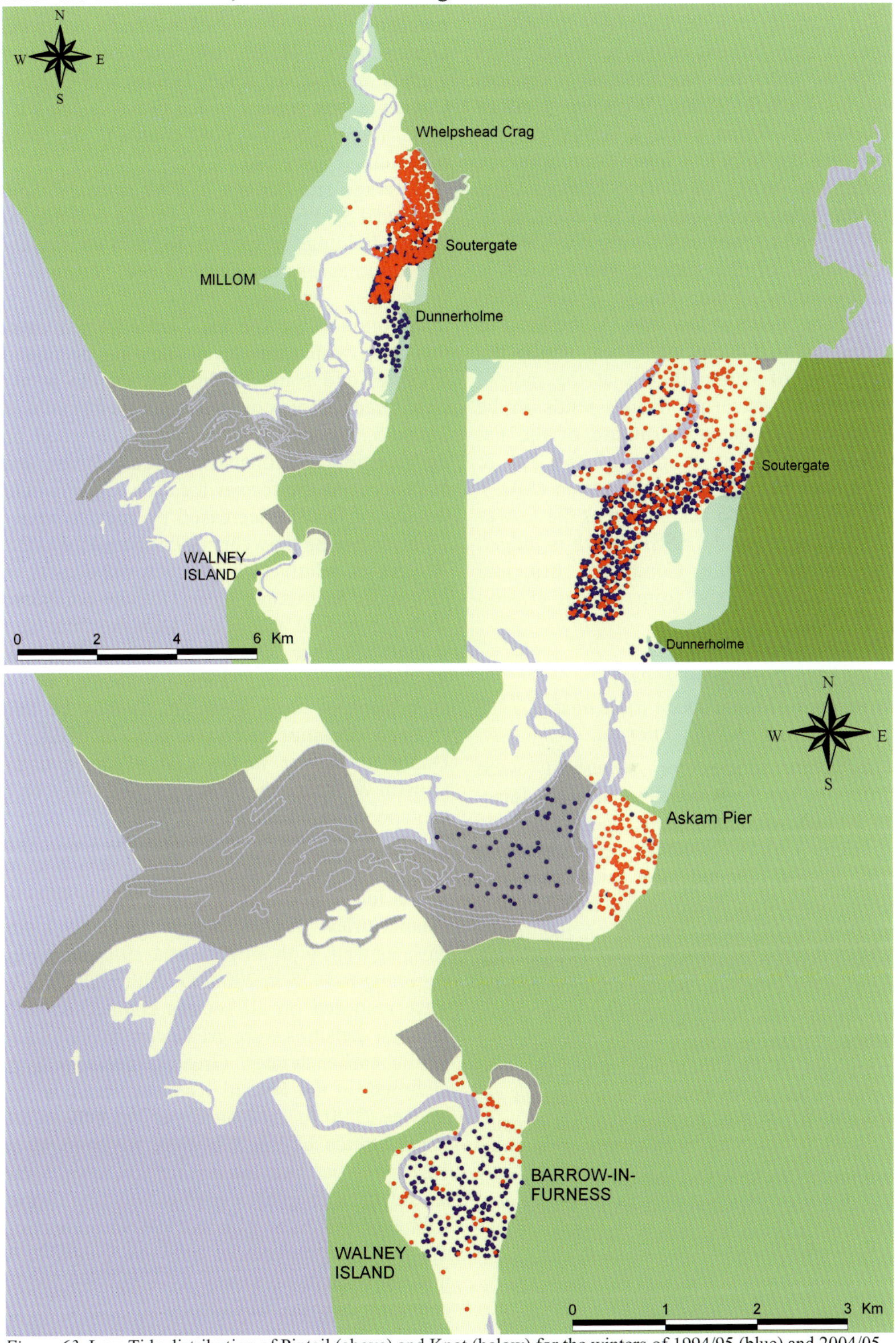

Figure 63. Low Tide distribution of Pintail (above) and Knot (below) for the winters of 1994/95 (blue) and 2004/05 (red); 1 dot = 2 birds. Inset shows Soutergate area in detail. Yellow = intertidal; pale green = nontidal; pale blue = subtidal. Grey areas not covered in later winter.

FAL COMPLEX

Site description

The Fal Complex in south Cornwall consists of a series of slow-current tidal creeks and rivers that empty into the main estuary channel known as Carrick Roads. With the Helford River, the Fal forms a ria: a drowned river valley, with typically low freshwater input and supporting a variety of marine habitats. The major intertidal areas occur in the network of creeks and river branches flowing into Carrick Roads. Much of the main channel itself is sided by rocky shore, which is not often surveyed for WeBS at low tide. At the head of the complex is the city of Truro, whilst the fishing town of Falmouth lies at the mouth; elsewhere, non-estuarine habitat tends to be agricultural. The habitat structure of the Fal Complex qualifies it as part of the Fal & Helford Special Area of Conservation (SAC). Part of the site is also designated as the Lower Fal & Helford Intertidal SSSI, which is judged to be in Favourable Condition. Principal threats to the site exist from coastal erosion and land claim.

General bird distribution 2004/05

Area covered 313 ha; Mean total birds 1,170; Mean bird density 3.7 birds per ha.

Twenty-seven species of waterbird were recorded on the Fal Complex, most in fairly small average winter numbers. Species such as Little Egret and Greenshank were thinly but widely spread around the site, whereas wildfowl including Mute Swan and Mallard were most highly concentrated on the lower Tresillian River. Mean Shelduck density was highest on the Tresillian River, with the nearby Truro River also holding notably high mean densities. The same area supported aggregations of Oystercatcher and Lapwing, and these species were also concentrated on the Percuil River. Black-tailed Godwit density was greatest on the Truro River and at Restronguet Creek, whilst the muddy sediment at the latter also attracted Redshank, found also at high density on the upper reaches of the Tresillian River.

Comparative bird distribution

The Fal Complex does not hold great numbers of waterbirds, but is notable for nationally important numbers of some relatively scarce species (*e.g.* Black-necked Grebe, Greenshank). Here, distributions of two waders are focused upon; Dunlin and Curlew. Coverage was similar in the two winters, with only three sectors not counted on both occasions (Figure 64).

Average site density of Dunlin in 1995/96 reached 1.73 birds per ha, with the majority of sectors holding the species. In 2004/05, density of Dunlin across the site had declined to 0.45 birds per ha, suggesting a change in distribution of the species. Figure 64 illustrates this case, with some areas showing decreased densities and some holding no Dunlin at all; only three sectors covered supported Dunlin in 2004/05. Greatest changes were recorded on the intertidal sections of the Truro River, and at Penryn. At the former location, mean winter density of Dunlin decreased by approximately 15 birds per ha between 1995/96 and 2004/05. At Penryn, relatively high mean site density of just under 5 birds per ha in 1995/96 compared with a complete absence of Dunlin in the later winter. Further changes were evident at Restronguet Creek, with evidence of a movement of birds upstream. It may be that reduced densities reflect a general trend for Dunlin to winter in the east of the country (Austin & Rehfisch 2004), or there may be a site-specific explanation.

Changes in Curlew distribution were also apparent at Restronguet Creek, with 1.5 birds per ha fewer recorded in 2004/05. Decreases in mean Curlew density were evident on a site basis as well, with a figure of 1.19 birds per ha declining to 0.78 in 2004/05. Although Curlew were recorded on all sectors counted in both winters, densities were greater on 75% of these in the earlier winter. Greatest declines were recorded on the Percuil River upstream of St. Mawes (2.90 to 1.46 birds per ha) and on the lower stretch of the Tresillian River (2.95 to 1.21 birds per ha). Interestingly, Curlew were recorded at higher densities on the upper stretch of this river than in 1995/96, indicating a within-site movement between years. It is unclear what factors may have contributed to changing densities of Curlew.

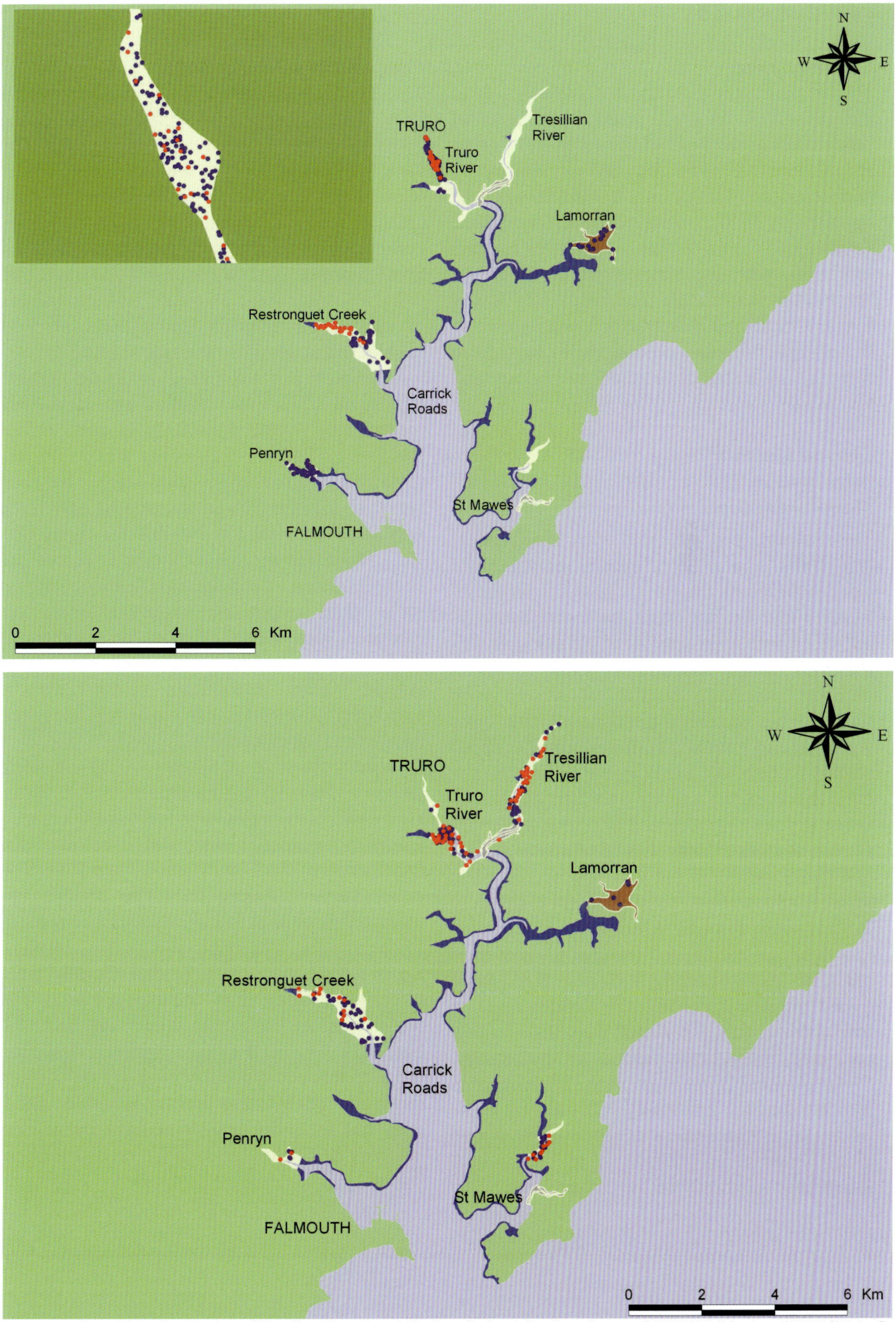

Figure 64. Low Tide distribution of Dunlin (above) and Curlew (below) for the winters of 1995/96 (blue) and 2004/05 (red); 1 dot = 3 birds. Inset shows Truro River in detail. Yellow = intertidal; pale green = nontidal; pale blue = subtidal. Grey areas not covered in earlier winter, brown in later winter; dark blue areas never covered.

FOWEY ESTUARY

Site description

The small Fowey Estuary in Cornwall lies to the east of St. Austell, with the villages of Lostwithiel at the head of the river and Fowey at the mouth. The estuary is extremely undeveloped, lying in a patchwork of agricultural fields, meadows and fragmented woodland. The estuary contains some saltmarsh, with the majority of the intertidal habitat on the middle reaches. The Fowey Estuary Partnership oversees management of the estuary. Pressures on waterbirds exist from recreational disturbance such as sailing and tourism.

General bird distribution 2004/05

Area covered 103 ha; Mean total birds 237; Mean bird density 2.3 birds per ha.

The Fowey does not hold vast numbers of waterbirds, and only 12 species of waterbird were recorded on the site in 2004/05. Of these, Mallard and Curlew were most abundant and are discussed in greater detail. Shelduck was the only other species to occur in mean numbers exceeding 30 for the winter, with fairly stable numbers throughout. The birds were almost exclusively found on the upper estuary, north of St Winnow Point at the confluence of the Fowey and Lerryn Rivers. Other species only counted on the upper count section included Little Grebe, Canada Goose and Oystercatcher, all at very low density. The lower count section, stretching approximately 2.5 km south from St Winnow Point, contains less intertidal habitat and features a wider river channel. Consequently, Cormorant, Mute Swan and Grey Heron were concentrated at low density here. Redshank were also largely restricted to the lower reaches of the estuary. Mean winter counts of nine Little Egrets indicate that the estuary is still used by these birds, their distribution fairly evenly spread along the river. Finally, an occasional Kingfisher was noted on the estuary.

Comparative bird distribution

The Fowey was last counted for the WeBS Low Tide Scheme in 1995/96 and thus the count in 2004/05 was a welcome update. The same two count sections were counted in both winters and so coverage of the major intertidal areas was complete. Overall species composition on the two counts showed no great changes, and two species - Mallard and Curlew - were most abundant in both winters of survey. Distribution across the estuary is considered in detail for the two species.

Although not exploiting intertidal habitat like some waterbirds, Mallard are commonly recorded on WeBS Low Tide Counts, and at the Fowey Estuary the winter site mean has remained remarkably consistent between surveys; in 1995/96, an average of 104 birds was recorded, whilst in 2004/05 the figure was 101. Densities for the two count sectors are slightly different for the two winters, but no more than might be expected from, for example, movements of birds in response to the tide (Figure 65). It therefore seems that current Mallard distribution on the Fowey is extremely similar to that when last surveyed nine winters previously.

Winter average numbers of Curlew were lower in 2004/05 than 1995/96, by 20 birds. However, distribution was very similar, with Curlew scarcely recorded below the confluence of the Fowey and Lerryn Rivers on either survey (Figure 65). The majority of intertidal habitat is found on the upper reaches of the estuary, where the river widens south of Lostwithiel, and it is here that Curlew were likely to be foraging on the exposed mud. The peak count in the earlier winter reached 103, but only 64 in the later winter. However, without this exceptionally high count, the site mean in both winters was comparable. There does not therefore appear to have been great changes in Curlew abundance or distribution.

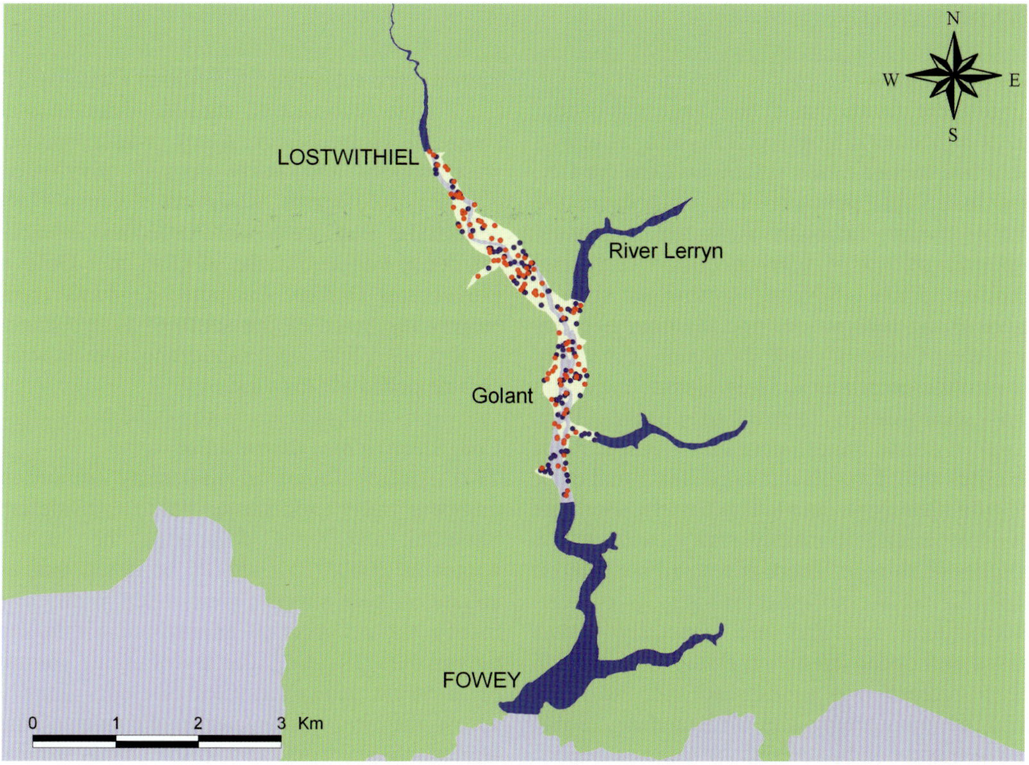

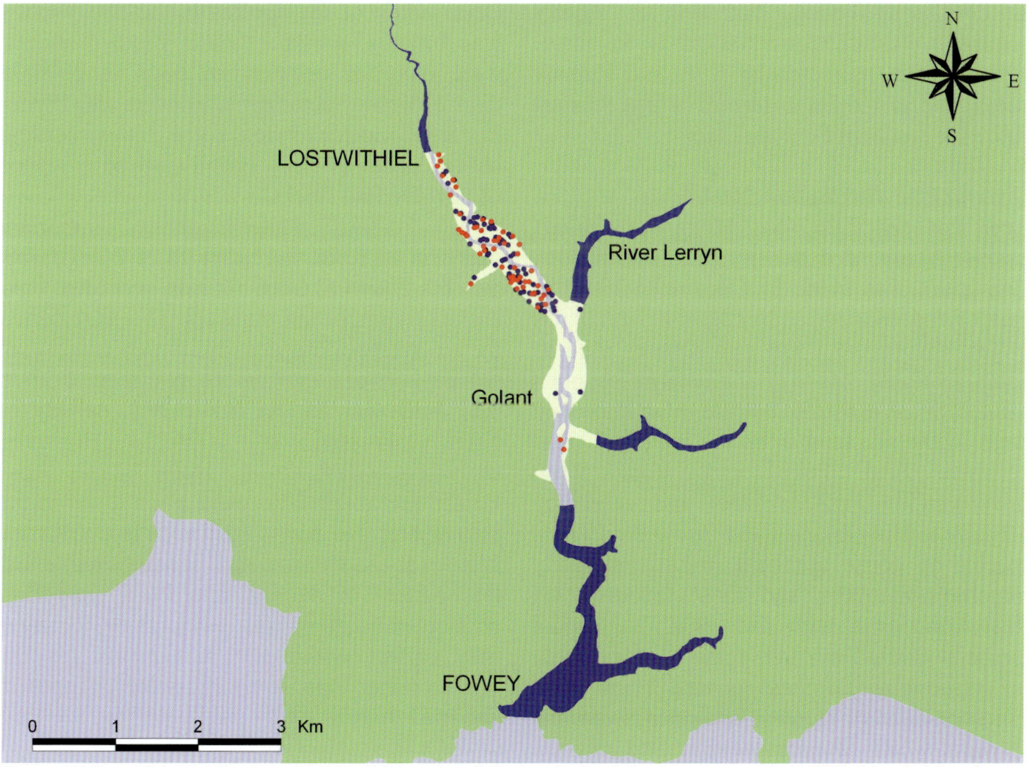

Figure 65. Low Tide distribution of Mallard (above) and Curlew (below) for the winters of 1995/96 (blue dots) and 2004/05 (red). Yellow = intertidal; pale blue = subtidal. Dark blue areas never covered.

LINDISFARNE

Site description

Lindisfarne forms one of the largest intertidal areas in northeast England. This site, as one of only two barrier beach systems within the UK, has an unusual structure. The majority of the site is sandy, although there are increasing amounts of silt in parts of Budle Bay and Fenham Flats. Several freshwater creeks traverse the flats at low tide. Saltmarsh exists between Goswick and Fenham, especially around the causeway to Holy Island, and along the southwestern shore of Budle Bay. Extensive sand dunes occur on several parts of the site, with dune slacks, dune heath and dune pasture also represented. The eastern shoreline of Holy Island is mainly rocky, with a few patches of shingle. There is a small harbour on Holy Island but no other industry is present. Recreational activities are generally water-based and occur mainly in Budle Bay, though beach recreation is widespread over the entire area, as are walking and birdwatching. Some grazing and hand-gathering of mussels occurs, as does wildfowling, but this is strictly licensed. Wildlife conservation is in force, with the area protected by SPA and Ramsar status, and in 1997 a waterbird refuge was set up on the southern Fenham Flats.

General bird distribution 2004/05

Area covered 2,965 ha; **Mean total birds** 28,094; **Mean bird density** 9.5 birds per ha. One count was made in December 2004, as part of the continuing Northumberland Atlas project (hence the 1 km grid squares used as count sections). Most of Lindisfarne supports some birds at low tide, but the most important areas of the site tend to be the intertidal areas between Beal Point and Ross Point (incorporating Fenham Flats and extending offshore to Holy Island), and Budle Bay. Three species of goose are held in internationally important numbers, and most favoured the Ross Point area, with additional Pink-footed Geese at Budle Bay. These species may occur on surrounding farmland as well as on more estuarine habitat, as may Whooper Swan and Wigeon, which favoured similar grid squares. Pintail were scattered off Fenham Flats, whilst Eider were recorded on subtidal areas, mostly south of Holy Island. Many wader species were widely and fairly densely distributed (Curlew, Redshank), whilst the Fenham Flats and Budle Bay areas were especially notable for dense concentrations of Grey and Golden Plover and Dunlin. Bar-tailed Godwits were largely restricted to the south of Holy Island, an area also featuring Sanderling, which were also recorded at Goswick.

Comparative bird distribution

Two species found in nationally important numbers at Lindisfarne are considered, Shelduck and Knot, comparing the current winter to 2000/01. It should be noted that as counts on both surveys were made in any one month, the prospect of anomalous within-site movements affecting the distributions cannot be ruled out. However, mean Shelduck density across the site was greater in the later of the winters, rising from 0.35 to 0.50 birds per ha. Distribution of the species in the two winters was broadly similar, though more Shelduck were recorded on Fenham Flats in 2004/05 than 2000/01. Sector densities were also higher in the second winter, and some of the intertidal habitat south of Holy Island showed evidence of new feeding flocks.

In 2000/01, internationally important numbers of Knot were present at Lindisfarne, but by 2004/05 Core Counts recorded only nationally important numbers. This is reflected by an average site density, which more than halved between the two counts, from 1.33 to 0.62 birds per ha. At Budle Bay, density of Knot increased in 2004/05. However, elsewhere (Beal Point, Holy Island, Ross Point), sector densities were much lower in the later winter. No Alerts have been issued for the species at Lindisfarne, so it seems possible that this pattern of decline may be a temporary phenomenon; likewise, distributional changes may reflect within-site movements for the winter rather than any large-scale changes in site use.

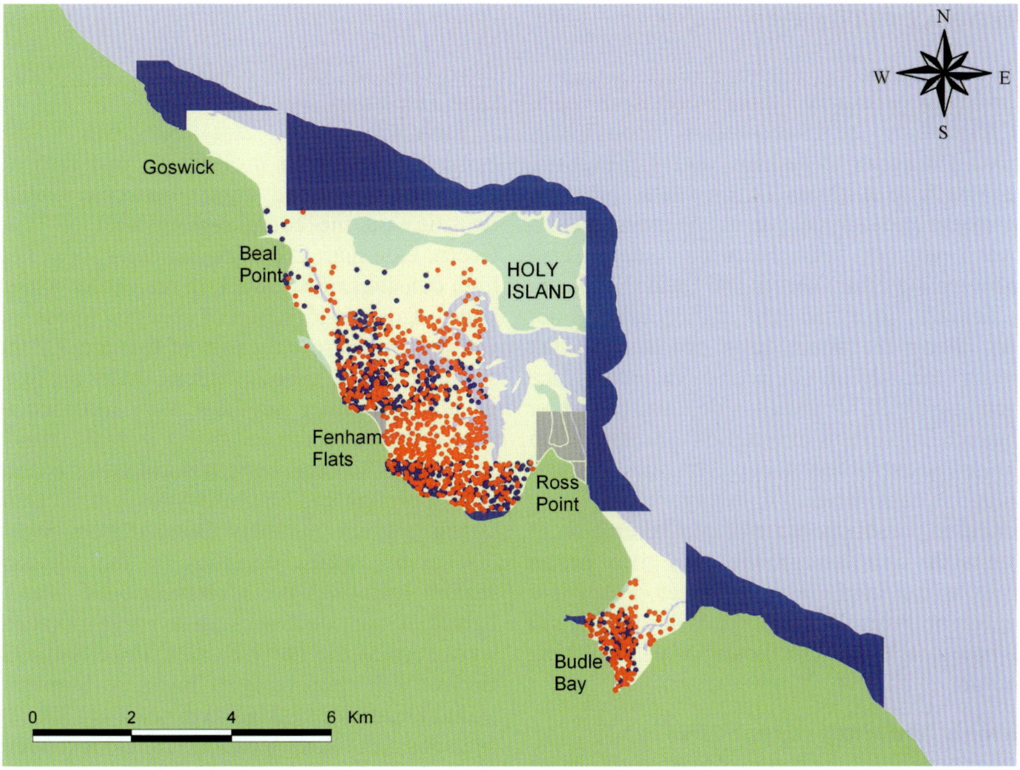

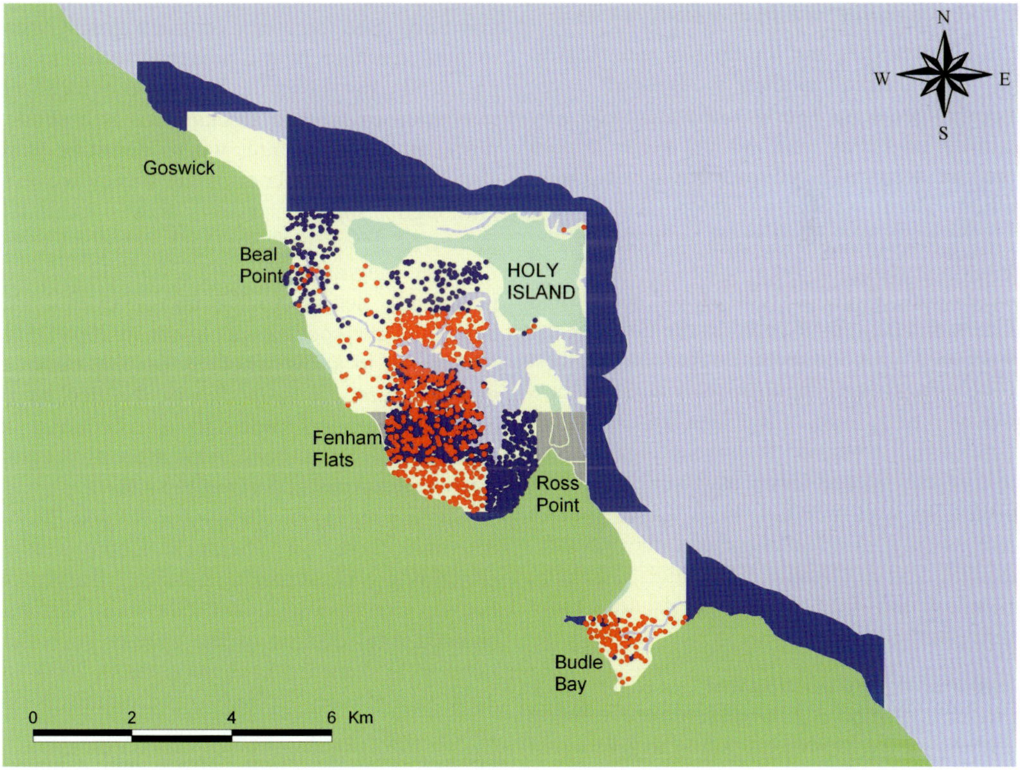

Figure 66. Low Tide distribution of Shelduck (above) and Knot (below) for the winters of 2000/01 (blue dots) and 2004/05 (red); 1 dot = 2 birds. Yellow = intertidal; pale blue = subtidal; pale green = nontidal. Grey area not covered in later winter; dark blue areas never covered.

MEDWAY ESTUARY

Site description

The Medway Estuary, on the North Kent coast, is formed by the river outflow upon which stand Rochester, Chatham and Gillingham, opening into a series of mudflats, brackish saltmarsh and subtidal islands. It is central to a local network of estuarine sites, also encompassing the Thames and Swale. The area is heavily urbanised and industrialised, not only along the south bank but also at the mouth: the Isle of Grain and Sheerness face each other on opposite sides of the river channel. The extent of intertidal and saltmarsh habitat is protected under SPA designation, though is considered in unfavourable declining condition, largely because of coastal erosion of saltmarsh. *Spartina* growth may also be an issue, whilst land claim for development is a persistent threat. Watersports, wildfowling and shipping activity are potentially disturbing factors.

General bird distribution 2004/05

Area covered 2,552 ha; Mean total birds 20,017; Mean bird density 7.8 birds per ha.

Barksore Marshes and the area of intertidal to the east called Bedlam's Bottom were important areas for a variety of waterbirds. The majority of Pintail and Avocet were located here, as were a large proportion of Dark-bellied Brent Geese, Wigeon and Knot. Moving eastward, count units around Chetney Marshes supported high densities of Brent Geese and Wigeon, with most Bar-tailed Godwit and Teal distributed at the north of the marsh area. The one sector counted in the west of the estuary at Gillingham was notable for a pocket of high Black-tailed Godwit aggregation, whereas Stoke Saltings to the west of the Isle of Grain also supported this species, as well as Knot. Five wader species were widely distributed across the estuary, Lapwing and Dunlin at high density, Oystercatcher, Curlew and Redshank more thinly spread.

Comparative bird distribution

A number of species have undergone declining numbers on the Medway, leading to a series of WeBS Alerts (Maclean *et al.* 2005). Here, the distributions of Shelduck and Ringed Plover (both with High Alerts over ten years) are examined for the latest winter, 2004/05, and the winter of 1996/97 when the last WeBS Low Tide Count was performed.

Coverage in 2004/05 was less comprehensive than in 1996/97, and thus no comparisons can be drawn between relative densities on the count sectors west of Ham Green. Otterham Creek in particular supported high densities of Shelduck in the earlier winter and so any consideration of changes across the site will relate to only part of the site. Of the sectors counted in both winters, major declines in average winter numbers were recorded at Ham Green and the low-lying marshy islands off shore to the east (Burntwick Island, Greenborough and Slayhills Marshes). At Ham Green, average numbers dropped from above 500 to 68, whilst densities on the adjacent sectors including the aforementioned islands halved. Other decreases in average numbers were seen at the north east of Stoke Saltings, Bedlam's Bottom (east of Barksore Marshes) and at Deadman's Island to the east of Chetney Marshes; at this location, Shelduck were virtually absent where previously an average 164 birds used the area over the winter. Owing to non-coverage of the adjacent sector to the east, it is not possible to rule out localised relocation of feeding flocks. This is a similar problem at Gillingham, where densities were also reduced in the later of the survey winters. Although some survey areas exhibited increased density of Shelduck (especially the Chetney Marshes area, with a six-fold increase in average density to the north), overall site density of the species was lower in 2004/05. It is difficult to know whether the distributional changes seen are reflective of the decline in Shelduck numbers known to have occurred, or whether there has been relocation of feeding birds to unsurveyed parts of the site.

Some patterns of change for Ringed Plover resemble those for Shelduck. Count sectors containing Deadman's Island, Burntwick Island, Greenborough and Slayhills Marshes showed total decline of the species such that none were recorded in these areas in 2004/05. This pattern also applied to the mouth of the estuary, along the south shore of the Isle of Grain. Between Ham Green and Chetney Marshes, overall density of Ringed Plover remained similar, with decreases at the latter and Bedlam's Bottom counterbalanced by increases at the former. At Stoke Saltings, declines in density from small average numbers were recorded. Overall mean site density was half the value in 1996/97 for the species, and whilst it may be possible that Ringed Plover have redistributed to count sectors not covered in 2004/05, other evidence suggests that the species may havc moved to neighbouring estuaries such as the Swale and Thames, at least at roost (Banks *et al.* 2005).

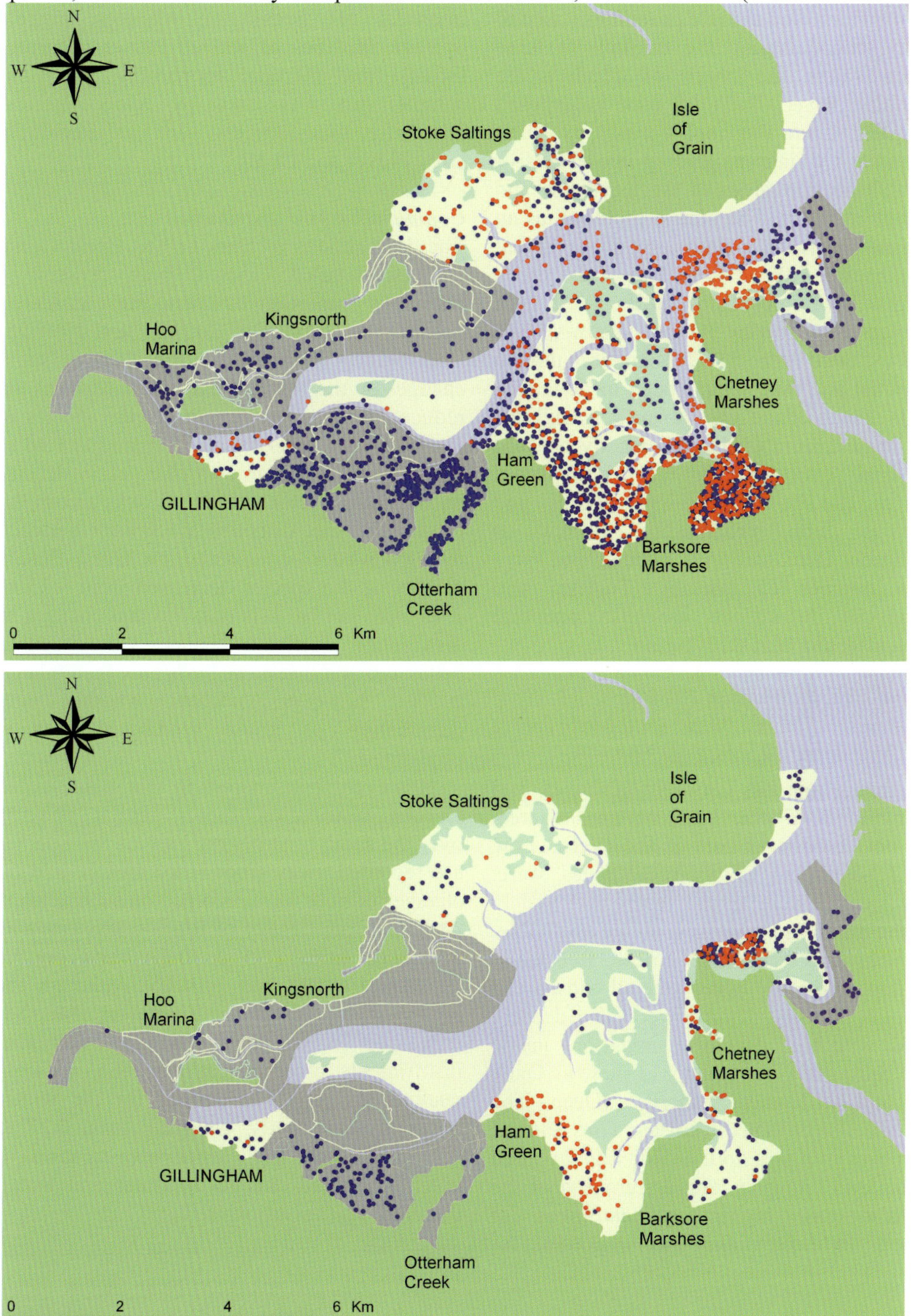

Figure 67. Low Tide distribution of Shelduck (above) (1 dot = 2 birds) and Ringed Plover (below) for the winters of 1996/97 (blue dots) and 2004/05 (red). Yellow = intertidal; pale blue = subtidal pale green = nontidal. Grey area not counted in later winter.

NORTH WEST SOLENT

Site description

The area of the North West Solent counted for WeBS stretches along the Hampshire coast from the Hurst Spit shingle promontory east to Sowley, encompassing the outflow of three running waters, the largest of which is the Lymington River. Intertidal mud is exposed principally inside the hook formed by Hurst Spit and at Lymington, grading into extensive saltmarsh on both sides of the Keyhaven and Lymington Rivers. The area is protected as an SSSI and forms the western end of the Solent & Southampton Water SPA. Part of the site is managed by Hampshire Wildlife Trust as a nature reserve. Sowley Pond is also an SSSI and is included in the SPA designation, though is nontidal and thus not included in Low Tide Counts. Much of the site habitat is considered to be in unfavourable decline, mostly because of coastal squeeze of saltmarsh against sea defences. The site borders the New Forest, and there is little urbanisation except at Lymington. Here, sailing is popular and there are a number of marinas. Tourism and recreational disturbance are also potential factors affecting bird distribution.

General bird distribution 2004/05

Area covered 753 ha; Mean total birds 7,433; Mean bird density 9.9 birds per ha.

The grazing marshes between Keyhaven and Lymington, and other sites within the area, were not counted as part of the WeBS Low Tide scheme, but all intertidal areas were. Many species were spread thinly and widely over the mudflats, suggesting all are profitable. Shelduck, Oystercatcher, Curlew, Redshank, Turnstone and Little Egret were all widespread at varying density. Dunlin were also densely spread, with high density aggregations inside Hurst Spit, off Pennington Marshes, at Oxey Lake and at Tanners Lane/Pitts Deep. The latter site also attracted Knot and Grey Plover, whilst Sowley Farm, to the east, was favoured by Teal and Lapwing. The eastern end of the site also saw highest concentrations of Wigeon. Pylewell Lake was important for Pintail, Grey Plover and Lapwing, whilst at the nearby mouth of Lymington River, additional Teal concentrations were found. On the intertidal off Pennington Marshes, highest average densities of Ringed Plover occurred. Finally, the intertidal formed by the Hurst shingle spit was used by many species, especially mud-foragers such as Shelduck, Grey Plover and Dunlin.

Comparative bird distribution

The distributions of Dark-bellied Brent Geese and Black-tailed Godwit, both present in nationally important numbers, are considered here for the winters of 2004/05 and 1997/98. Dark-bellied Brent Geese have shown some changes on the North West Solent. Most of the reduced average site density for the species in 2004/05 (0.76 birds per ha compared to 1.18 in 1997/98) is attributable to changes in density on the count sector at Hurst Beach; across the rest of the site, densities are similar between the two winters considered. At Hurst Beach, density of the species decreased from 9.33 to 2.45 birds per ha. It is possible that changes in distribution reflect changes in feeding behaviour and that more birds now feed inland. It is, however, unclear whether any degradation of goose feeding habitat is implicated.

However, it is interesting that Black-tailed Godwit distribution shows similar patterns over the same time period. Few of this species are recorded away from Hurst Beach, and no changes of note were recorded on such sectors. At Hurst Spit, density of Black-tailed Godwits decreased from 1.95 to 0.23 birds per ha between 1997/98 and 2004/05. Although the count sector abutting Keyhaven Marshes witnessed increased site density, this was insufficient to counterbalance the apparent decline. Average site density for the winter declined accordingly, changing from 0.27 to 0.11 birds per ha. Closer investigation of potential habitat changes at Hurst Spit would therefore seem worthwhile.

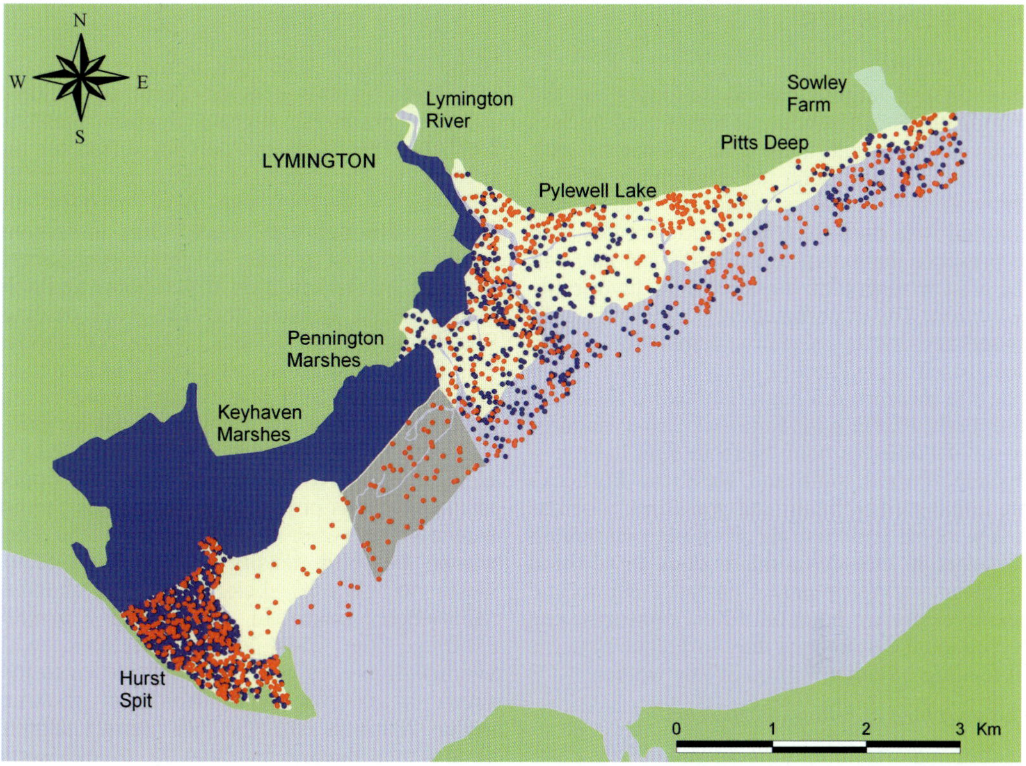

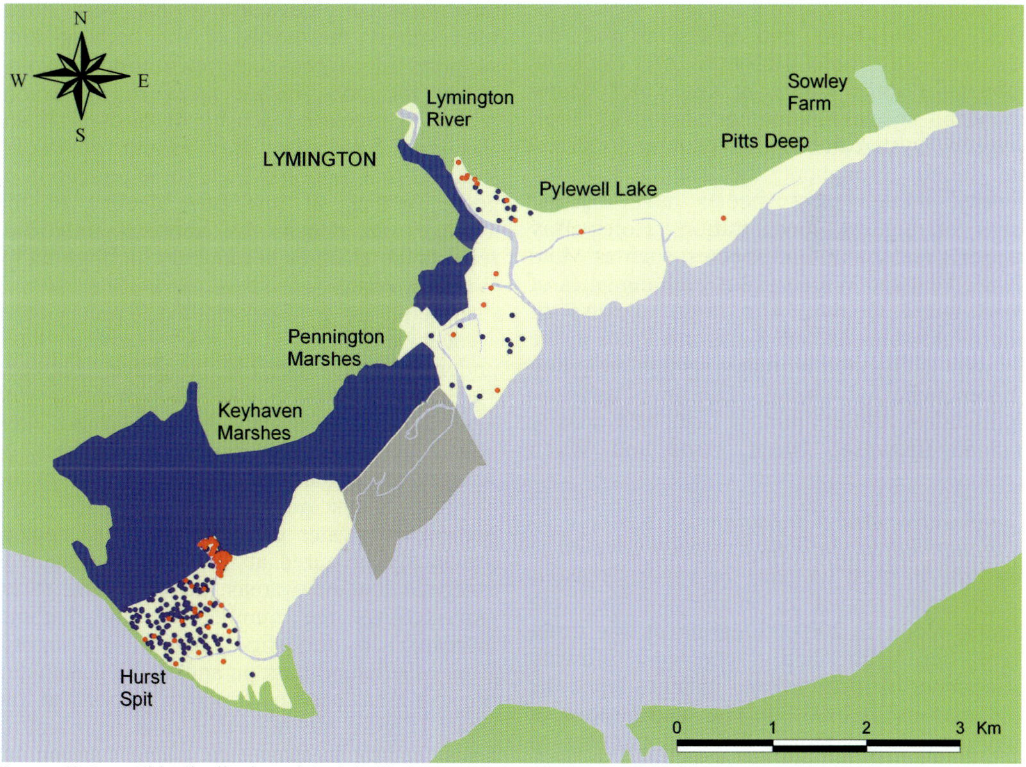

Figure 68. Low Tide distribution of Dark-bellied Brent Goose (above) and Black-tailed Godwit (below) for the winters of 1997/98 (blue dots) and 2004/05 (red). Yellow = intertidal; pale blue = subtidal; pale green = nontidal. Grey area not covered in earlier winter; dark blue areas never covered.

POOLE HARBOUR

Site description

One of the largest natural harbours in the world, Poole Harbour, Dorset, comprises extensive intertidal flats and saltmarsh, with nontidal refuges such as Brownsea Island. Unusual tidal patterns and low tidal range coupled with poor flushing characteristics help to preserve the extent of intertidal habitat, and the vast majority of the site is in favourable condition. The importance of Poole Harbour is underlined by its protective legislation, being designated as a Ramsar site, SPA and SSSI. The site incorporates National Nature Reserves and both RSPB and National Trust manage areas of Poole Harbour. Surrounding habitat is diverse. To the south and west, the Purbecks are dominated by heath and grassland, whereas to the north and east the Poole-Bournemouth conurbation includes a major dock in addition to urban development. Coastal erosion and sea-level rise are potential future concerns, and recreational disturbance may influence patterns of feeding bird distribution at low water.

General bird distribution 2004/05

Area covered 1,563 ha; Mean total birds 15,693; Mean bird density 10.0 birds per ha.

Most of the bays and inlets around the perimeter of Poole Harbour offer suitable habitat for some species at low water. Many abundant species including Shelduck, Teal, Wigeon, Oystercatcher, Redshank and Curlew are found throughout the site at varying density, though most species have discrete areas of highest concentration. Holes Bay supports many wildfowl species, such as Mute Swan, Teal and Wigeon, plus high densities of Black-tailed Godwit. The large expanse of intertidal habitat around Kesworth Point in the far west is favoured by many waders including Oystercatcher, Lapwing, Dunlin, Redshank and Curlew. Other areas of high bird density include Brownsea Island, Arne and Wych Lake.

Comparative bird distribution

Although Poole Harbour has been covered at low tide in most winters, for various reasons counts have tended to be analysed for Core Counts alone. In 2004/05, counts at low water were fed into both schemes, allowing assessment of distributional changes since the last analysis in 1993/94. The distributions of two species undergoing different patterns of change, Dark-bellied Brent Goose and Avocet, are investigated.

In 1993/94, Dark-bellied Brent Goose were present in nationally importance numbers in Poole Harbour. The site remains nationally important for this species, currently holding average numbers slightly higher than in 1993/94. The Medium Alert identified for the species at the site (Maclean *et al.* 2005) indicates a decline that may involve site factors, although the species has undergone general national decline. An interesting pattern emerges in the low tide distribution at Poole Harbour that reflects the site-level decline. Previously favoured areas, where the species was found at high density, show evidence of greatly reduced bird density, and in one case a complete absence. Of these areas, all lie along the south of the harbour, from Brand's Bay to west of Arne. At the former, density dropped from 0.84 birds per ha to just 0.1 birds per ha. Newton Bay was not surveyed in the later winter, but bird density on the count sector west to Wych Lake roughly halved. Middlebere Lake, in the Wych Channel, exhibited the most severe declines, with average counts of over 400 birds down to just seven in 2004/05; to the northeast of Arne, lower numbers in 1993/94 again showed almost total decline by 2004/05, though some movement to Kesworth Point was clear. Interestingly the north of the harbour now supports more geese, compensating to some extent the now vacant areas in the south. Although increased density of birds at Holes Bay and Sandbanks Bay (south of Poole) suggests that there has been some relocation of geese feeding at low tide, decline of the average site density by approximately half reflects the changes in numbers highlighted by WeBS Alerts.

In contrast to Dark-bellied Brent Geese, Avocet numbers at Poole Harbour have undergone a threefold increase between 1993/94 and 2004/05, and although the species has increased nationally, Poole Harbour now holds internationally important numbers of Avocet, more than any other single site in the country. Consequently, Avocets are now present in greater densities on those sectors previously favoured at low tide, and there is evidence that new areas are beginning to be exploited as the numbers in the harbour increase. In the first winter of survey, Brownsea Island Lagoon and Wych Lake were the sole areas used by foraging Avocets at low water (Figure 69: note that dots representing birds feeding in Brownsea Island Lagoon are arbitrarily placed on intertidal areas of the island). Bird density at these areas has risen sharply: from 12.4 to 29.3 birds per ha on Brownsea Island, and from 2.1 to 3.8 birds per ha at Wych Lake. Another area of the Wych Channel, Middlebere Lake, contains previously unprecedented concentrations of Avocet (3.9 birds per ha), suggesting that this area is an overspill feeding site for increasing numbers of the species. Furthermore, signs of expansion into the north of the harbour (at Lytchett and Holes Bays) suggest that the carrying capacity of the site has not yet been reached. It will be interesting to see whether further colonisation of these areas occurs in future.

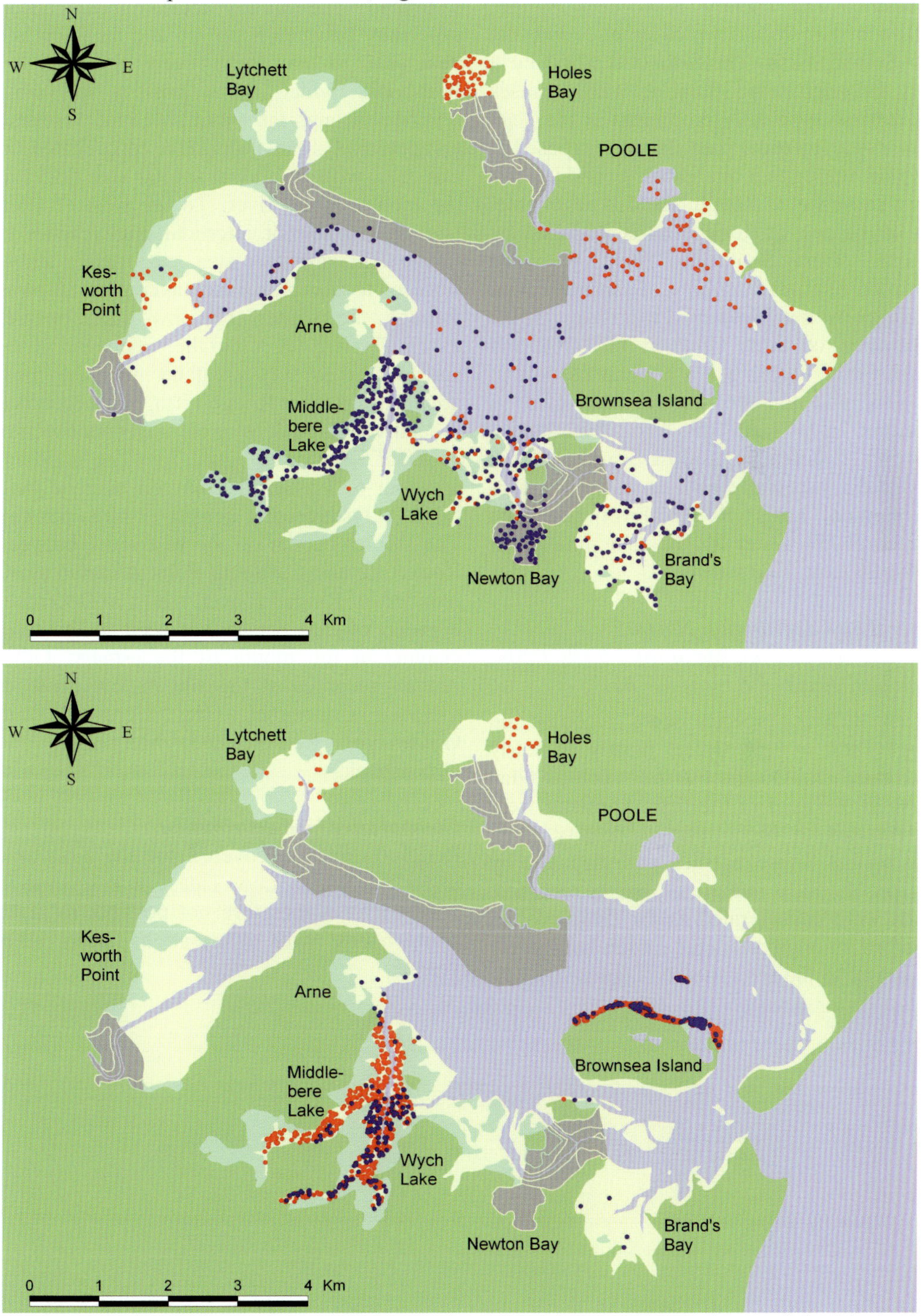

Figure 69. Low Tide distribution of Dark-bellied Brent Geese (above) (1 dot = 2 birds) and Avocet (below) for the winters of 1993/94 (blue dots) and 2004/05 (red). Yellow = intertidal; pale blue = subtidal; pale green = nontidal. Grey areas not counted in later winter.

ROUGH FIRTH

Site description

Rough Firth is a small inlet on the Dumfries & Galloway coast, 6 km to the south of Dalbeattie. The estuary is formed by the outflow of Urr Water, which empties into the west Solway Firth. The firth is typified by a rocky and steep-rising shoreline, with expanses of saltmarsh especially prominent at the head of the river. Intertidal mud flats are exposed at low tide throughout the estuary, and at this time a causeway to Rough Island is negotiable. Part of the site, at Rockcliffe, is managed by National Trust Scotland. There is some indication of coastal squeeze of saltmarsh, and cockling in the Solway Firth may affect movements of birds between the area and Rough Firth. Yachting is a popular pastime at the site, and tourism, though restricted in location, may lead to bird disturbance. However, disturbance is most heavily created by walkers, especially with unrestrained dogs on intertidal areas. Rough Firth has been counted at low water for the first time under the WeBS Low Tide scheme.

General bird distribution 2004/05

Area covered 539 ha; Mean total birds 1,509; Mean bird density 2.8 birds per ha.

Fifteen different species were recorded on Rough Firth at low water, mostly in fairly small numbers. On the marshes such as Glen Marsh at the head of the firth, Lapwing were concentrated at high average density (2.83 birds per ha). These marshes were also used by Wigeon, the most numerous species recorded, which were concentrated as far downriver as Kippford, though scattered individuals were present up to the river mouth (Figure 70). The species was at greatest average winter density on Glen Marsh, where just under 8 birds per ha were recorded grazing or loafing on the saltmarsh. The muddy creeks in the north, especially between Kippford and the Glen Isle peninsula to the west, were favoured by Redshank, which were present in stable numbers through the winter. Mallard were also comparatively most densely congregated in this area, with additional similar size flocks at the mouth of the river. To the west of Glen Isle, an expanse of intertidal mud reaches to Castle Hill, and this area supported a stable flock of foraging Shelduck averaging 88 birds through the winter, but few other species except Oystercatcher and the widely and thinly spread Curlew. Oystercatcher was the most abundant wader at Rough Firth, being recorded at relatively high density on most count sectors south of Kippford (Figure 70). The intertidal area out to Castlehill Point on average held highest Oystercatcher density (2 birds per ha). The stonier habitat on the east shore between Rockcliffe and Castlehill Point attracted small numbers of Ringed Plover, whilst in December a group of 200 Scaup was seen in the river channel at the mouth of the estuary. Other species recorded at the site in low densities included Cormorant, Grey Heron, Pintail, Goldeneye, Red-breasted Merganser and Greenshank.

Comparative bird distribution

Rough Firth was covered for the first time under the WeBS Low Tide Count scheme in 2004/05; it is therefore not possible to make comparisons with distributions from other years.

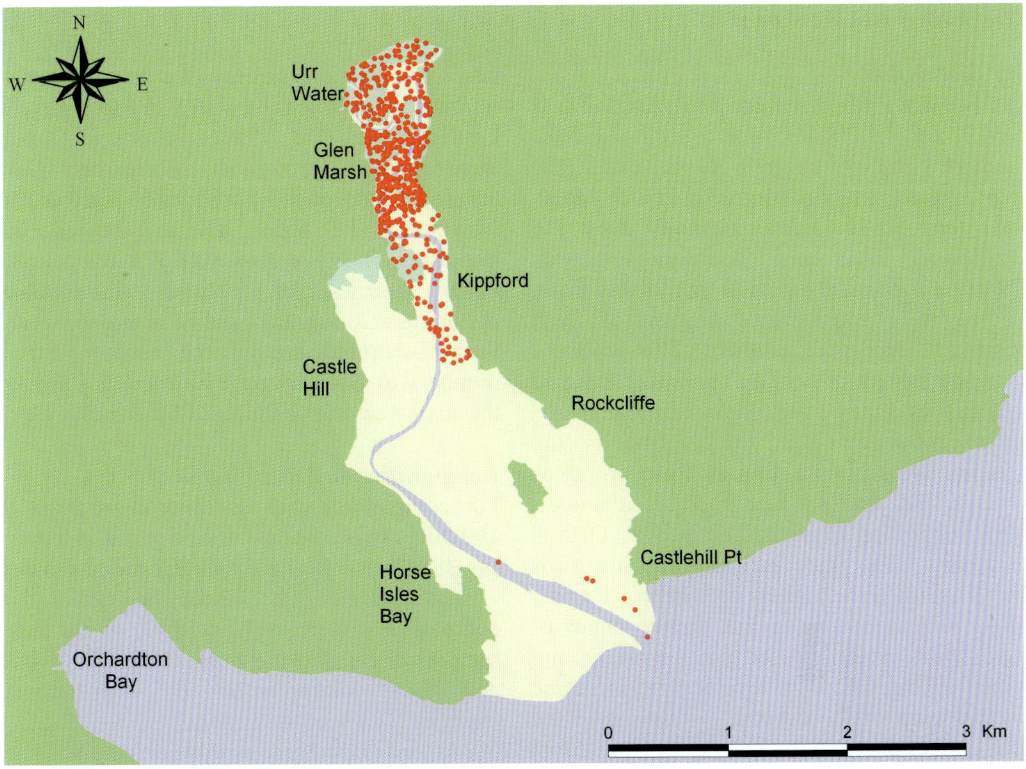

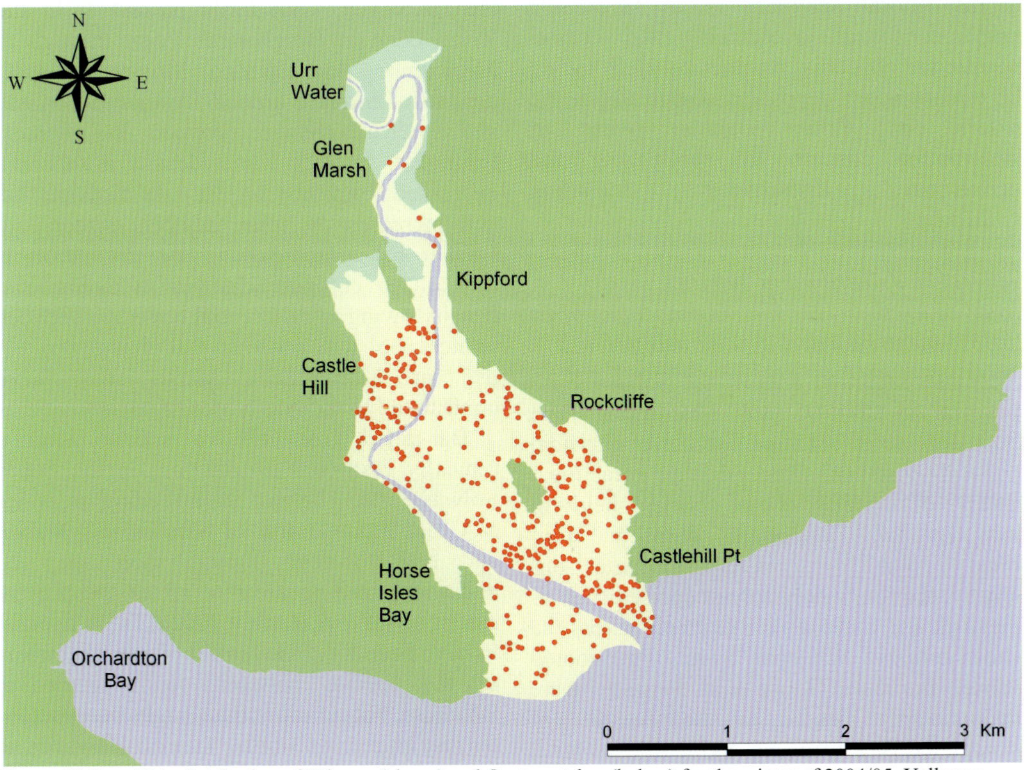

Figure 70. Low Tide distribution of Wigeon (above) and Oystercatcher (below) for the winter of 2004/05. Yellow = intertidal; pale blue = subtidal; pale green = nontidal.

SOLWAY FIRTH (SOUTH)

Site description

The Solway Firth demarcates the Anglo-Scots border, and in 2004/05 count sectors on the southern (English) shore were surveyed. The area covered extended from Rockcliffe Marsh and the River Eden channel west to Cardurnock Flats and Skinburness Marsh. There are nine main inputs to the Solway Firth, leading to a considerable amount of fresh water and deposited sediment. This produces an active system of shifting channels, flats and saltmarshes, many of the latter only inundated on exceptionally high tides. Sand is the predominate substrate, though there are areas of mud and fine silt and rockier substrates, some supporting mussels. The Solway Firth is of utmost importance for non-breeding birds and is afforded the highest protection (Ramsar, SPA, SSSI), forming a link in the chain of west coast estuaries used by migratory and wintering birds. Urban development is scarce on the south of the estuary, with most of the area comprising pasture, marsh or merse. Most of the littoral sediment is in favourable condition, though overgrazing and coastal squeeze are concerns. Another problem facing the estuary is silt deposition, especially at the eastern end. Other potential issues of conservation concern are shellfishing and recreational disturbance including wildfowling.

General bird distribution 2004/05

Area covered 5,492 ha; Mean total birds 26,485; Mean bird density 4.8 birds per ha.

The most important area of the Solway Firth surveyed in 2004/05 was Moricambe Bay. This large inlet contains extensive intertidal flats and substantial saltmarsh, thus attracting a variety of species. Few species held in (inter)nationally important numbers were found in notable densities east of Bowness-on-Solway, though widely spread species such as Shelduck and Curlew were scattered across the whole estuary. Whooper Swan, Pink-footed Goose, Barnacle Goose and Wigeon were predominantly limited to the marshes of Moricambe Bay. Of these, Skinburness Marsh was also noteworthy for high densities of Pintail and Teal; Whitrigg and Newton Marshes for Golden Plover. Ringed Plover and Redshank were also distributed at highest density in Moricambe Bay, both also scattered on intertidal habitat elsewhere. Waders such as Oystercatcher, Grey Plover, Knot and Dunlin were present throughout Moricambe Bay, across Cardurnock Flats and upriver to Bowness. The latter species showed the highest average site density (1.37 birds per ha). Other species of national importance, including Cormorant and Turnstone, were thinly scattered around the estuary. Small numbers of Scaup were also recorded around the scars near the channels of Moricambe Bay.

Comparative bird distribution

For comparison with distribution of birds in 1999/00, two species issued with Medium WeBS Alerts - Mallard and Bar-tailed Godwit - were selected. It should be noted that comparisons were only possible with those sectors covered on the south shore in 2004/05 and thus interpretation of distributional changes is subject to caution.

Mallard were fairly evenly spread across the area surveyed, with birds present from the River Eden at Burghmarsh Point through to Moricambe Bay. Within Moricambe Bay, density of Mallard around Skinburness Marsh shows clear change between the winters examined. Average winter density in this area decreased from 0.61 to just 0.08 birds per ha, and increases on some adjacent sectors are smaller by comparison to the decline. It is difficult to compare changes in overall site density owing to differential coverage between winters, but it is possible that changes in site use at Skinburness Marsh may have contributed to the Alerts identified.

Bar-tailed Godwit is a species with a typically restricted distribution on the South Solway Firth. The two main feeding areas are at Grune Point, north of Skinburness Marsh, and off Bowness (though there may be additional feeding areas south of the present extent of survey: Hartley 2002). There appears to have been local movement of small numbers of the species between count sectors at Bowness, but no more than might be expected by disturbance or similar factors. However, at Grune Point, sector density declined from 0.68 to 0.07 birds per ha. It is possible that there have been counteractive movements to areas beyond the survey area in 2004/05, or it may be that site-level declines of the species, as identified by WeBS Alerts (Maclean *et al.* 2005), are reflected in the differing distributions.

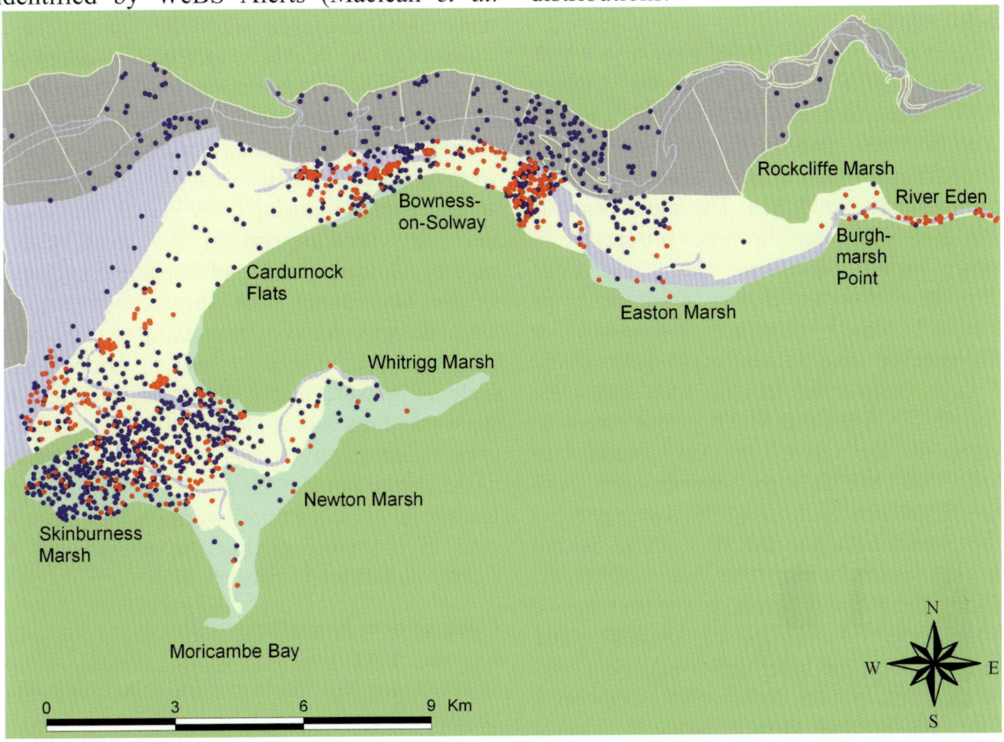

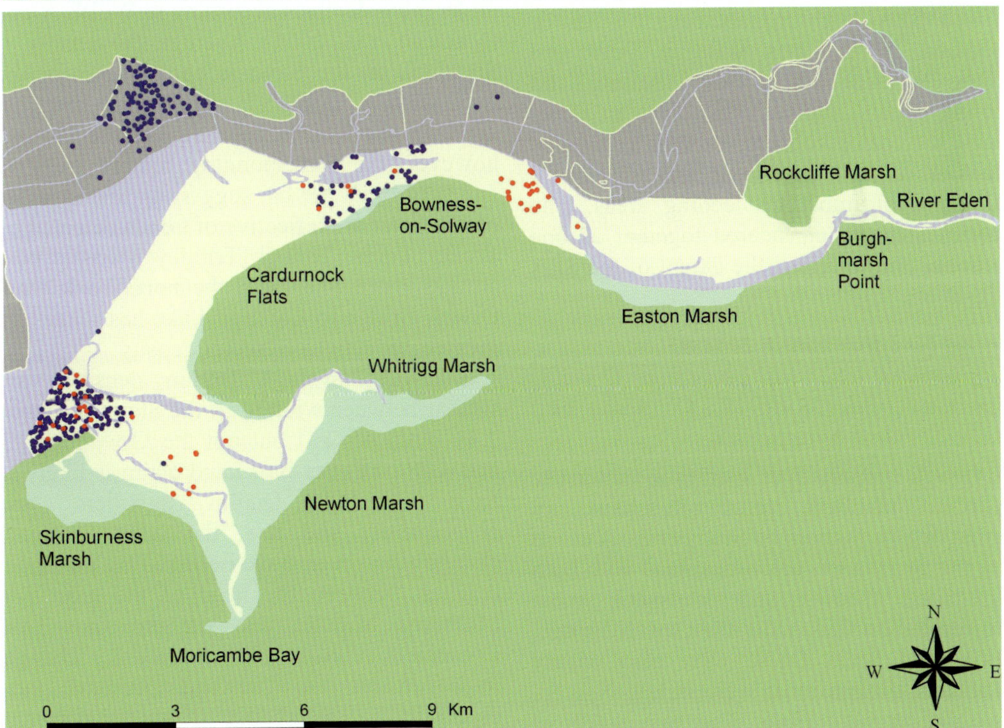

Figure 71. Low Tide distribution of Mallard (above) and Bar-tailed Godwit (below) for the winters of 2000/01 (blue dots) and 2004/05 (red). Yellow = intertidal; pale blue = subtidal; pale green = nontidal. Grey areas not counted in later winter.

STOUR AND ORWELL ESTUARIES

Site description

The Stour is a long and straight estuary, which forms the eastern end of the border between Suffolk and Essex. The estuary's mouth converges with that of the Orwell, which extends from Ipswich to Felixstowe, as the two rivers enter the North Sea. The outer Stour is sandy and substrates become progressively muddier further upstream. There are seven shallow bays along the estuary and much of its length is bordered by sharply rising land or cliffs, covered with ancient coastal woodland and agricultural land, leaving little room for saltmarsh development. Much of the intertidal substrate of the Orwell is fairly muddy. In mitigation for the latest port development, both the north and south shores of the lower reaches of the estuary have had soft silts placed behind stiff clay bunds within the intertidal areas, changing the substrate again. Long stretches of farmland and wet meadow are situated along the mid-estuary, the latter providing roost sites for waterbirds. Nature conservation in the area includes the Stour & Orwell Estuaries Ramsar site and SPA, with management by RSPB, Woodland Trust, Essex Wildlife Trust and Suffolk Wildlife Trust. Some sailing and shooting occurs, and disturbance is an issue, though the major concern remains continued expansion of dock operations and subsequent land claim of important feeding areas. The estuaries are here considered together as one functional unit to reflect the extent of the SPA designation.

General bird distribution 2004/05

Areas covered 1,627/1,227 ha (Stour/Orwell); Mean total birds 38,495/17,747; Mean bird density 23.7/14.5 birds per ha.

The Orwell supports four species in nationally important numbers (plus Redshank in internationally important numbers), and of these the two waders are widespread with both Black-tailed Godwit and Redshank most concentrated on flats under the Orwell Bridge. Dark-bellied Brent Geese were also widely distributed, though most were below Nacton, especially at Jill's Hole. The two wildfowl species, Gadwall and Pintail, favoured Trimley Marshes; Loompit Lake and Mulberry Middle were also preferred by the two species respectively.

The Stour has very different characteristics, with more sheltered intertidal habitat in a series of bays. A wider variety of waders are attracted in important numbers. Most are found throughout the site, with each bay supporting concentrations of waders. Black-tailed Godwit were distributed mostly west of Holbrook Bay, an area also harbouring Grey Plover. Redshank too were predominant in the west of the estuary. Golden Plover and Turnstone occurred in the bays along both shores, whilst Knot were densely packed throughout the estuary. By contrast, Great Crested Grebe and Goldeneye were thinly distributed, tending to the west of the Stour. Dark-bellied Brents were present on many count sectors, especially those on the north bank around Erwarton Bay. Conversely, most Pintail were on the opposite side of the river channel in areas such as Copperas Bay and Mistley.

Comparative bird distribution

Counts from the winter of 1996/97 were selected for comparison, and Shelduck and Dunlin have been selected as species undergoing declines according to WeBS Alerts (Maclean *et al.* 2005). Average site density of Shelduck on the Stour is comparable for the two winters (1996/97: 0.66 birds per ha; 2004/05: 0.63 birds per ha). On the Orwell, however, the corresponding figures are 0.42 and 0.24 birds per ha, suggesting that changes on the latter are reflective of trends detected by WeBS Alerts. Though Trimley Marshes now attracts more Shelduck, the north bank from Orwell Bridge to Loompit Lake has witnessed greatest reductions in sector density.

Likewise, average Dunlin densities are similar between winters on the Stour (7.53 and 7.41 birds per ha) but not the Orwell (10.02 and 4.78 birds per ha). Land claim is likely to have affected Dunlin numbers and consequently site use; the most profound changes, however, have occurred on the upper reaches, between the head of the river and Mulberry Middle, and on the flats near Loompit Lake. Substrate change, improvements to sewage outputs and especially disturbance are pertinent explanatory factors. The latter is partially from dog-walkers and bait-diggers, causing high disturbance on the upper Orwell (Ravenscroft

2005); potentially this may contribute to changes in Dunlin distribution.

The Stour & Orwell Estuaries are counted by Suffolk Wildlife Trust under contract to Harwich Haven Authority. These data are generously made available to The Wetland Bird Survey.

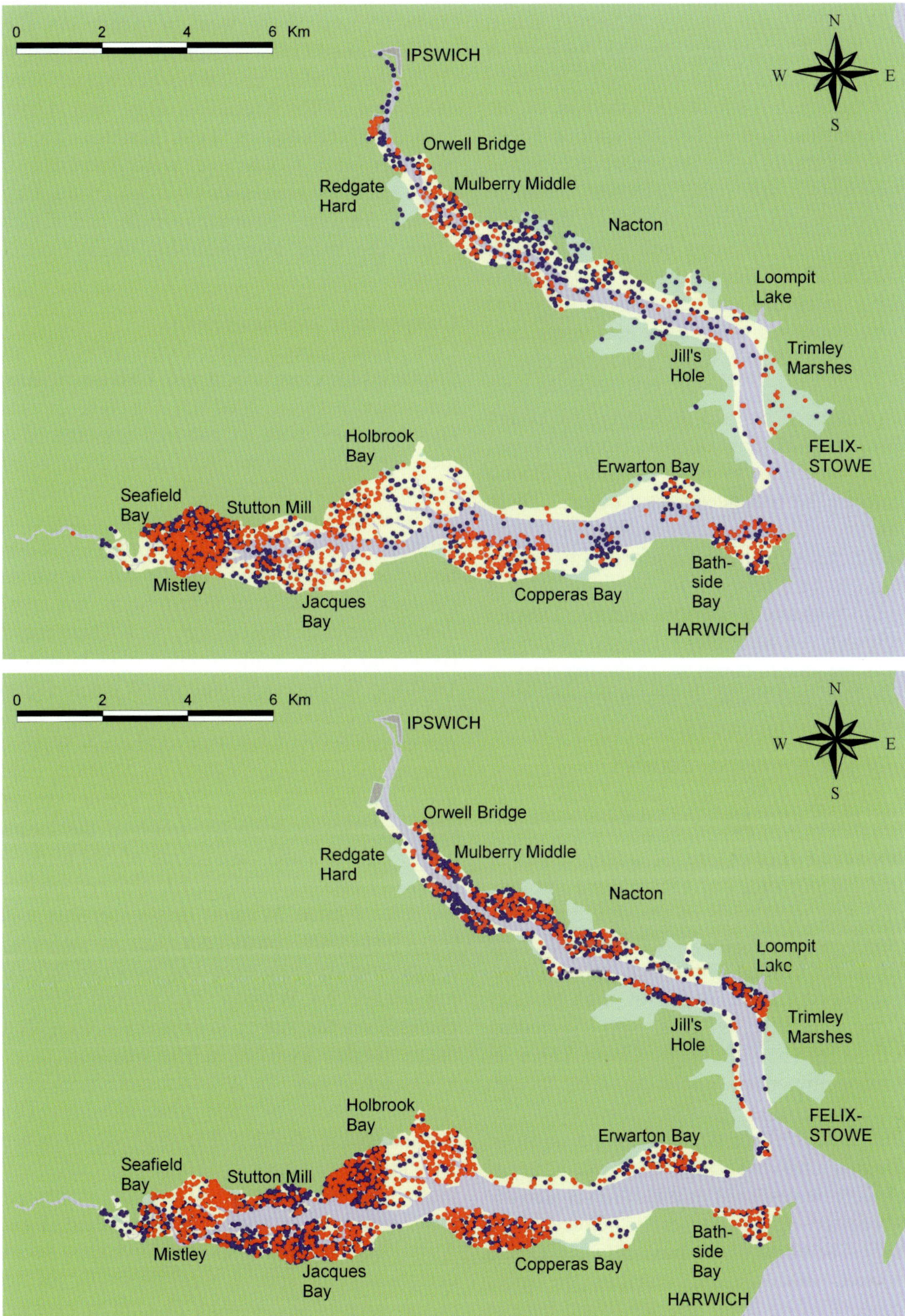

Figure 72. Low Tide distribution of Shelduck (above) (1 dot = 2 birds) and Dunlin (below) (1 dot = 10 birds) for the winters of 1996/97 (blue dots) and 2004/05 (red). Yellow = intertidal; pale blue = subtidal; pale green = nontidal. Grey areas not counted in earlier winter.

STRANGFORD LOUGH

Site description

Strangford Lough is a large shallow sea lough on the east coast of Northern Ireland, protected as a SPA, a Marine Nature Reserve, and a Ramsar Site. The site includes the Narrows, a deep rocky channel to the Irish Sea. The main body of the lough is sheltered to the east by the Ards Peninsula, and is fed by various rivers and tributaries. Downpatrick and Newtownards are the largest human habitations nearby. Within the lough there are numerous rocky outcrops and small islands. The north of the lough in particular holds extensive intertidal mud and sand flats and there are countless other bays and inlets, and large expanses of open water, providing a wide diversity of habitat. Since 2001, mobile gear fishing has been banned in Strangford Lough to allow populations of the Horse Mussel *Modiolus modiolus* to recover. Static fishing and catching of crustaceans still occurs. There is some recreational activity within the lough, including sailing, and Strangford Narrows is proposed to be a test site for a marine current turbine to generate renewable energy. Despite the enormity of Strangford Lough, dedicated counters are able to count along the majority of its shoreline, and do so at low tide annually - an impressive achievement.

General bird distribution 2004/05

Area covered 4,335 ha; Mean total birds 43,182; Mean bird density 10.0 birds per ha.

Twenty-seven species are of international or national importance at Strangford Lough, meaning that most parts of the site, especially the bays and expanses of intertidal, are important for at least one species at low tide. Several species were found at high density over the majority of the site, including Nearctic Light-bellied Brent Geese (here at their most important wintering site, though typically in smaller numbers than in earlier months), Shelduck (especially in the north of the lough), Lapwing and Redshank, whereas others were widespread at much lower density (Red-breasted Merganser, Greenshank, Curlew). The north of the bay, between Castle Espie and Newtownards, contains the most profitable feeding flats. Species found here at high density were typically waders such as Ringed Plover (also around the east coast), Golden Plover (also in dense pockets elsewhere), Grey Plover (restricted here), Knot (restricted to the area), Dunlin (also densely distributed in smaller bays), Black-tailed Godwit (most north west of Ardmillan Bay and Bar-tailed Godwit (from Ardmillan Bay to Greyabbey). Mute and Whooper Swans were associated with Ardmillan Bay and Castle Espie, and clumps of Teal, Mallard, Pintail, Shoveler and Goldeneye were recorded in areas such as Danes Point, Castleward Bay, Mount Stewart, Ardmillan Bay and Castle Espie, the latter also featuring Great Crested Grebe.

Comparative bird distribution

Low Tide Count data from 1994/95 are displayed for comparison with bird distribution ten years later in 2004/05, for Wigeon and Oystercatcher, both of national importance at the site. The former has undergone some apparent within-site movements over the past ten winters; these could be due to changes in habitat suitability, or simply reflect behavioural patterns. The most profound changes have occurred in two areas; the small islands around Ballymoran and Quarterland Bays, and in the south around Gores Island. In 1994/95, the average size of Wigeon flocks was over 150 birds at a density of 0.37 birds per ha off Ballymoran Bay. Although Wigeon were recorded close to shore in 2004/05, none were present on the count sector further offshore. However, in the south of the lough, below Gores Island, an average of 247 birds, at a density of 3.92 birds per ha, was recorded in the later winter. Wigeon density in this area was just 0.06 birds per ha in 1994/95, suggesting that the area has assumed greater importance for Wigeon. The overall effect of changes on the site is to produce a consistent mean site density (0.07 in both winters).

Oystercatcher in 2004/05 were present in second highest average site densities (the highest being Golden Plover). Most sectors around the site supported some Oystercatchers, though the intertidal expanse in the north was most favoured in both winters studied. From Ardmillan Bay north to Newtownards, sector densities of the species were generally greater in the later of the winters or similar between winters. However, on the east coast, from Newtownards to Greyabbey, density in the later winter was considerably lower. Why this area should now be unfavourable is unclear, but it may be that a lower number of Oystercatcher using the site can feed profitably on the main northern mudflats and elsewhere around the lough (average site density has dropped from 1.72 to 1.25 birds per ha).

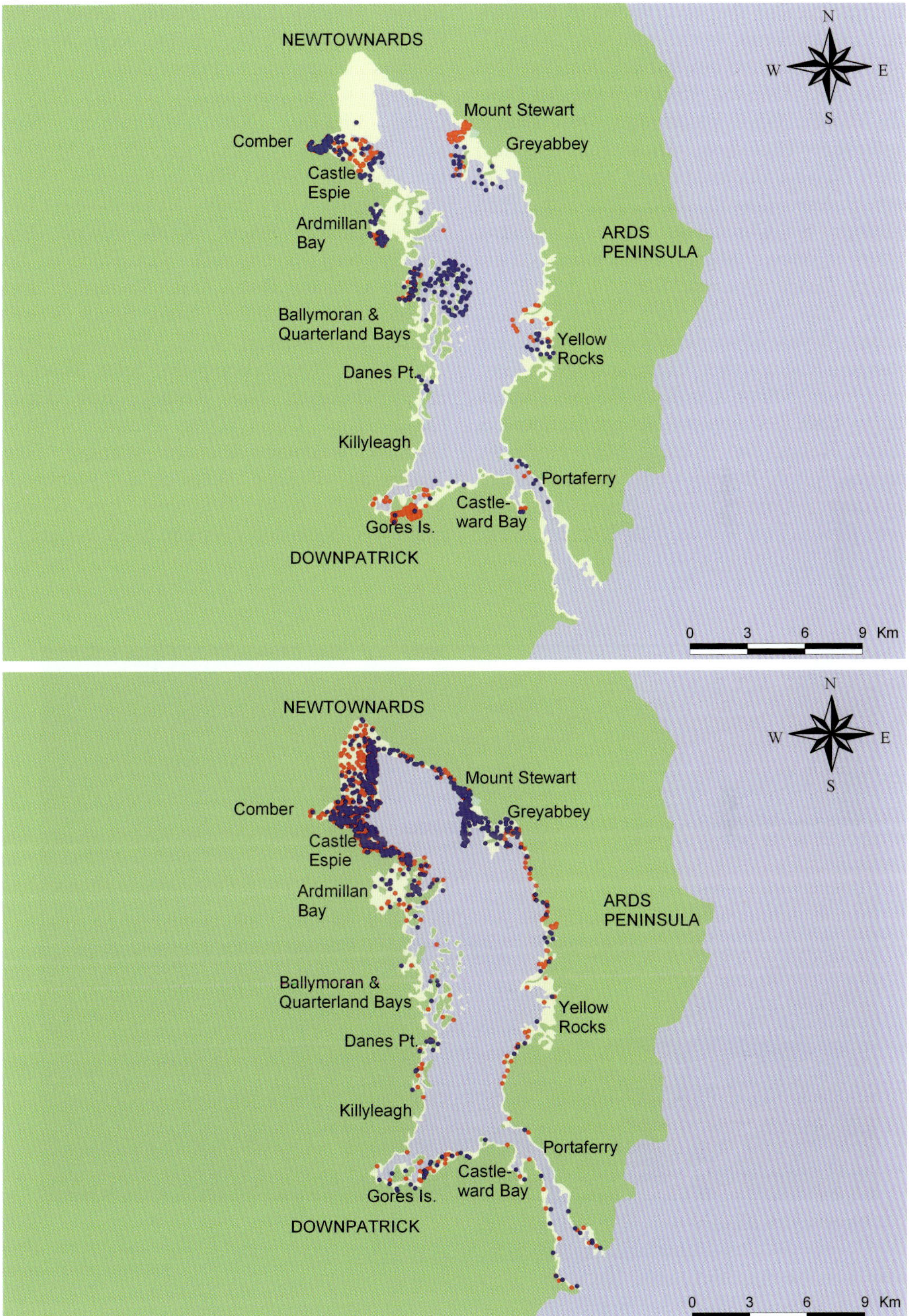

Figure 73. Low Tide distribution of Wigeon (above) (1 dot = 2 birds) and Oystercatcher (below) (1 dot = 10 birds) for the winters of 1994/95 (blue dots) and 2004/05 (red). Yellow = intertidal; pale blue = subtidal; pale green = nontidal. Grey area not counted in later winter.

ACKNOWLEDGEMENTS

We are very grateful to the following people and organisations that contributed to the Low Tide Count scheme in the winter of 2004/05. Apologies to anyone omitted accidentally from the list. J.R. Alderton, Rachel Bain, Jill Bale, Judy Baxter, Dave Blackledge, Dot Blakey, Christopher Bradshaw, Dick Burt, Ronald Butler, David Callahan, Geoff Campbell, Mike Carrier, Alex Carroll, Bob Chapman, Paul Charlton, S. Christmas, Chris Cockburn, Declan Coney, Mark Constantine, Paul Corbett, David Cousins, Kevin Crisp, Jason Crook, Peter Davidson, Jeff Delve, Frances Donnan, Bryan Edwards, Michael Ellison, Ian Enlander, R.R. Farrant, Colin Gay, Stan Gay, Bob Glover, Mhairi Gordon , Bobbie Hamill, Frances Hall, J. Hampton, George Henderson, Graham Hobin, Jim Hobson, Mike Hodgson, Norman Holton, Bob Howells, Andy Humber, Ed Hunter, David James, Philip Johnston, Geoff Kelso, Paul Kemp, Bill Kendall, Mike Lawson, Russell Leavett, Jon Lees, Chris Lewis, Harold Lilley, Paddy Livingstone, Ralph Loughlin, Paddy Mackie, Kerry Mackie, Seamus Magouran, Peter Mason, Frank Mawby, Kevin Mawhinney , Hilary Mayne, Craig McCoy, Neil McCulloch, Niall McCutcheon, Roy McGregor, James McNair, Ian Morrison, John O'Boyle, Geoffrey Orton, Andy Parfitt, Terry Paton, Brian Pavey, Colin Peake, George Peters, Keith Powrie, Eric Rainey, Neil Ravenscroft, Nick Robson, Jim Rowe, Graham Rutt, Alan Shearring, Jack Sheldon, John Shillitoe, Mark Smart, Martin Smith, Len Stewart, G.H. Thomas, David Thompson, Mert Thompson, John Threlfall, Hugh Thurgate, Jack Torney, Bob Treen, Derek Tutt, Chris Tyas, Andrew Upton, Stephen Wadman, Richard Waring, Tony Waterman, Gregor Watson, J. Webb, Jo Whatmough, Colin Williams, Jim Wilson, Eddie Wiseman, John Wright, Mick Wright.

Dunlin (John Bowers)

References

Banks, A.N., Bolt, D., Bullock, I., Haycock, B., Musgrove, A., Newson, S., Fairney, N., Sanderson, W., Schofield, R., Smith, L., Taylor, R. & Whitehead, S. 2005. *Ground and aerial monitoring protocols for in shore Special Protection Areas*. Countryside Council for Wales Marine Monitoring Report No: 11, 89pp.

Beekman, J.H., Dirksen, S. & Slagboom, T.H. 1985. Population size and breeding success of Bewick's Swans wintering in Europe in 1983-4. *Wildfowl* 36: 5-12.

Bibby, C.J., Burgess, N.D., Hill, D.A. & Mustoe, S. 2000. *Bird Census Techniques. Second Edition.* Academic Press, London.

Bill, D.I. & Hollins, J.R. 1989. Report on Portsmouth Harbour low water waterbird counts - Winter 1988/89. Unpublished.

BOURC. 1999. British Ornithologists' Union Records Committee: 25th Report (October 1998). *Ibis* 141: 175-180.

Buckland, S.T., Anderson, D., Burnham, K., Laake, J., Borchers, D. & Thomas, L. 2001. *Introduction to distance sampling: estimating abundance of biological populations*. Oxford University Press, Oxford.

Buckland, S.T., Anderson, D.R., Burnham, K.P., Laake, J.L., Borchers, D.L. & Thomas, L. (editors). 2004. *Advanced Distance Sampling.* Oxford University Press.

Burton, N.H.K., Musgrove, A.J., Rehfisch, M.M., Sutcliffe, A. & Waters, R.J. 2003. Numbers of wintering gulls in the United Kingdom, Channel Islands and Isle of Man: a review of the 1993 and previous Winter Gull Roost Surveys. *British Birds* 96: 376-401.

Clark, J.M. & Eyre, J.A. 1993. *Birds of Hampshire.* Hampshire Ornithological Society, Hampshire.

Collier, M.P., Banks, A.N., Austin, G.E., Girling, T., Hearn, R.D. & Musgrove, A.J. 2005. *The Wetland Bird Survey 2003/04: Wildfowl and Wader Counts.* BTO/WWT/RSPB/JNCC, Thetford.

Cranswick, P.A., Kirby, J.S., Salmon, D.G., Atkinson-Willes, G.L., Pollitt, M.S. & Owen, M. 1997. A history of wildfowl counts by The Wildfowl & Wetlands Trust. *Wildfowl* 47: 217-230.

Crowe, O. 2005. *Ireland's Wetlands and their Waterbirds: Status and Distribution.* BirdWatch Ireland, Newcastle, Co. Wicklow.

Danielsen, F., Skov, H. & Durnick, J. 1993. Estimates of the wintering population of Red-throated Diver *Gavia stellata* and Black-throated Diver *Gavia arctica* in northwest Europe. *Proceedings of the 7th Nordic Congress of Ornithology, 1990.* pp18-24.

Dean, B.J., Webb, A., McSorley, C.A., Schofield, R.A. & Reid, J.A. 2004. *Surveillance of wintering seaducks, divers and grebes in UK inshore areas: Aerial surveys and shore-based counts 2003/04*. JNCC report 357.

Fox, A.D. & Francis, I. 2004. *Report of the 2003/2004 national census of Greenland White-fronted Geese in Britain.* Greenland White-fronted Goose Study, Kalø.

Gilbert, G., Gibbons, D.W. & Evans, J. 1998. *Bird Monitoring Methods*. RSPB, Sandy.

Griffin, L.R. & Mackley, E.R. 2004. *WWT Svalbard Barnacle Goose Project Report 2003-04.* WWT Internal report, Slimbridge.

Hall, C., Smith, L.E., Trinder, M.N. & Cranswick, P.A. 2005. *All Wales Common Scoter survey: report on 2003/04 work programme.* WWT Wetlands Advisory Service report to Countryside Council for Wales, CCW Contract Science Report No. 709.

Hastie, T. & Tibshirani, R. 1990. *Generalized Additive Models*. Chapman & Hall, London.

Heubeck, M. & Mellor, M. 2005. *SOTEAG ornithological monitoring programme: 2004 summary report.* SOTEAG, Aberdeen.

Holmes, J.S., Marchant, J., Bucknell, N., Stroud, D.A. & Parkin, D.T. 1998. The British List: new categories and their relevance to conservation. *British Birds*, 92: 2-11.

Holmes, J.S. & Stroud, D.A. 1995. Naturalised birds: feral, exotic, introduced or alien? *British Birds*, 92: 2-11.

Kershaw, M. & Cranswick, P.A. 2003. Numbers of wintering waterbirds in Great Britain, 1994/1995-1998/1999: I. Wildfowl and selected waterbirds. *Biological Conservation* 111: 91-104.

Kirby, J.S., Salmon, D.G., Atkinson-Willes, G.L. & Cranswick, P.A. 1995. Index numbers for waterbird populations, III. Long-term trends in the abundance of wintering wildfowl in Great Britain, 1966/67 to 1991/92. *Journal of Applied Ecology* 32: 536-551.

Knox, A.G., Collinson, M., Helbig, A.J., Parkin. D.T. & Sangster, G. 2002. Taxonomic recommendations for British birds. *Ibis* 144: 707-710.

Laubek, B., Nilsson, L., Wieloch, M., Koffijberg, K., Sudfelt, C. & Follested, A. 1999. Distribution, numbers and habitat choice of the NW European Whooper Swan *Cygnus cygnus* population: results of an international census in January 1995. *Vogelwelt* 120: 141-154.

Maclean, I.M.D., Austin, G.E., Mellan, H.J. and Girling, T. 2005. *WeBS Alerts 2003/2004: Changes in numbers of wintering waterbirds in the United Kingdom, its Constituent Countries, Special Protection Areas (SPAs) and Sites of Special Scientific Interest (SSSIs).* BTO Research Report No. 416 to the WeBS partnership. BTO, Thetford. Available: http://blx1.bto.org/webs/alerts/index.htm.

Mitchell, P.I., Newton, S.F., Ratcliffe, N. & Dunn, T.E. 2004. *Seabird Populations of Britain and Ireland Results of the Seabird 2000*. T & AD Poyser, London.

Musgrove, A.J., Langstone, R.H.W., Baker, H. & Ward, R.M. 2003. *Estuarine Waterbirds at Low Tide: the WeBS Low Tide Counts 1992/93 to 1998/99.* WSG/BTO/WWT/RSPB/JNCC, Thetford.

Newson, S.E., Ekins, G.R., Marchant, J.H., Rehfisch, M.M. & Sellers, R.M. 2006. *The status of inland and coastal breeding Great Cormorants Phalacrocorax carbo in England.* BTO Research Report 433. British Trust for Ornithology, Thetford.

Prŷs-Jones, R.P., Underhill, L.G. & Waters, R.J. 1994. Index numbers for waterbird populations. II Coastal wintering waders in the United Kingdom, 1970/71-1990/91. *Journal of Applied Ecology* 31: 481-492.

Ramsar Convention Bureau. 1988. *Convention on Wetlands of International Importance especially as Waterfowl Habitat.* Proceedings of the third meeting of the Conference of the Contracting Parties, Regina, Canada, 1987. Ramsar, Switzerland.

Raven, M.J., Noble, D.G., & Baillie, S.R. 2005. *The Breeding Bird Survey 2004.* BTO Research Report 403. British Trust for Ornithology, Thetford.

Ravenscroft, N. 2005. *Pilot study into disturbance of waders and wildfowl on the Stour-Orwell SPA: analysis of 2004-05 data.* Era report to Suffolk Coast & Heaths Unit. Era: Woodbridge, Suffolk.

Rehfisch, M.M, Austin, G.E., Armitage, M.J.S., Atkinson, P.W., Holloway, S.J., Musgrove, A.J. & Pollitt, M.S. 2003. Numbers of wintering waterbirds in Great Britain and the Isle of Man (1994/1995-1998/1999): II. Coastal waders (Charadrii). *Biological Conservation* 112: 329-341.

Rogers, N.J., & the Rarities Committee. 2004. Report on rare birds in Great Britain in 2003. *British Birds* 2004, 558-625

Rose, P.M. & Scott, D.A. 1997. *Waterfowl Population Estimates - Second Edition.* Wetlands International Publ. 44, Wageningen, The Netherlands.

Rose, P.M. & Stroud, D.A. 1994. Estimating international waterfowl populations: current activity and future directions. *Wader Study Group Bulletin* 73: 19-26.

Rowell, H.E. & Hearn, R.D. 2005. *The 2003 Icelandic-breeding Goose Census*. The Wildfowl & Wetlands Trust / Joint Nature Conservation Committee, Slimbridge.

Sangster, G., Collinson, M.J., Helbig, A.J., Knox, A.G. & Parkin, D.T. 2005. Taxonomic recommendations for British birds: third report. Ibis 147; 821-826.

Simpson, J. & Maciver, A. 2005. *Population and distribution of Bean Geese in the Slamannan area 2004/2005*. Report to Bean Goose Action Group.
Stroud, D.A., Chambers, D., Cook, S., Buxton, N., Fraser, B., Clement, P., Lewis, P., McLean, I., Baker, H. & Whitehead, S. (2001) *The UK SPA network: its scope and content.* JNCC, Peterborough, UK.

Underhill, L.G. 1989. *Indices for waterbird populations*. BTO Research Report 52.

Underhill, L.G. & Prŷs-Jones, R. 1994. Index numbers for waterbird populations. I. Review and methodology. *Journal of Applied Ecology*, 31: 463-480.

Vinicombe, K., Marchant, J. & Know, A. 1993. Review of status and categorization of feral birds on the British List. *British Birds*, 75: 1-11.

Waltho, C.M. 2004. *Firth of Clyde Eider News: No 5. August 2004*. Private report

Way, L.S., Grice, P., MacKay, A., Galbraith, C.A., Stroud, D.A. & Pienkowski, M.W. 1993. *Ireland's internationally important bird sites: a review of sites for the EC Special Protection Area network*. JNCC, Peterborough, 231 pp.

Webb, A., McSorley, C.A., Dean, B.J., O'Brien, S., Reid, J.B., Cranswick, P.A., Smith L.E. & Hall, C. In prep. *An assessment of the numbers and distribution of aggregations of inshore waterbirds using the Greater Thames during the non-breeding season*. Unpublished JNCC Report, Joint Nature Conservation Committee, Peterborough.

Wetlands International. 2002. *Waterbird Population Estimates - Third Edition. Wetlands International Global Series No. 12*, Wageningen, the Netherlands.

Wilson, L.J., Dean, B.J., Webb, A., McSorley, C.A. and Reid, J.B. 2006. *Wintering seaducks, divers and grebes in UK inshore areas: Aerial surveys and shore-based counts 2004/05. JNCC Report No. 371.* Joint Nature Conservation Committee, Peterborough.

WWT Wetlands Advisory Service. 2005. *Aerial Surveys of Waterbirds in Strategic Windfarm Areas: 2004/05 Final Report.* WWT Wetlands Advisory Service Report to Department of Trade and Industry.

Glossary

The terms listed below are generally restricted to those that have been adopted specifically for use within WeBS or more widely for monitoring.

Autumn For waders, autumn comprises July to October inclusive. Due to differences in seasonality between species, a strict definition of autumn is not used for wildfowl.

British Trust for Ornithology (BTO) The BTO is a well-respected organisation, combining the skills of professional scientists and volunteer birdwatchers to carry out research on birds in all habitats and throughout the year. Data collected by the various surveys form the basis of extensive and unique databases, which enable the BTO to objectively advise conservation bodies, government agencies, planners and scientists on a diverse range of issues involving birds.

Complex site A *WeBS site* that consists of two or more *sectors*.

Core Counts The fundamental WeBS counts that monitor all types of wetlands throughout the UK once per month on priority dates. Used to determine population estimates and trends and identify important sites.

Local Organiser Person responsible for coordinating counters and counts at a local level, normally a county or large estuary, and the usual point of contact with the *WeBS office*.

Incomplete counts When presenting counts of an individual species, a large proportion of the number of birds was suspected to have been missed, *e.g.* due to part coverage of the site or poor counting conditions, or when presenting the total number of birds of all species on the site, a significant proportion of the total number was missed.

I-WeBS An independent but complementary scheme operating in the Republic of Ireland to monitor non-breeding waterbirds, organised by the IWC BirdWatch Ireland, the National Parks and Wildlife Service (Ireland) and The Wildfowl & Wetlands Trust.

Joint Nature Conservation Committee (JNCC) JNCC is the statutory body constituted by the Environmental Protection Act 1990 to be responsible for research and advice on nature conservation at both UK and international levels. The committee is established by English Nature, Scottish Natural Heritage and the Countryside Council for Wales, together with independent members and representatives from the Countryside Commission and Northern Ireland, and is supported by specialist staff.

Low Tide Counts (LTC) WeBS counts made at low tide to assess the relative importance of different parts of individual estuaries as feeding areas for intertidal waterbirds.

Royal Society for the Protection of Birds (RSPB) The RSPB is the charity that takes action for wild birds and the environment in the UK. The RSPB is the national BirdLife partner in the UK.

Spring For waders, spring comprises April to June inclusive. Due to differences in seasonality between species, a strict definition of spring is not used for wildfowl.

Waterbirds WeBS follows the definition adopted by Wetlands International. This includes a large number of families, those occurring regularly in the UK being divers, grebes, cormorants, herons, storks, ibises and spoonbills, wildfowl, cranes, rails, waders and gulls and terns.

WeBS count unit The area/boundary within which a count is made. The generic term for *sites*, *sub-sites* and *sectors*.

WeBS Office Main administrative centre for the day-to-day running of WeBS and main point of contact for information or data pertaining to WeBS (see *Contacts* section).

WeBS sector The unit of division of large *sites* into areas that can be counted by one person in a reasonable time period. They are often demarcated by geographic features to facilitate recognition of the boundary by counters. The finest level at which data are recorded.

WeBS site A biologically meaningful area that represents a discrete area used by waterbirds such that birds regularly move within but only occasionally between sites. The highest level at which count data are stored.

WeBS sub-site A grouping of *sectors* within a *site* to facilitate coordination. In most cases, sub-sites also relate to biologically meaningful units for describing waterbird distribution.

WeBS Year Defined as July to June inclusive the WeBS Year is centred on the time when most waterbird species are present in their largest number, during winter. Counts during autumn passage and spring passage the following calendar year are logically associated with the intervening winter.

Wildfowl & Wetlands Trust (WWT) Founded by Sir Peter Scott in 1946, WWT is the largest wildlife conservation charity specialising in wetlands and the wildlife they support in the UK. It has pioneered the bringing together of people and wildlife for the benefit of both and seeks to raise awareness of the value of wetlands, the threats they face and the actions needed to save them. To this end, WWT has eight centres throughout the UK and is dedicated to saving wetlands for wildlife and people.

Winter For waders, winter comprises November to March inclusive. Due to differences in seasonality between species, a strict definition of winter is not used for wildfowl.

1% criterion The Ramsar Convention has established site selection criteria. Criterion 6 states that ". . . a wetland should be considered internationally important if it regularly supports 1% of the individuals in a population of one species or subspecies of waterbird".

1% threshold This logically derives from the *1% criterion* and relates to the number of birds that are used as the nominal 1% of the population for the purposes of site selection. Thus, an international population of 75,215 Shelduck has a derived 1% threshold (adopting rounding conventions) of 750.

APPENDIX 1. INTERNATIONAL AND NATIONAL IMPORTANCE

Any site recognised as being of international ornithological importance is considered for classification as a Special Protection Area (SPA) under the EC Directive on the Conservation of Wild Birds (EC/79/409), whilst a site recognised as an internationally important wetland qualifies for designation as a Ramsar site under the Convention on Wetlands of International Importance especially as Waterfowl Habitat. Criteria for assessing the international importance of wetlands have been agreed by the Contracting Parties to the Ramsar Convention on Wetlands of International Importance (Ramsar Convention Bureau 1988). Under criterion 6, a wetland is considered internationally important if it regularly holds at least 1% of the individuals in a population of one species or subspecies of waterbird, while criterion 5 states that any site regularly supporting 20,000 or more waterbirds also qualifies. Britain and Ireland's wildfowl belong, in most cases, to the northwest European population and the waders to the east Atlantic flyway population (Wetlands International 2002).

A wetland in Britain is considered nationally important if it regularly holds 1% or more of the estimated British population of one species or subspecies of waterbird, and in Northern Ireland important in an all-Ireland context if it holds 1% or more of the estimated all-Ireland population.

The 1% thresholds for British, all-Ireland and international waterbird populations, where known, are listed in Table A1. Thus, any site regularly supporting at least this number of birds potentially qualifies for designation under national legislation, or the EC Birds Directive or Ramsar Convention. The international population for each species and subspecies is also specified in the table. However, it should be noted that, where 1% of the national population is less than 50 birds, 50 is normally used as a minimum qualifying threshold for the designation of sites of national or international importance.

It was agreed at the meeting of the Ramsar Convention in Brisbane that population estimates will be reviewed by Wetlands International every three years and 1% thresholds revised every nine years (Rose & Stroud 1994; Ramsar Resolution VI.4). 1% thresholds have not been derived for introduced species since protected sites would not be identified for these birds.

Sources of qualifying levels represent the most up-to-date figures following recent reviews: for wildfowl in Britain see Kershaw & Cranswick (2003); for waders in Britain see Rehfisch *et al.* (2003); for gulls in Britain see Burton *et al.* (2003); for all-Ireland importance for divers see Danielsen *et al.* (1993) and for other waterbirds see Whilde (*in prep.*) cited in Way *et al.* (1993). International criteria follow Wetlands International (2002).

It should be noted that for some populations, where the British total is the international total, the precise figure given for the estimates may differ because of different rounding conventions applied in the relevant publications.

Table A1. 1% thresholds for national and international importance

	Great Britain	all-Ireland	International	Subspecies/Population
Red-throated Diver	49	*10	10,000	NW Europe (non-br)
Black-throated Diver	*7	*1	10,000	*arctica*
Great Northern Diver	*30	?	50	NW Europe (non-br)
Little Grebe	78	?	3,400	*ruficollis*
Great Crested Grebe	159	*30	4,800	*cristatus*
Red-necked Grebe	*2	?	1,000	*grisegena*, NW Europe (non-br)
Slavonian Grebe	*7	?	35	*auritus*, NW Europe (large billed)
Black-necked Grebe	*1	?	2,800	*nigricollis*, Europe, N Africa
Cormorant	230	?	1,200	*carbo*, NW Europe
Shag	?	?	2,400	*aristotelis*
Little Egret	?	?	1,300	*garzetta*, W Europe, NW Africa
Grey Heron	?	?	2,700	*cinerea*, W Europe, NW Africa (br)

Table A1. continued

	Great Britain	all-Ireland	International	Subspecies/Population
Mute Swan: *British*	375	n/a	380	Britain
Irish	n/a	100	100	Ireland
Bewick's Swan	81	*25	290	*bewickii*, NW Europe (non-br)
Whooper Swan	57	100	210	Iceland (br)
Bean Goose: *Taiga*	*4	+	1,000	*fabalis*
Pink-footed Goose	2,400	+	2,400	Greenland, Iceland (br)
European White-fronted Goose	58	+	10,000	*albifrons*, Baltic-North Sea
Greenland White-fronted Goose	209	140	300	*flavirostris*
Greylag Goose: *Iceland*	819	*40	1,000	*anser*, Iceland (br)
Hebrides/N Scotland	90	n/a	90	*anser*, NW Scotland
Barnacle Goose: *Greenland*	450	75	540	E Greenland (br)
Svalbard	220	+	230	Svalbard (br)
Dark-bellied Brent Goose	981	+	2,200	*Bernicla*, W Siberia (br)
Light-bellied Brent Goose: *Canada*	+	200	200	*hrota*, Ireland (non-br)
Svalbard	*30	+	50	*hrota*, Svalbard, N Greenland (br)
Shelduck	782	70	3,000	NW Europe (br)
Wigeon	4,060	1,250	15,000	NW Europe (non-br)
Gadwall	171	+	600	*strepera*, NW Europe (br)
Teal	1,920	650	4,000	NW Europe (non-br)
Mallard	3,520	500	**20,000	*platyrhynchos*, NW Europe (non-br)
Pintail	279	60	600	NW Europe (non-br)
Garganey	+	+	**20,000	W Africa (non-br)
Shoveler	148	65	400	NW & C Europe (non-br)
Red-crested Pochard	+	+	500	C Europe & W Mediterranean
Pochard	595	400	3,500	NE & NW Europe (non-br)
Tufted Duck	901	400	12,000	NW Europe (non-br)
Scaup	76	*30	3,100	*marila*, W Europe (non-br)
Eider	730	*20	15,500	*mollissimma*, NW Europe[1]
Long-tailed Duck	160	+	**20,000	W Siberia, N Europe (br)
Common Scoter	500	*40	16,000	*nigra*
Velvet Scoter	*30	+	10,000	*fusca*, Baltic, W Europe (non-br)
Goldeneye	249	110	4,000	*clangula*, NW & Central Europe (non-br)
Smew	*4	+	400	NW & C Europe (non-br)
Red-breasted Merganser	98	*20	1,700	NW & C Europe (non-br)
Goosander	161	+	2,700	*merganser*, NW Europe[2]
Moorhen	7500	?	**20,000	*chloropus*, Europe, N Africa (br)
Coot	1,730	250	17,500	*atra*, NW Europe (non-br)
Oystercatcher	3,200	500	10,200	*ostralegus*, Europe, NW Africa
Avocet	*35	+	730	W Europe (br)
Ringed Plover: *winter*	330	125	730	*hiaticula*, Europe & N Africa (non-br)
passage	300			
Golden Plover	2,500	2,000	9,300	*altifrons*, Iceland & Faeroes, E Atlantic[3]
Grey Plover	530	*40	2,500	E Atlantic (non-br)
Lapwing	**20,000	2,500	**20,000	Europe (br)
Knot	2,800	375	4,500	*islandica*
Sanderling: *winter*	210	*35	1,200	E Atlantic, W & S Africa (non-br)
passage	300			
Purple Sandpiper	180	*10	750	*maritima*, E Atlantic
Dunlin: *winter*	5,600	1,250	13,300	*alpina*, W Europe (non-br)[4]
passage	2,000			
Ruff	*7	+	?	W Africa (non-br)
Jack Snipe	?	250	?	NE Europe (br)
Snipe	?	?	**20,000	*gallinago*, Europe (br)
Woodcock	?	?	**20,000	Europe (br)
Black-tailed Godwit	150	90	350	*islandica*
Bar-tailed Godwit	620	175	1,200	*lapponica*
Whimbrel	+	+	6,100	*islandicus*
Curlew	1,500	875	4,200	*arquata*
Spotted Redshank	+	+	1,000	Europe (br)
Redshank	1,200	245	1,300	*brittanica*[5]
Greenshank	*6	*9	3,100	Europe (br)
Green Sandpiper	?	?	14,500	Europe (br)
Common Sandpiper	?	?	17,000	N, W & C Europe (br)
Turnstone	500	225	1,000	*interpres*, NE Canada, Greenland (br)

Table A1. continued

	Great Britain	all-Ireland	International	Subspecies/Population
Little Gull	?	?	840	N, C & E Europe (br)
Black-headed Gull	19,000	?	**20,000	N & C Europe (br)
Common Gull	9,000	?	17,000	*canus*
Lesser Black-backed Gull	500	?	5,300	*graellsii*
Herring Gull	4,500	?	13,000	*argentatus*[6]
Great Black-backed Gull	400	?	4,700	NE Atlantic
Kittiwake	?	?	**20,000	*tridactyla*, E Atlantic (br)
Sandwich Tern	?	?	1,700	*sandvicensis*, W Europe (br)
Common Tern	?	?	1,900	*hirundo*, S, W Europe (br)
Little Tern	?	?	340	*albifrons*, W Europe (br)
Black Tern	?	?	4,000	*niger*

? *Population size not accurately known.*
\+ *Population too small for meaningful figure to be obtained.*
* *Where 1% of the British or all-Ireland wintering population is less than 50 birds, 50 is normally used as a minimum qualifying level for national or all-Ireland importance respectively.*
** *A site regularly holding more than 20,000 waterbirds qualifies as internationally important by virtue of absolute numbers.*

1 The degree of interchange of UK Eiders with birds on the continent is unclear, and although Wetlands International (2002) has recommended that birds in Britain and Ireland should be treated as a separate biogeographical population, a recent review of available data by DEFRA's SPA and Ramsar Scientific Working Group has found limited evidence to support this conclusion, and recommended that for site-selection purposes, British Eider continue to be considered as a component of the four groups of the Northwest European groups of the race *mollissima* with an international 1% threshold of 15,500. It is hoped that future genetic studies will help clarify the situation.

2 Although Wetlands International (2002) considers Goosanders breeding in Scotland, northern England and Wales to be a discrete population, a recent review of available data by DEFRA's SPA and Ramsar Scientific Working Group has found limited evidence to support this conclusion for the time being, and recommended that for site-selection purposes, British Goosanders continue to be considered as a component of the NW and C European population of Goosander, with an international 1% threshold of 2,700.

3 Three populations of Golden Plover listed by Wetlands International (2002) overlap in the UK in winter. Draft guidelines from Ramsar suggest that the largest of the three thresholds (*i.e.* that for *altifrons,* Iceland & Faeroes, E Atlantic) should be used for site-selection purposes.

4 Whilst several populations of Dunlin occur in the UK at different times of the year, most wintering birds are referable to the listed population.

5 Three populations of Redshank listed by Wetlands International (2002) overlap in the UK in winter: *totanus* E Atlantic (non-br), *robusta* and *brittanica*. Most *totanus* winter outside the UK but the other populations are known to occur widely. Draft guidelines from Ramsar suggest that the larger of the two thresholds (*i.e.* that for *brittanica*) should be used for site-selection purposes.

6 Two populations of Herring Gull overlap in the winter in the UK; *argentatus* and *argenteus*. Draft guidelines from Ramsar suggest that the larger of the two thresholds, *i.e.* that for *argentatus*, should be used for site-selection purposes.

APPENDIX 2. LOCATIONS OF WeBS COUNT SITES MENTIONED IN THIS REPORT

Table A2 provides details of all WeBS sites that are mentioned in this report. Sites are listed alphabetically, with details of the Ordnance Survey 1-km square that the centre of the sites falls into and the region also provided. Principal Core Count sites, as listed in Table 6, are highlighted in **bold**. Numbers following Principal Core Count sites refer to the sites' location in Figure A1.

Note that this is not an exhaustive list of sites counted during 2004/05, simply those mentioned in this report.

Table A2. Details of sites mentioned in this report. Numbers following Principal Core Count sites refer to the sites' location in figure A1.

Site	1-km sq	Region	
Abberton Reservoir	**TL9717**	**Essex**	95
Aberarth	SN4764	Dyfed	
Aberdeen Bay offshore	NK0323	Grampian	
Aberlady Bay	NT4581	Lothians	
Acre Nook Sand Quarry	SJ8272	Cheshire	
Adur Estuary	TQ2006	W Sussex	
Aignish Bay (Lewis)	NB4832	Western Isles	
Alde Complex	**TM4257**	**Suffolk**	89
Allasdale Bay to Borve (Barra)	NF6503	Western Isles	
Allerton Bywater	SE4127	W Yorkshire	
Allington Gravel Pit	SU4717	Hampshire	
Alnmouth to Boulmer	NU2511	N`thumberland	
Alresford Pond	SU5933	Hampshire	
Alt Estuary	**SD2903**	**Merseyside**	70
Alton Water	TM1536	Suffolk	
Amberswood Cmn	SD6003	Gtr Manchester	
Ampton Water	TL8770	Suffolk	
Angler`s CP Lake	SE3716	West Yorkshire	
Anstruther Harbour	NO5603	Fife	
Aqualate Mere	SJ7720	Staffordshire	
Ardivachar Point	NF7345	Western Isles	
Ardrossan	NS2045	Ayrshire & Arran	
Arlington Reservoir	TQ5307	E Sussex	
Arnot Park Lake	SK5844	Nottinghamshire	
Arran	NR9535	Ayrshire & Arran	
Arun Valley	**TQ0314**	**W Sussex**	109
Attenborough GP	SK5234	Nottinghamshire	
Auchenharvie	NS2541	Ayrshire & Arran	
Avon Estuary	SX6745	Devon	
Avon Valley: Salisbury-Fordingbridge	SU1619	Wiltshire	
Axe Estuary	SY2590	Devon	
Ayr to North Troon	NS3427	Ayrshire & Arran	
Bainton Pits	TF1107	Cambridgeshire	
Bala Lake	SH9033	Merioneth	
Baleshare	NF7760	Western Isles	
Ballycastle: Fair Head	ID1442	Antrim	
Ballysaggart Lough	IH7961	Tyrone	
Balnakeil Bay	NC3869	Highland	
Balranald (RSPB)	**NF7169**	**Western Isles S**	10
Balta Sound	HP6308	Shetland	
Bann Estuary	IC7935	Londonderry	
Barcombe Mills Res	TQ4314	E Sussex	
Bardolf Wtr Meadows	SY7796	Dorset	
Barmston Ponds	NZ3356	Tyne & Wear	
Barn Elms Reservoirs	TQ2277	Gtr London	
Barons Folly	NT6426	Roxburgh	

Site	1-km sq	Region	
Barrow Gurney Res	ST5367	Avon	
Bartley Reservoir	SP0081	W Midlands	
Barton Broad	TG3621	Norfolk	
Barton Pits	SK2017	Staffordshire	
Baston & Langtoft GPs	TF1212	Lincolnshire	
Beaulieu Estuary	**SZ4297**	**Hampshire**	115
Beauly Firth	NH5848	Highland	
Bedfont & Ashford GPs	TQ0872	Gtr London	
Beesands Ley	SX8141	Devon	
Belfast Lough	**IJ3983**	**Down**	55
Belhus Woods CP	TQ5782	Gtr London	
Bellflask	SE2977	N Yorkshire	
Belvide Reservoir	SJ8610	Staffordshire	
Benacre Broad	TM5383	Suffolk	
Benbecula	NF8152	Western Isles	
Benston Loch	HU4653	Shetland	
Besthorpe & Girton GPs and Fleet	SK8165	Nottinghamshire	
Bewl Water	TQ6733	E Sussex	
Bickershaw Colliery	SD6300	Gtr Manchester	
Big Waters Seaton Burn	NZ2273	Tyne & Wear	
Birns Farm Gravel Pit	NO2911	Fife	
Bittell Reservoirs	SP0275	Hereford & Worcs	
Black Cart Water (Gryfe-White Cart)	NS4767	Renfrew	
Blackwater Estuary	**TL9307**	**Essex**	96
Blagdon Lake	ST5159	Avon	
Blatherwyke Lake	SP9796	N`thamptonshire	
Blenheim Park Lake	SP4316	Oxfordshire	
Blithfield Reservoir	SK0524	Staffordshire	
Blucks Pool to Freshwater West	SR8898	Dyfed	
Blyth Estuary	**TM4675**	**Suffolk**	88
Blyth Estuary	NZ3082	N`thumberland	
Boat of Garten Pools	NH9319	Highland	
Boddam Voe	HU3915	Shetland	
Bolton-on-Swale GPs	SE2498	N Yorkshire	
Bornish & Ormiclate Machairs SSSI Coast	NF7129	Western Isles	
Bough Beech Res	TQ4947	Kent	
Boulmer to Howick	NU2615	N`thumberland	
Bourne Hall Ponds	TQ2162	Surrey	
Bourton-on-the-Water Gravel Pits	SP1720	Glos	
Brading Harbour	SZ6388	Isle of Wight	
Bradley Pools	SK2245	Derbyshire	
Braewick Loch	HU2478	Shetland	
Bramshill Park Lake	SU7560	Hampshire	
Bramshill Plantation Lakes	SU7462	Hampshire	

Site	1-km sq	Region	
Branahuie Banks	NB4731	Western Isles	
Branahuie Saltings	**NB4631**	**Western Isles**	**8**
Bredon`s Hardwick Gravel Pits	SO9035	Hereford & Worcs	
Brent Reservoir	TQ2287	Gtr London	
Bressay Sound	HU4741	Shetland	
Breydon Water and Berney Marshes	**TG4706**	**Norfolk**	**86**
Bridge of Crathies	NO2745	Perth & Kinross	
Broad Bay (Lewis)	NB4733	Western Isles	
Broadford Bay	NG6523	Skye	
Broadwater Lake: South Harefield	TQ0589	Surrey	
Broubster Leans	ND0360	Highland	
Buckden & Stirtloe Pits	TL2066	Cambridgeshire	
Burghfield Gravel Pits	SU6870	Berkshire	
Burra and Trondra	HU3633	Shetland	
Burrator Reservoir	SX5568	Devon	
Burry Inlet	**SS5096**	**Dyfed**	**78**
Burton Gravel Pits	SK9474	Lincolnshire	
Busbridge Lakes	SU9742	Surrey	
Bute	NS0761	D`barton/Argyll	
Buxton Pavilion Gdns	SK0573	Derbyshire	
Caerphilly Castle	ST1587	Glamorgan	
Caistron Quarry	NU0001	N`thumberland	
Caithness Lochs	ND1859	Highland	
Caldey Island	SS1496	Dyfed	
Cambois to Newbiggin	NZ3084	N`thumberland	
Camel Estuary	SW9275	Cornwall	
Cameron Reservoir	**NO4711**	**Fife**	**28**
Capesthorne Hall Pls	SJ8472	Cheshire	
Cardigan Bay	SN5293	Merioneth	
Carlingford Lough	**IJ1814**	**Down**	**53**
Carmarthen Bay	**SN2501**	**Dyfed**	**77**
Carr Vale Flash and Reserve Pond	SK4570	Derbyshire	
Carsebreck and Rhynd Lochs	**NN8609**	**Perth & Kinross**	**32**
Carsington Water	SK2451	Derbyshire	
Cassington & Yarnton Gravel Pits	SP4710	Oxfordshire	
Castle Loch (Lochmaben)	NY0881	D & Galloway	
Castle Park Lochan	NS7894	Central	
Ceann a Bhaigh	NF7567	Western Isles	
Cemlyn Bay & Lgn	SH3393	Anglesey	
Chard Reservoir	ST3309	Somerset	
Chasewater	SK0307	W Midlands	
Chat Moss	SJ7196	Gtr Manchester	
Cheddar Reservoir	ST4454	Somerset	
Chelmarsh Reservoir	SO7387	Shropshire	
Chetwynd Pool	SJ7420	Shropshire	
Chew Valley Lake	**ST5659**	**Avon**	**123**
Chichester GPs	SU8703	W Sussex	
Chichester Harbour	**SU7700**	**W Sussex**	**111**
Chilham & Chartham Gravel Pits	TR0954	Kent	
Chillington Hall Pool	SJ8505	Staffordshire	
Chilton Foliat	SU3270	Wiltshire	
Christchurch Harbour	SZ1791	Dorset	
Church Wilne Res	SK4632	Derbyshire	
Clachan & Whitehouse	NR7959	Argyll	
Clashnessie Bay	NC0631	Highland	
Clatto Reservoir	NO3607	Fife	

Site	1-km sq	Region	
Cleddau Estuary	**SN0005**	**Dyfed**	**76**
Clifford Hill GPs	SP7859	N`thamptonshire	
Clifton House Pond	SD7704	Gtr Manchester	
Clumber Park Lake	SK6374	Nottinghamshire	
Clwyd Estuary	SJ0079	Clwyd	
Clyde Estuary	NS2778	Renfrew	
Cochrage Loch	NO1549	Perth & Kinross	
Coll & Tiree offshore	NM0654	Coll/Tiree	
College Lake Reserve	SP9314	Bucks	
Colliford Reservoir	SX1772	Cornwall	
Colne Estuary	**TM0614**	**Essex**	**94**
Colne Valley GPs	TQ0391	Hertfordshire	
Colney Gravel Pits	TG1708	Norfolk	
Colonsay/Oronsay	NR3793	Islay/Jura/Colonsay	
Colt Crag Reservoir	NY9478	N`thumberland	
Colwick Country Park	SK6039	Nottinghamshire	
Combermere	SJ5844	Cheshire	
Connaught Water	TQ4095	Essex	
Coombe CP	SP3979	Warwickshire	
Coombe Hill Canal	SO8626	Glos	
Copgrove Lake	SE3463	N Yorkshire	
Coquet Island	NU2804	N`thumberland	
Corby Loch	NJ9214	Grampian	
Cors Caron (Cors Tregaron)	SN6863	Dyfed	
Cosmeston Lakes	ST1769	Glamorgan	
Cotswold Water Park (East)	SU1999	Glos	
Cotswold Water Park (West)	**SU0595**	**Glos**	**121**
Cransley Reservoir	SP8278	N`thamptonshire	
Cranwich Gravel Pits	TL7795	Norfolk	
Cray Reservoir	SN8821	Powys	
Cresswell Pond	NZ2894	N`thumberland	
Cromarty Firth	**NH7771**	**Highland**	**15**
Crookfoot Reservoir	NZ4331	Durham	
Cropston Reservoir	SK5410	Leicestershire	
Crouch-Roach Est	**TQ9895**	**Essex**	**98**
Crowdy Reservoir	SX1483	Cornwall	
Croxall Pits	SK1914	Staffordshire	
Cuckmere Estuary	TV5197	E Sussex	
Cuttmill Ponds	SU9145	Surrey	
Dagenham Chase GP	TQ5186	Gtr London	
Dart Estuary	SX8456	Devon	
Darwell Reservoir	TQ7121	E Sussex	
Deben Estuary	**TM2942**	**Suffolk**	**90**
Dee Estuary (England & Wales)	**SJ2675**	**Cheshire**	**73**
Dee Estuary (Scotland)	NJ9505	Grampian	
Dee Flood Meadows	**SJ4059**	**Cheshire**	**72**
Dee Mth to Don Mth	NJ9507	Grampian	
Deene Lake	SP9492	N`thamptonshire	
Dengie Flats	**TM0302**	**Essex**	**97**
Derwent Reservoir	NZ0251	Durham	
Derwent Water	NY2520	Cumbria	
Dingle Marshes and Walberswick NNR	TM4872	Suffolk	
Dinton Pastures	SU7872	Berkshire	
Diss Mere	TM1179	Norfolk	
Ditchford Gravel Pits	SP9468	N`thamptonshire	
Doddington Pool	SJ7146	Cheshire	
Doffcocker Lodge	SD6810	Gtr Manchester	
Dolydd Hafren	SJ2000	Powys	
Don Mth to Ythan Mth	NJ9819	Grampian	
Dorchester GPs	SU5795	Oxfordshire	

Site	1-km sq	Region	
Dornoch Firth	**NH7384**	**Highland**	**14**
Doxey Marshes SSSI	SJ9024	Staffordshire	
Drakelow Gravel Pit	SK2320	Derbyshire	
Draycote Water	SP4469	Warwickshire	
Drift Reservoir	SW4329	Cornwall	
Drumbowie Reservoir	NS7981	Central	
Drumgay Lough	IH2348	Fermanagh	
Druridge Pool	NZ2796	N`thumberland	
Duddon Estuary	**SD2081**	**Cumbria**	**60**
Dundrum Bay	**IJ4135**	**Down**	**58**
Dungeness GPs	**TR0619**	**Kent**	**106**
Duns Dish	NO6460	Angus	
Dupplin Lochs	NO0320	Perth & Kinross	
Durham Coast	NZ4349	Durham	
Dyfi Estuary	**SN6394**	**Dyfed**	**75**
Earls Barton GPs	SP8863	N`thamptonshire	
East Aberthaw Quarry Pool	ST0367	Glamorgan	
East Chevington Pls	NZ2799	N`thumberland	
East Fenton Farm Reservoir	NT5281	Lothians	
East Mains Flood	NJ5602	Grampian	
East Unst	HP6506	Shetland	
East Wretham Meres	TL9088	Norfolk	
Eccup Reservoir	SE3041	W Yorkshire	
Eden Estuary	**NO4719**	**Fife**	**27**
Edington Lake	ST9253	Wiltshire	
Ellesmere Lakes	SJ4133	Shropshire	
Endrick Water: Council Boundary to Drymen Bridge	NS4787	Central	
Entrance to Deer and Shapinsay Sounds	HY5612	Orkney	
Erme Estuary	SX6249	Devon	
Eversley Cross and Yateley Gravel Pits	SU8061	Hampshire	
Exe Estuary	**SX9883**	**Devon**	**126**
Exeter River Vly Pk & Matford Marshes	SX9390	Devon	
Eyebrook Reservoir	SP8595	Leicestershire	
Fairburn Ings	SE4627	N Yorkshire	
Fairfield SSSI	TQ9626	Kent	
Fal Complex	SW8541	Cornwall	
Fala Flow	NT4258	Lothians	
Farmland near Monymusk	NJ6816	GrampianS.E.	
Farne Islands	NU2136	N`thumberland	
Fen Drayton GPs	TL3470	Cambridgeshire	
Fernworthy Reservoir	SX6684	Devon	
Fiddlers Ferry Power Station Lagoons	SJ5585	Cheshire	
Filey Bay	TA1279	N Yorkshire	
Findhorn Bay	NJ0462	Grampian	
Firth of Clyde	NS2778	D`barton/Argyll	
Fisherwick & Elford GPs	SK1710	Staffordshire	
Fleet and Wey	**SY6976**	**Dorset**	**120**
Fleet Pond	SU8255	Hampshire	
Folkestone: Copt Pt & East Wear Bay	TR2436	Kent	
Fonthill Lake	ST9331	Wiltshire	
Foreland	SZ6587	Isle of Wight	
Forest Mere	SU8230	W Sussex	
Forest of Dean Ponds	SO6010	Glos	
Fort Henry Ponds and Exton Park Lake	SK9412	Leicestershire	
Forth Estuary	**NT2080**	**Lothians**	**31**
Foryd Bay	SH4459	Caernarvon	
Fowey Estuary	SX1254	Cornwall	
Frampton Pools	SO7507	Glos	
Frensham Great Pd	SU8440	Surrey	
Gadloch	NS6471	Lanarkshire	
Gannel Estuary	SW8060	Cornwall	
Gare Loch	NS2486	D`barton/Argyll	
Garths Loch	HU3809	Shetland	
Gatton Park	TQ2753	Surrey	
Gerrans Bay	SW8937	Cornwall	
Girvan to Turnberry	NS2003	Ayrshire & Arran	
Gladhouse Reservoir	NT2953	Lothians	
Goring	TQ1001	W Sussex	
Gourock to Largs	NS1968	Ayrshire & Arran	
Grafham Water	TL1568	Cambridgeshire	
Grnd Western Canal: Greenway Bridge - N. Devon Lk Rd	ST0213	Devon	
Grt Pl Westwood Pk	SO8763	Hereford & Worcs	
Grimley New Wkings	SO8359	Hereford & Worcs	
Grimsthorpe Lake	TF0222	Lincolnshire	
Grouville Marsh	WV6949	Channel Islands	
Gruinard Bay	NG9590	Highland	
Gualan and Balgarva	NF7648	Western Isles	
Guernsey Shore	WV2575	Channel Islands	
Hacosay & Bluemull & Colgrave Sounds	HU5595	Shetland	
Haddo House Lakes	**NJ8734**	**Grampian**	**20**
Hagnaby Lock Fen	TF3459	Lincolnshire	
Hallington Reservoir	NY9776	N`thumberland	
Hamford Water	**TM2225**	**Essex**	**93**
Hamilton Low Parks & Strathclyde Park	NS7257	Lanarkshire	
Hampton & Kempton Reservoirs	TQ1269	Gtr London	
Hanningfield Res	TQ7398	Essex	
Hardley Flood	TM3899	Norfolk	
Harewood Lake	SE3144	W Yorkshire	
Harrow Lodge Park	TQ5286	Gtr London	
Hastings to Bexhill	TQ7708	E Sussex	
Hauxley Haven	NU2802	N`thumberland	
Haweswater Res	NY4713	Cumbria	
Hayle Estuary	SW5436	Cornwall	
Headley Mill Pond	SU8138	Hampshire	
Heath Pond	SU7522	Hampshire	
Heaton Park Res	SD8205	Gtr Manchester	
Heigham Holmes	**TG4420**	**Norfolk**	**85**
Helford Estuary	SW7526	Cornwall	
Henfield Brooks	TQ1913	W Sussex	
Heritage Park Loch	NS6255	Lanarkshire	
Herne Bay	TR1768	Kent	
Hetton Bogs	NZ3448	Tyne & Wear	
Hetton Lyons Park	NZ3647	Tyne & Wear	
Hickling Broad	TG4121	Norfolk	
High Batts	SE3173	N Yorkshire	
Hightae Loch	NY0880	D & Galloway	
Hilfield Park Res	TQ1596	Hertfordshire	
Hill Ridware Lake	SK0717	Staffordshire	
Hillsborough Lake	IJ2458	Down	
Hogganfield Loch	NS6467	Lanarkshire	
Holbeach St Matthew	TF4334	Lincolnshire	
Holburn Moss	**NU0536**	**N`thumberland**	**41**
Holkham Bay	TF8845	Norfolk	
Holland Marshes	TQ8696	Essex	

Site	1-km sq	Region	
Hollowell Reservoir	SP6872	N`thamptonshire	
Holme Pierrepont GP	SK6239	Nottinghamshire	
Holmethorpe	TQ2951	Surrey	
Holy Loch-Toward Pt	NS1672	D`barton/Argyll	
Hornsea Mere	**TA1947**	**Humberside**	**64**
Horsey Mere	**TG4422**	**Norfolk**	**84**
Houb of Urafirth	HU3078	Shetland	
Houghton Green Pool	SJ6292	Cheshire	
Hoveringham & Bleasby Gravel Pits	SK7047	Nottinghamshire	
Hoveton Great Broad	TG3116	Norfolk	
Howmore Estuary SSSI Coast	NF7436	Western Isles	
Hule Moss	**NT7149**	**Berwickshire**	**37**
Humber Estuary	**TA2020**	**Humberside**	**65**
Hunterston Lagoon	NS2054	Ayrshire & Arran	
Hunterston Sands	NS1852	Ayrshire & Arran	
Hurleston Reservoir	SJ6255	Cheshire	
Ifield Mill Pond	TQ2435	W Sussex	
Ingrebourne Valley	TQ5485	Gtr London	
Beddmanarch Bay & Alaw Estuary	SH2779	Anglesey	
Inner Firth of Clyde	**NS3576**	**D`barton/Argyll**	**34**
Inner Firth of Tay	NO3225	Fife	
Inner Loch Fyne	NR9894	Argyll	
Inner Moray and Inverness Firth	**NH6752**	**Highland**	**16**
Irvine & Garnock Est	NS3038	Ayrshire & Arran	
Irvine Bay	NS3036	Ayrshire & Arran	
Island of Egilsay	HY4730	Orkney	
Island of Islay	NR3060	Islay/Jura/Colonsay	
Island of Papa Westray	HY4952	Orkney	
Islands South Of Barra	NF6393	Western Isles	
Isle of Coll	NM2055	Coll	
Isle of Colonsay	NR3896	Colonsay	
Isle Of Jura	NR5880	Jura	
Isle of Lismore	NM8441	Lismore	
Jersey Shore	WV6249	Channel Islands	
Joe`s Ponds	NZ3248	Tyne & Wear	
Kedleston Park Lake	SK3141	Derbyshire	
Keills Peninsula & Isle of Danna	NR7385	Argyll	
Kemerton Lake	SO9436	Hereford & Worcs	
Kenfig NNR	SS7981	Glamorgan	
Kenfig Pool	SS7981	Glamorgan	
Kentra Moss & Lower Loch Shiel	NM6668	Highland	
Kilconquhar Loch	**NO4801**	**Fife**	**29**
Killantringan Bay	NW9857	D & Galloway	
Killimster Loch	ND3056	Highland	
Killington Reservoir	SD5991	Cumbria	
Killough Harbour	**IJ5337**	**Down**	**59**
Kilmardinny Loch	NS5472	D`barton/Argyll	
Kilsyth Cemetery Pd	NS7177	Lanarkshire	
King George V Res	TQ3796	Gtr London	
King George VI Res	TQ0473	Surrey	
King`s Bromley GPs	SK1116	Staffordshire	
King`s Dyke Pits Whittlesey	TL2397	Cambridgeshire	
Kingsbridge Estuary	SX7441	Devon	
Kinnordy Loch	NO3654	Angus	
Kirkabister & Dury Voe	HU5159	Shetland	

Site	1-km sq	Region	
Kirkabister to Wadbister Ness	HU4453	Shetland	
Kirkby-on-Bain GPs	TF2360	Lincolnshire	
Knepp Castle Lake	TQ1521	W Sussex	
Knight & Bessborough Res	TQ1268	Surrey	
Kyle of Durness	NC3765	Highland	
Lackford GPs	TL7971	Suffolk	
Lade Sands	TR0921	Kent	
Lake of Menteith	**NN5700**	**Central**	**33**
Lakenheath Fen	TL7186	Suffolk	
Land Gate Pond	SD5800	Gtr Manchester	
Langford Lowfields GPs	SK8259	Nottinghamshire	
Langstone Harbour	**SU6902**	**Hampshire**	**112**
Larne Lough	**ID4100**	**Antrim**	**54**
Lavan Sands	**SH6474**	**Caernarvon**	**74**
Lee Valley GPs	**TL3807**	**Hertfordshire**	**104**
Leven Cut	NO2000	Fife	
Leybourne & New Hythe GPs	TQ6959	Kent	
Liden Lagoon	SU1983	Wiltshire	
Linacre Reservoirs	SK3372	Derbyshire	
Lindisfarne	**NU1041**	**N`thumberland**	**40**
Linga Beach	HU5599	Shetland	
Little Loch Broom	NH0491	Highland	
Little Mollands Fm Pts	TQ5982	Essex	
Little Paxton GPs	TL1963	Cambridgeshire	
Livermere	TL8771	Suffolk	
Llandegfedd Res	ST3298	Gwent	
Llangorse Lake	SO1326	Powys	
Llanon & Llansantffraid	SN5166	Dyfed	
Llyn Alaw	SH3986	Anglesey	
Llyn Maelog	SH3272	Anglesey	
Llyn Traffwll	SH3276	Anglesey	
Llynnau Y Fali	SH3076	Anglesey	
Llys-y-fran Reservoir	SN0324	Dyfed	
LNER Ballast Pits	SK9571	Lincolnshire	
Loans of Tullich	NH8677	Highland	
Loch A Chairn Bhain	NC1934	Highland	
Loch a` Phuill	**NL9541**	**Tiree**	**12**
Loch Arnol (Lewis)	NB3048	Western Isles	
Loch Ashie	**NH6234**	**Highland**	**17**
Loch Bee	**NF7743**	**Western Isles**	**11**
Loch Bee SSSI Coast	NF7542	Western Isles	
Loch Branahuie	NB4732	Western Isles	
Loch Broom	**NH0988**	**Highland**	**7**
Loch Clash	NC2156	Highland	
Loch Coulter	NS7687	Central	
Loch Croispol	NC3967	Highland	
Loch Doine	NN4719	Central	
Loch Eaval & Loch Hosta	NF7272	Western Isles	
Loch Ewe	NG8486	Highland	
Loch Eye & Cromarty Firth	NH8379	Highland	
Loch Fleet Complex	**NH7896**	**Highland**	**13**
Loch Flemington	NH8152	Highland	
Loch Gairloch	NG7674	Highland	
Loch Garten	**NH9718**	**Highland**	**23**
Loch Gelly	NT2092	Fife	
Loch Gruinart	NR2971	Islay	
Loch Heilen	ND2568	Highland	
Loch Hempriggs	ND3447	Highland	
Loch Indaal	NR3261	Islay	

Site	1-km sq	Region	
Loch Innis Na Ba Buidhe	NC2256	Highland	
Loch Insh and Spey Marshes	NH8304	Highland	
Loch Inver	NC0922	Highland	
Loch Katrine	NN4309	Central	
Loch Ken	**NX7168**	**D & Galloway**	**47**
Loch Kirkaig	NC0719	Highland	
Loch Kishorn	NG8240	Highland	
Loch Leven	**NO1401**	**Perth & Kinross**	**30**
Loch Linnhe - Camas Shallachain	NM9862	Highland	
Loch Lomond	NS3599	D`barton/Argyll	
Loch Long & Loch Goil	NS2191	D`barton/Argyll	
Loch Moraig	NN9066	Perth & Kinross	
Loch Mullion	NN9833	Perth & Kinross	
Loch nan Capull	NF7516	Western Isles	
Loch Nan Gabhar	NM9663	Highland	
Loch of Beith	HU3789	Shetland	
Loch of Boardhouse	**HY2725**	**Orkney**	**1**
Loch of Brockan	HY3919	Orkney	
Loch of Brow	HU3815	Shetland	
Loch of Clumlie	HU4017	Shetland	
Loch of Harray	**HY2915**	**Orkney**	**2**
Loch of Hillwell	HU3713	Shetland	
Loch of Lintrathen	**NO2754**	**Angus**	**25**
Loch of Mey	ND2773	Highland	
Loch of Skaill	HY2418	Orkney	
Loch of Skene	**NJ7807**	**Grampian**	**21**
Loch of Spiggie	HU3716	Shetland	
Loch of Stenness	**HY2812**	**Orkney**	**3**
Loch of Strathbeg	**NK0758**	**Grampian**	**19**
Loch of Swannay	HY3127	Orkney	
Loch of Tankerness	**HY5109**	**Orkney**	**4**
Loch of Vatster	HU3723	Shetland	
Loch of Wester	**ND3259**	**Highland**	**5**
Loch Paible (North Uist)	**NF7268**	**Western Isles**	**9**
Loch Roag	NB1234	Western Isles	
Loch Ryan	NX0565	D & Galloway	
Loch Sarclet	ND3442	Highland	
Loch Shawbost and Loch A` Bhaile	NB2547	Western Isles	
Loch Shieldaig	NG8156	Highland	
Loch Slapin	NG5719	Skye	
Loch Spynie	**NJ2366**	**Grampian**	**18**
Loch Stiapavat	NB5264	Western Isles	
Loch Tuamister	NB2645	Western Isles	
Loch Tullybelton	NO0034	Perth & Kinross	
Loch Urrahag	NB3148	Western Isles	
Lochs Davan and Kinord	**NO4499**	**Grampian**	**22**
Loe Pool	SW6424	Cornwall	
Long Eaton GPs	SK5032	Derbyshire	
Longnewton Res	NZ3616	Cleveland	
Longueville Marsh	WV6748	Channel Islands	
Looe Estuary	SX2553	Cornwall	
Lossie Estuary	NJ2470	Grampian	
Lost & Golding Hill & Baldwins Hill Pds	TQ4297	Essex	
Lough Foyle	**IC5925**	**Londonderry**	**49**
Lough McNean Lower	IH1038	Fermanagh	
Lough Money	IJ5345	Down	
Loughs Neagh and Beg	**IJ0475**	**Armagh**	**50**

Site	1-km sq	Region	
Lower Derwent Ings	**SE6938**	**Humberside**	**66**
Lower Loch Long	NS2084	D`barton/Argyll	
Lower Lough Erne	**IH0960**	**Fermanagh**	**51**
Lower Teviot Valley	**NT6725**	**Roxburgh**	**38**
Lower Windrush Valley Gravel Pits	SP4004	Oxfordshire	
Lunan Bay	NO6950	Angus	
Lunda Wick	HP5704	Shetland	
Lunning & Lunna Holm	HU5070	Shetland	
Lynemouth Ash Lgns	NZ3089	N`thumberland	
Lynford Gravel Pit	TL8194	Norfolk	
Machrihanish	NR6522	Argyll	
Maer Lake	SS2007	Cornwall	
Malltraeth Cob & Pls	SH4066	Anglesey	
Malltraeth RSPB	SH4471	Anglesey	
Marlee Loch	NO1444	Perth & Kinross	
Marloes Mere	SM7708	Dyfed	
Marsh Lane GPs Hemingford Grey	TL3069	Cambridgeshire	
Martin Mere & Ribble Estuary	SD4015	Lancashire	
Maxey Pits	TF1207	Cambridgeshire	
Meadow Lane Gravel Pits St Ives	TL3270	Cambridgeshire	
Medway Estuary	**TQ8471**	**Kent**	**100**
Melbost / Tong / Broad Bay	NB4534	Western Isles North	
Mere Sands Wood Nature Reserve	SD4415	Lancashire	
Mersehead RSPB	**NX9255**	**D & Galloway**	**45**
Mersey Estuary	**SJ4578**	**Cheshire**	**71**
Messingham Sand Quarries	SE9103	Humberside	
Middle Tame Valley GPs	SP2096	Warwickshire	
Middle Yare Marshes	**TG3504**	**Norfolk**	**87**
Milldam & Balfour Mains Pools	HY4817	Orkney	
Minsmere	TM4666	Suffolk	
Monikie Reservoirs	NO5038	Angus	
Monk Myre	NO2042	Perth & Kinross	
Montrose Basin	**NO6958**	**Angus**	**24**
Moore Nature Res	SJ5785	Cheshire	
Moray Coast	NJ2868	Grampian	
Moray Firth	NH8060	Highland	
Morecambe Bay	**SD4070**	**Lancashire**	**61**
Mount Castle Quarry	NO3012	Fife	
Munlochy Bay	NH6752	Highland	
Nafferton Mere	TA0558	Humberside	
Needingworth Quarry Lakes	TL3672	Cambridgeshire	
Nene Washes	**TF3300**	**Cambridgeshire**	**80**
Netherfield GPs	SK6339	Nottinghamshire	
Nevern Estuary	SN0639	Dyfed	
Newsham Park	SJ3791	Merseyside	
Newton Pool & Sea	NU2325	N`thumberland	
Newtown Estuary	SZ4291	Isle of Wight	
Norham West Mains	NT9147	N`thumberland	
North Bay	NF7546	Western Isles	
North Bressay	HU5143	Shetland	
North Cave Wetlands	SE8833	Humberside	
North Fetlar	HU6592	Shetland	
North Norfolk Coast	**TF8546**	**Norfolk**	**83**
North Sutherland	NC5363	Highland	

Site	1-km sq	Region	
North Third Reservoir	NS7588	Central	
North Uist	NF8070	Western Isles	
North Warren & Thorpeness Mere	TM4658	Suffolk	
North West Solent	**SZ3395**	**Hampshire**	**116**
NE Glamorgan Moorland Pools	SO0808	Glamorgan	
NW Yell Sound	HU3886	Shetland	
Nosterfield GPs	SE2779	N Yorkshire	
Nunnery Lakes	TL8781	Norfolk	
Ogmore Estuary	SS8675	Glamorgan	
Ogston Reservoir	SK3760	Derbyshire	
Old Moor	**SE4302**	**S Yorkshire**	**67**
Old Romney	TR0025	Kent	
Old Romney	TR0323	Kent	
Orchardton and Auchencairn Bays	**NX8151**	**D & Galloway**	**46**
Orkney	HY2915	Orkney	
Orwell Estuary	**TM2238**	**Suffolk**	**91**
Osterley Park Lakes	TQ1478	Gtr London	
Otmoor	SP5614	Oxfordshire	
Otter Estuary	SY0782	Devon	
Ouse Washes	**TL5394**	**Cambridgeshire**	**81**
Outer Ards Shoreline	**IJ6660**	**Down**	**57**
Outer Tay and St Andrews Bay offshore	NO5525	Fife	
Outwood Swan Sanctuary	TQ3246	Surrey	
Over Water and Chapelhouse Res	NY2535	Cumbria	
Pagham Harbour	**SZ8796**	**W Sussex**	**110**
Palmarsh Gravel Pit	TR1333	Kent	
Panshanger Estate	TL2812	Hertfordshire	
Papil Water Fetlar	HU6090	Shetland	
Par Sands Pools and St Andrews Road	SX0853	Cornwall	
Passfield Pond	SU8234	Hampshire	
Paultons Bird Park	SU3116	Hampshire	
Peasholme Park Lake	TA0389	North Yorkshire	
Pegwell Bay	**TR3563**	**Kent**	**103**
Pembroke Power Station Ponds	SM9202	Dyfed	
Pitsford Reservoir	SP7669	N`thamptonshire	
Point of Ayre GP	NX4504	Isle of Man	
Pontllyfni to Aberdesach	SH4252	Caernarvon	
Pool of Virkie	HU3911	Shetland	
Pool of Virkie	HU3911	Shetland	
Poole Harbour	**SY9988**	**Dorset**	**119**
Port Stewart: Portrush	IC8339	Antrim	
Porth Reservoir	SW8662	Cornwall	
Portsmouth Harbour	**SU6204**	**Hampshire**	**113**
Portworthy Mica Dam	SX5660	Devon	
Potteric Carr	SE5900	S Yorkshire	
Pugneys CP Lakes	SE3218	W Yorkshire	
Pulfin Bog	TA0544	Humberside	
Purfleet Chalk Pit	TQ5778	Essex	
Queen Elizabeth II Res	TQ1167	Surrey	
Queen Mary Res	TQ0769	Surrey	
Queen Mother Res	TQ0076	Berkshire	
Queens Valley Res	WV5652	Channel Islands	
Quendale Bay	HU3712	Shetland	
Quendale to Virkie	HU3709	Shetland	

Site	1-km sq	Region	
R Clyde: Carstairs to Thankerton	**NS9842**	**Lanarkshire**	**35**
Radnor Mere	SJ8475	Cheshire	
Ramsbury Lake	SU2671	Wiltshire	
Ranworth and Cockshoot Broads	TG3515	Norfolk	
Ravensthorpe Res	SP6770	N`thamptonshire	
Ravenstruther	NS9343	Lanarkshire	
Red Moss SSSI	SD6309	Gtr Manchester	
Red Point to Port Henderson	NG7470	Highland	
Reedham Water	TG3618	Norfolk	
Rescobie Loch	NO5151	Angus	
Rhunahaorine	NR7049	Argyll	
Ribble Estuary	**SD3825**	**Lancashire**	**68**
Ripon RC GP	SE3269	N Yorkshire	
Rispond Bay	**NC4565**	**Highland**	**6**
River Avon: Fordingbridge to Ringwood	**SU1410**	**Hampshire**	**117**
River Avon: Ringwood to Christchurch	**SZ1499**	**Hampshire**	**118**
River Avon: Salisbury	SU1430	Wiltshire	
River Cam: Kingfishers Bridge	TL5473	Cambridgeshire	
River Cam: Owlstone Rd to Baits Bite Lock	TL4760	Cambridgeshire	
River Devon: Kersiepow Ponds	NS8996	Central	
River Earn: Lawhill Oxbows	NN9517	Perth & Kinross	
River Eden: Warcop-Ltle Salkeld	**NY6424**	**Cumbria**	**63**
River Fort: Meiklewood	NS7295	Central	
River Hul: Wansford to Whinhill	TA0556	Humberside	
River Kennet: Ramsbury to Chilton Foliat	SU2871	Wiltshire	
River Nith: Keltonbank-Nunholm	**NX9774**	**D & Galloway**	**44**
River Severn & River Vyrnwy Confluence	SJ3215	Shropshire	
River Tay - Haughs of Kercock	NO1439	Perth & Kinross	
River Test - Fullerton to Stockbridge	SU3535	Hampshire	
River Test - Romsey	SU3420	Hampshire	
River Thames - Lechlade	SU2199	Glos	
River Thames at Staines Bridge	TQ0371	Surrey	
River Trent - Burton Joyce to Stoke Ferry	SK6541	Nottinghamshire	
River Trent & Pastures	SK6743	Nottinghamshire	
River Tweed: Kelso to Coldstream	NT7737	Roxburgh	
River Wye: Bakewell to Haddon	SK2267	Derbyshire	
River Wye: Putson	SO5138	Hereford & Worcs	
Rivers Eamont and Eden: Honeypot to Edenhall	NY5831	Cumbria	
Roadford Reservoir	SX4291	Devon	

Site	1-km sq	Region	
Roath Park Lake	ST1879	Glamorgan	
Romney Sands	TR0823	Kent	
Rossie Bog	NO2610	Fife	
Rossie Bog	NO2610	Fife	
Rostherne Mere	SJ7484	Cheshire	
Roswell Gravel Pits	TL5580	Cambridgeshire	
Rother Valley CP	SK4582	S Yorkshire	
Rough Firth	NX8453	D & Galloway	
Roughrigg Reservoir	NS8164	Lanarkshire	
Rova Head to Wadbister Ness	HU4546	Shetland	
Roxton Gravel Pits	TL1553	Bedfordshire	
RSPB Hanson Wetland Creation	TL3672	Cambridgeshire	
Rutland Water	**SK9207**	**Leicestershire**	**79**
Ryde Pier- Puckpool Pt	SZ6092	Isle of Wight	
Rye Harbour and Pett Level	**TQ9418**	**E Sussex**	**108**
Salhouse Broad	TG3115	Norfolk	
Saltwell Park Lake	NZ2561	Tyne & Wear	
Sandbach Flashes	SJ7259	Cheshire	
Scaling Dam Res	NZ7412	Cleveland	
Scalloway Islands	HU3641	Shetland	
Scapa Flow	HY4101	Orkney	
Scarp to Vatersay	NF8187	Western Isles	
Scolt Head	TF8046	Norfolk	
Scorton Quarry	NZ2300	N Yorkshire	
Sea Bank Clay Pits	TF5477	Lincolnshire	
Seahouses to Budle Point	NU1934	N`thumberland	
Seaton & Murton Pds	NZ3848	Durham	
Seaton Gravel Pits and River	TR2258	Kent	
Sedgeletch & Lambton	NZ3150	Durham	
Selset Reservoir	NY9121	Durham	
Sennowe Park Lk	TF9825	Norfolk	
Severn Estuary	**ST5084**	**Glos**	**122**
Severn Hams	SO8426	Glos	
Shipton On Cherwell Quarry	SP4717	Oxfordshire	
Siblyback Reservoir	SX2370	Cornwall	
Sizewell Belts	TM4663	Suffolk	
Skinflats	NS9284	Central	
Slamannan Area	NS8475	Central	
Snetterton Gravel Pits	TL9991	Norfolk	
Snettisham	TF6430	Norfolk	
Solway Estuary	**NY1060**	**Cumbria**	**42**
Solway Firth	NY0555	Cumbria	
Somerset Levels	**ST4137**	**Somerset**	**124**
Somersham GP	TL3779	Cambridgeshire	
Sonning Eye & Henley Road GPs	SU7375	Oxfordshire	
Sound of Barra	NF7509	Western Isles	
Sound of Gigha	NR6950	Argyll	
Sound of Harris	NF9781	Western Isles	
Sound of Harris (NW)	NF9788	Western Isles	
Sound of Taransay	NG0798	Western Isles North	
South Fetlar	HU6387	Shetland	
South Ford	NF7948	Western Isles	
South Havra	HU3726	Shetland	
South Medwin Pools	NT0547	Lanarkshire	
South Studdock Pond	SM8502	Dyfed	
South Uist	NF7930	Western Isles	
South Uist West Coast	NF7030	Western Isles	

Site	1-km sq	Region	
South Unst	HU5999	Shetland	
South Walls (Hoy)	ND3090	Orkney	
South Yell	HU5178	Shetland	
South Yell Sound	HU4575	Shetland	
Southampton Water	**SU4507**	**Hampshire**	**114**
Southeast Yell	HU5382	Shetland	
Southfield Reservoir	SE6519	Humberside	
Southwest Lancashire	SD4015	Lancashire	
Southwold Sole Bay	TM5177	Suffolk	
Spade Oak GP (Little Marlow)	SU8887	Bucks	
Spittal-Cocklawburn	NU0250	N`thumberland	
St Andrews Bay	NO5121	Fife	
St Benet`s Levels	TG3815	Norfolk	
St Mary`s Island	NZ3476	N`thumberland	
St Mary`s Island	NZ3575	Tyne & Wear	
Staines Reservoirs	TQ0573	Surrey	
Stanford Reservoir	SP6080	Leicestershire	
Stanford	TL8695	Norfolk	
Stanley Park Lakes	SD3235	Lancashire	
Stanton Lake	SU1790	Wiltshire	
Stanwick GPs	SP9670	N`thamptonshire	
Steeple Langford GP	SU0437	Wiltshire	
Steinish Canal	NB4533	Western Isles	
Stibbington GP	TL0899	Cambridgeshire	
Stockgrove CP	SP9229	Bedfordshire	
Stodmarsh NNR and Collards Lagoon	TR2061	Kent	
Stour Estuary	**TM1732**	**Essex**	**92**
Strangford Lough	**IJ5460**	**Down**	**56**
Stranraer Lochs	NX1161	D & Galloway	
Strathearn (West)	NN8819	Perth & Kinross	
Strathearn South Kinkell	NN9516	Perth & Kinross	
Strawberry Hill Ponds	TQ4196	Essex	
Sullom Voe	HU3772	Shetland	
Sutton & Lound GPs	SK6985	Nottinghamshire	
Sutton Place	TQ0153	Surrey	
Swale Estuary	**TQ9765**	**Kent**	**101**
Swanbourne Lake	TQ0108	W Sussex	
Swansea Bay	SS6391	Glamorgan	
Swarkestone GPs	SK3627	Derbyshire	
Swillington Ings	SE3828	W Yorkshire	
Swithland Reservoir	SK5513	Leicestershire	
Tabley Mere	SJ7276	Cheshire	
Talkin Tarn	NY5458	Cumbria	
Tamar Complex	SX4453	Cornwall	
Tarbat Ness	NH9487	Highland	
Tattershall Pits	TF2057	Lincolnshire	
Taw-Torridge Est	**SS4731**	**Devon**	**125**
Tay and Isla Valley	NO1638	Perth & Kinross	
Tay Estuary	**NO4828**	**Fife**	**26**
Tees Estuary	**NZ5528**	**Cleveland**	**63**
Teign Estuary	SX8772	Devon	
Teignmouth to Berry Head	SX9463	Devon	
Testwood Lakes	SU3514	Hampshire	
Thames and Suffolk Offshore	TR1791	Essex	
Thames Estuary	**TQ7880**	**Essex**	**99**
Thanet Coast	**TR2669**	**Kent**	**102**
The Houb (Whalsay)	HU5565	Shetland	
The Ouse & Lairo Wtr	HY5019	Orkney	
The Wash	**TF5540**	**Lincolnshire**	**82**
The Wash Strategic Area	TF7768	Norfolk	

Site	1-km sq	Region	
Theale Gravel Pits	SU6570	Berkshire	
Thoresby Lake	SK6370	Nottinghamshire	
Thornham	TF7344	Norfolk	
Thorpe Water Park	TQ0268	Surrey	
Thrapston Gravel Pits	SP9979	N`thamptonshire	
Threave Estate	NX7362	D & Galloway	
Threipmuir and Harlaw Reservoirs	NT1764	Lothians	
Thurso Bay	ND1169	Highland	
Thwaite Flat & Roanhead Ponds	SD2174	Cumbria	
Tiree	NL9741	Tiree	
Tongue End Nature Reserve (Baston Fen)	TF1417	Lincolnshire	
Tongwell Lake	SP8642	Bucks	
Tophill Low Res	TA0748	Humberside	
Traeth Dulas	SH4888	Anglesey	
Traeth Melynog	SH4463	Anglesey	
Traigh Luskentyre	NG0798	Western Isles	
Tring Reservoirs	SP9113	Hertfordshire	
Trinity Broads	TG4614	Norfolk	
Troon Meikle Craigs	NS3228	Ayrshire & Arran	
Tweed Estuary	**NT9853**	**N`thumberland**	**39**
Twyford Gravel Pits	SU7875	Berkshire	
Tyne Estuary	NZ2464	Tyne & Wear	
Tyninghame Estuary	NT6379	Lothians	
Ugie Estuary	NK1247	Grampian	
Ullswater	NY4420	Cumbria	
Upper Loch Torridon	NG8955	Highland	
Upper Lough Erne	**IH3131**	**Fermanagh**	**52**
Upper Quoile River	IJ4745	Down	
Upper River Avon: Wester Jaw	NS8573	Central	
Upper Tay	NN9557	Perth & Kinross	
Upton Warren LNR	SO9367	Hereford & Worcs	
Usk Reservoir	SN8126	Powys	
Uyea Sound	HP5900	Shetland	
Vaila Sound & Gruting Voe	HU2747	Shetland	
Vale of Coustry	NN7300	Central	
Vasa Loch Shapinsay	HY4718	Orkney	
Vyne Floods	SU6356	Hampshire	
Walland Marsh	**TQ9824**	**Kent**	**107**
Walmore Common	SO7415	Glos	
Walthamstow Res	TQ3589	Gtr London	
Wanlip Gravel Pits	SK6011	Leicestershire	
Warkworth Lane Pds	NZ2793	N`thumberland	
Water Sound	ND4795	Orkney	
Watermead GPs	SK6008	Leicestershire	

Site	1-km sq	Region	
Waulkmill Glen & Littleton Reservoirs	NS5257	Renfrew	
Weald Park	TQ5694	Gtr London	
Wellington CP	SU7362	Hampshire	
Wellington GPs	SO5047	Hereford & Worcs	
Wemyss Bay to Fairlie	NS2059	Ayrshire & Arran	
West Coast (Benbecula)	NF7652	Western Isles	
West Mull offshore	NM4132	Mull	
West Water Res	**NT1252**	**Tweeddale area**	**36**
West Whalsay & Sds	HU5264	Shetland	
West Yell	HU4485	Shetland	
Weybread Pits	TM2481	Suffolk	
Whiteadder Reservoir	NT6563	Lothians	
Whitemoor Haye	SK1813	Staffordshire	
Whiteness to Skelda Ness	HU3547	Shetland	
Whitlingham CP	TG2508	Norfolk	
Whitrig Bog	NT6235	Tweeddale area	
Widewall Bay	ND4292	Orkney	
Wigan Flashes	SD5803	Gtr Manchester	
Wigtown Bay	**NX4456**	**D & Galloway**	**48**
Wilderness Pond	SS8277	Glamorgan	
Willen Lake	SP8741	Bucks	
William Girling Res	TQ3694	Gtr London	
Willington	TL1050	Bedfordshire	
Wimbleball Lake	SS9631	Somerset	
Wintersett and Cold Hiendley Reservoirs	SE3714	W Yorkshire	
Woburn Park Lakes	SP9632	Bedfordshire	
Woodhorn Flashes	NZ2988	N`thumberland	
Woolmer Cmn Pds	SU7832	Hampshire	
Woolston Eyes	SJ6588	Cheshire	
Wootton Creek	SZ5592	Isle of Wight	
Wraysbury GPs	**TQ0074**	**Berkshire**	**105**
Wraysbury Pond	TQ0073	Berkshire	
Wraysbury Reservoir	TQ0274	Surrey	
WWT Caerlaverock	**NY0565**	**D & Galloway**	**43**
WWT Martin Mere	**SD4015**	**Lancashire**	**69**
Wynyard Lake	NZ4225	Cleveland	
Yar Estuary	SZ3588	Isle of Wight	
Yeadon Tarn	SE2141	W Yorkshire	
Yealm Estuary	SX5449	Devon	
Young`s Park: Goodrington	SX8959	Devon	
Ythan Estuary	NK0026	Grampian	
Ythan Estuary and Slains Lochs	NK0126	Grampian	

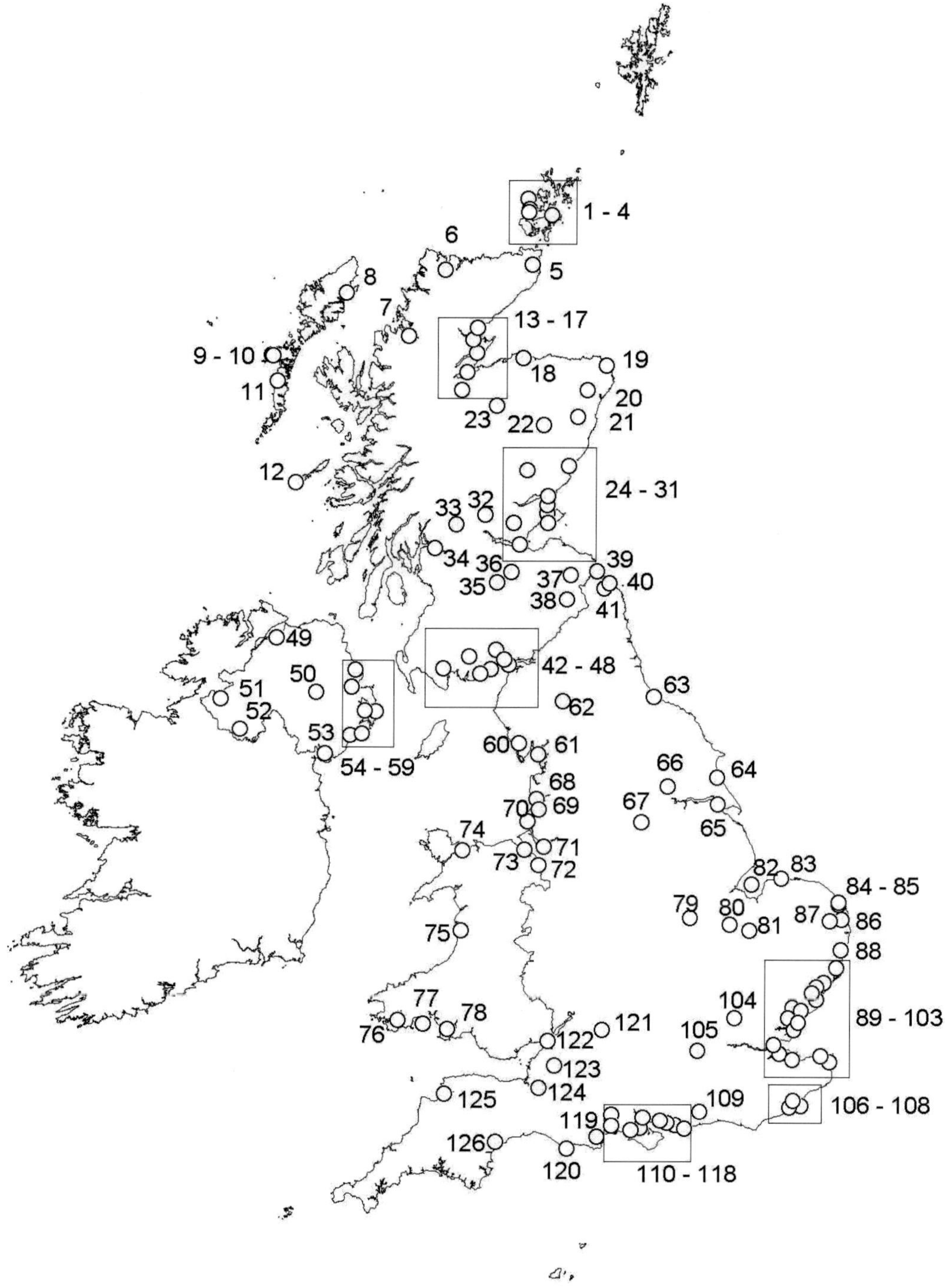

Figure A1. Locations of Core WeBS sites supporting more than 10,000 waterbirds or which support internationally important numbers of one or more waterbird species (see *PRINCIPAL SITES*). Numbers refer to sites listed in Table A2.

APPENDIX 3. WATERBIRDS RECORDED BY WeBS BETWEEN APRIL AND JUNE 2004

As a result of the change in the WeBS Year from April-March, to July-June, a short summary of key findings from April-June 2004 is presented here. If further information is required, please contact the WeBS Office at the BTO.

The total numbers of waterbirds recorded by WeBS between April and June 2004 are given in Tables A3 and A4 for Great Britain (including the Isle of Man, but excluding the Channel Islands) and Northern Ireland, respectively.

A single Bean Goose was recorded at Hornsea Mere in April, whilst another was present at Hauxley Haven. A large number of Pink-footed Geese was still present in April with flocks of 5,400 at Loch of Strathbeg, 3,500 at Morecambe Bay and about 1,500 each at the Ribble and Solway Estuaries. Similarly, Barnacle Geese were still present in considerable numbers in April with 9,338 on the Solway Estuary and 2,750 at Mersehead RSPB. By May there were still 2,250 on the Solway Estuary, although these had all left the site by June.

More than 7,500 Dark-bellied Brent Geese were counted on The Wash during April. By May this number had fallen to just over 6,100 and by June only nine birds remained at the site. Up to 57 Light-bellied Brent Geese were recorded at Lough Foyle in April, after which time no Brents were recorded in Northern Ireland until the following September.

Over 600 Wigeon were present at Breydon Water and Berney Marshes in April. This figure had dropped to 151 in May and only three in June. Additionally, 969 Teal were present at this site in April. Over 2,000 Common Scoter remained off the North Norfolk Coast during May and June.

A total of 126 Red-throated Divers were recorded at the Inner Firth of Clyde during April. During the same month 19 Black-throated and five Great Northern Divers were at Gerrans Bay. A total of nine Red-necked and 19 Slavonian Grebes were recorded at the Forth Estuary in April; this was also the top month for Black-necked Grebe with 23 at Woolston Eyes and 17 at a further known breeding site. Bittern were recorded at a total of six sites between April and June; all records were of single birds except for six at Minsmere in April. There was a single Glossy Ibis on the Exe Estuary in April and up to 12 Spoonbill on the Orwell Estuary in June.

Avocet numbers peaked at 591 at the North Norfolk Coast in April. The May total of Ringed Plovers at the Humber Estuary totalled 1,004. Dotterel were recorded at the Duddon Estuary in April (two) and The Wash in May (46). Grey Plover numbers peaked in May at the Wash with a total of 12,442 recorded; by June only 39 remained. Between April and May Knot numbers fell from over 20,000 to around 7,000 at two of the key sites for this species, the Ribble Estuary and The Wash; by June this figure had fallen to around 4,000. Numbers of Sanderling at the Ribble Estuary fell from 4,477 in April to 2,270 in May before rising slightly to 2,908 in June. Three Little Stints were at Toft Newton Reservoir in May. Temminck's Stints were recorded at four sites, all singles except for two at East Chevington Pools in May. Ruff numbers peaked in April with 57 at the North Norfolk Coast, 19 at the Ouse Washes and 18 at Breydon Water and Berney Marshes. Bar-tailed Godwit numbers fell from 4,662 in April to 2,199 in May. By June the summed site maximum was 756, over half of which were recorded at Humber Estuary, the North Norfolk Coast and The Wash. The highest single-site Core Count total of Whimbrel was 298 at the Exe Estuary in May. Up to 31 Spotted Redshank and 34 Greenshank were at Breydon Water and Berney Marshes and The Wash, respectively, in May. Green Sandpiper were recorded at just five sites during May; compared to 27 in April; all records were of single birds. By June this species was recorded from 40 sites, the highest single count was of six at Beddington Sewage Farm. Wood Sandpiper were recorded at nine sites between April and June; all were singles except for three at Loch of Strathbeg in May and two at Minsmere in April. A single Red-necked Phalarope was present at the Swale Estuary in May.

Table A3. Total numbers of waterbirds recorded by WeBS Core Counts in Great Britain between April and June 2004

	Species	Apr	May	Jun
Number of sites visited		*868*	*746*	*731*
MS	Mute Swan	8831	7995	9057
AS	Black Swan	17	21	21
WS	Whooper Swan	148	28	20
HN	Chinese Goose	4	9	1
BE	Bean Goose	2	0	0
PG	Pink-footed Goose	13757	273	26
EW	European White-fronted Goose	3	1	0
NW	Greenland White-fronted Goose	3	0	0
LC	Lesser White-fronted Goose	1	0	0
GJ	Greylag Goose	7438	8803	14228
JE	Re-established Greylag Goose	22	0	0
HD	Bar-headed Goose	12	5	8
SJ	Snow Goose	8	4	13
RJ	Ross's Goose	1	1	2
EM	Emperor Goose	0	4	0
CG	Greater Canada Goose	12254	12647	25226
BY	Barnacle Goose	12343	2404	314
DB	Dark-bellied Brent Goose	11973	7636	40
PB	Light-bellied Brent Goose	16	0	0
EB	Red-breasted Goose	2	2	2
ZL	Feral/hybrid Goose	268	194	205
EG	Egyptian Goose	160	169	246
UB	Paradise Shelduck	1	1	1
UD	Ruddy Shelduck	0	2	4
SU	Shelduck	18954	15772	22514
MY	Muscovy Duck	27	25	22
DC	Wood Duck	4	3	1
MN	Mandarin	97	79	111
WN	Wigeon	3199	420	283
HL	Chiloe Wigeon	2	1	0
GA	Gadwall	2575	1903	2221
T.	Teal	6244	1154	649
TA	Green-winged Teal	3	1	0
MA	Mallard	25028	25814	37208
ZF	Feral/hybrid mallard type	176	185	292
PT	Pintail	216	25	9
GY	Garganey	26	22	14
SV	Shoveler	1679	689	441
ZR	Hybrid Anas	1	0	0
IE	Ringed Teal	0	1	0
RQ	Red-crested Pochard	25	20	22
PO	Pochard	1121	745	981
NG	Ring-necked Duck	1	1	1
TU	Tufted Duck	14781	7671	8773
SP	Scaup	218	1	1
AY	Lesser Scaup	1	0	0
ZD	Aythya hybrid	2	0	0
E.	Eider	11003	9092	11141
LN	Long-tailed Duck	163	2	1
CX	Common Scoter	5969	3035	2269
FS	Surf Scoter	2	2	0
VS	Velvet Scoter	1011	134	32
GN	Goldeneye	982	71	83
SY	Smew	1	1	1
RM	Red-breasted Merganser	1593	416	443
GD	Goosander	295	138	369
RY	Ruddy Duck	933	500	558
OI	Argentine Bluebill	1	0	0
WQ	White-headed Duck	0	1	4
RH	Red-throated Diver	323	50	9
BV	Black-throated Diver	25	9	4
ND	Great Northern Diver	16	15	1
LG	Little Grebe	1193	855	872
GG	Great Crested Grebe	3628	2889	2697
RX	Red-necked Grebe	10	0	5
SZ	Slavonian Grebe	20	4	0
BN	Black-necked Grebe	66	28	18
CA	Cormorant	6074	4880	5244
SA	Shag	423	231	151
BI	Bittern	10	3	3
ET	Little Egret	579	497	558
H.	Grey Heron	2051	1841	1871

	Species	Apr	May	Jun
OR	White Stork	1	2	0
IB	Glossy Ibis	1	0	0
IS	Sacred Ibis	0	1	1
NB	Spoonbill	3	5	23
FM	Chilean Flamingo	1	0	0
WA	Water Rail	113	56	48
MH	Moorhen	4479	3400	3465
CO	Coot	16928	12968	22807
KF	Kingfisher	80	63	80
	TOTAL WILDFOWL	**199621**	**135920**	**175715**

	Species	Apr	May	Jun
Number of sites visited		*868*	*746*	*731*
OC	Oystercatcher	49454	31413	25018
IT	Black-winged Stilt	1	1	1
AV	Avocet	1888	1608	1371
TN	Stone-curlew	2	0	0
LP	Little Ringed Plover	225	201	223
RP	Ringed Plover	4492	7482	1871
KP	Kentish Plover	1	1	0
DO	Dotterel	2	46	0
GP	Golden Plover	4168	645	152
GV	Grey Plover	17183	21445	580
L.	Lapwing	6320	5389	11351
KN	Knot	53252	16929	11210
SS	Sanderling	8207	5175	4581
LX	Little Stint	1	5	0
TK	Temminck's Stint	0	5	0
PP	Pectoral Sandpiper	0	2	0
CV	Curlew Sandpiper	3	14	5
PS	Purple Sandpiper	549	28	0
DN	Dunlin	71451	72285	4166
RU	Ruff	135	17	7
JS	Jack Snipe	6	0	0
SN	Snipe	481	104	122
WK	Woodcock	2	2	1
BW	Black-tailed Godwit	10822	3628	2490
BA	Bar-tailed Godwit	4615	2195	754
WM	Whimbrel	1627	1068	61
CU	Curlew	9382	3142	7741
DR	Spotted Redshank	50	38	11
RK	Redshank	23129	4339	3967
GK	Greenshank	229	128	31
GE	Green Sandpiper	48	5	66
OD	Wood Sandpiper	4	7	1
CS	Common Sandpiper	319	311	254
TT	Turnstone	4986	1737	488
NK	Red-necked Phalarope	0	1	0
	TOTAL WADERS	**273034**	**179396**	**76523**

	Species	Apr	May	Jun
Number of sites visited		*723*	*606*	*609*
MU	Mediterranean Gull	56	34	33
LU	Little Gull	355	39	22
ON	Bonaparte's Gull	0	0	1
BH	Black-headed Gull	39725	32404	35059
IN	Ring-billed Gull	0	1	0
CM	Common Gull	3035	4462	1781
LB	Lesser Black-backed Gull	31735	42836	42359
YG/YM	Yellow-legged Gull	0	7	5
HG	Herring Gull	28987	27775	26151
GB	Great Black-backed Gull	1455	1256	1771
KI	Kittiwake	314	597	6948
	TOTAL GULLS	**105662**	**109411**	**114130**

	Species	Apr	May	Jun
Number of sites visited		*741*	*627*	*620*
AF	Little Tern	74	609	661
BJ	Black Tern	0	1	1
TE	Sandwich Tern	2440	5276	7565
CN	Common Tern	910	2622	2861
RS	Roseate Tern	0	2	0
AE	Arctic Tern	26	436	336
UI	Common/Arctic Tern	0	34	3
	TOTAL TERNS	**3450**	**8980**	**11427**

Table A4. Total numbers of waterbirds recorded by WeBS Core Counts in Northern Ireland between April and June 2004

	Species	**Apr**	**May**	**Jun**
Number of sites visited		*3*	*3*	*2*
MS	Mute Swan	68	51	14
WS	Whooper Swan	15	0	3
NW	Greenland White-fronted Goose	1	0	0
GJ	Greylag Goose	60	57	0
BY	Barnacle Goose	1	0	0
DB	Dark-bellied Brent Goose	1	0	0
PB	Light-bellied Brent Goose	57	0	0
SU	Shelduck	197	111	134
WN	Wigeon	0	2	4
T.	Teal	16	0	0
MA	Mallard	110	88	178
PO	Pochard	4	1	0
TU	Tufted Duck	27	36	0
SP	Scaup	11	0	0
E.	Eider	4	0	267
GN	Goldeneye	3	0	0
RM	Red-breasted Merganser	14	0	31
RH	Red-throated Diver	6	0	0
LG	Little Grebe	11	2	14
GG	Great Crested Grebe	5	0	0
GX	Gannet	1	0	0
CA	Cormorant	75	17	69
H.	Grey Heron	15	4	28
MH	Moorhen	6	21	3
CO	Coot	30	33	0
	TOTAL WILDFOWL	**738**	**423**	**745**

	Species	**Apr**	**May**	**Jun**
Number of sites visited		*3*	*3*	*2*
OC	Oystercatcher	381	282	351
RP	Ringed Plover	291	120	43
GP	Golden Plover	1068	0	1
L.	Lapwing	128	0	93
KN	Knot	5	0	0
SS	Sanderling	265	18	0
DN	Dunlin	330	118	117
BW	Black-tailed Godwit	116	1	0
BA	Bar-tailed Godwit	43	0	2
WM	Whimbrel	33	0	0
CU	Curlew	89	12	430
RK	Redshank	526	0	28
GK	Greenshank	0	0	12
TT	Turnstone	13	0	0
	TOTAL WADERS	**3288**	**551**	**1077**

	Species	**Apr**	**May**	**Jun**
Number of sites visited		*3*	*3*	*2*
BH	Black-headed Gull	176	42	196
CM	Common Gull	70	17	140
LB	Lesser Black-backed Gull	17	45	4
HG	Herring Gull	4	85	12
GB	Great Black-backed Gull	134	16	28
KI	Kittiwake	2	1	3
	TOTAL GULLS	**403**	**206**	**383**

	Species	**Apr**	**May**	**Jun**
Number of sites visited		*2*	*3*	*2*
TE	Sandwich Tern	13	22	34
CN	Common Tern	0	4	0
	TOTAL TERNS	**13**	**26**	**34**